History, Philosophy and Theory of the Life Sciences

Volume 27

Hans Metz, Leiden University, Leiden, The Netherlands

Roberta L. Millstein, Department of Philosophy, University of California, Davis, Davis, CA, USA

Staffan Müller-Wille, Dept of History & Philosophy of Science, University of Cambridge, Cambridge, UK

François Munoz, Université Montpellier 2, Montpellier, France

Dominic Murphy, School of History & Philosophy of Sci, University of Sydney, Camperdown, NSW, Australia

Stuart A. Newman, Dept Cell Bio & Anat, Basic Sci Bld, New York Medical College, Valhalla, NY, USA

Frederik Nijhout, Duke University, Durham, NC, USA

Samir Okasha, Philosophy, University of Bristol, Bristol, UK

Susan Oyama, The City University of New York, New York, USA

Kevin Padian, University of California, Berkeley, CA, USA

David Queller, Washington University in St. Louis, St. Louis, MO, USA

Stephane Schmitt, Archives Poincaré, Nancy, France

Phillip Sloan, Program of Liberal Studies, University of Notre Dame, Notre Dame, IN, USA

Jacqueline Sullivan, Western University, London, ON, Canada

Giuseppe Testa, University of Milan, Milan, Italy

J. Scott Turner, SUNY College of Environmental Science and Forestry, Syracuse, NY, USA

Denis Walsh, University of Toronto, Toronto, ON, Canada

Marcel Weber, Department of Philosophy, University of Geneva, Geneva, Switzerland

History, Philosophy and Theory of the Life Sciences is a space for dialogue between life scientists, philosophers and historians – welcoming both essays about the principles and domains of cutting-edge research in the life sciences, novel ways of tackling philosophical issues raised by the life sciences, as well as original research about the history of methods, ideas and tools, which constitute the genealogy of our current ways of understanding living phenomena.

The series is interested in receiving book proposals that • are aimed at academic audience of graduate level and up • combine historical and/or philosophical and/or theoretical studies with work from disciplines within the life sciences broadly conceived, including (but not limited to) the following areas: • Anatomy & Physiology • Behavioral Biology • Biochemistry • Bioscience and Society • Cell Biology • Conservation Biology • Developmental Biology • Ecology • Evolution & Diversity of Life • Genetics, Genomics & Disease • Genetics & Molecular Biology • Immunology & Medicine • Microbiology • Neuroscience • Plant Science • Psychiatry & Psychology • Structural Biology • Systems Biology • Systematic Biology, Phylogeny Reconstruction & Classification • Virology The series editors aim to make a first decision within 1 month of submission. In case of a positive first decision the work will be provisionally contracted: the final decision about publication will depend upon the result of the anonymous peer review of the complete manuscript. The series editors aim to have the work peer-reviewed within 3 months after submission of the complete manuscript. The series editors discourage the submission of manuscripts that contain reprints of previously published material and of manuscripts that are below 150 printed pages (75,000 words). For inquiries and submission of proposals prospective authors can contact one of the editors: Charles T. Wolfe: ctwolfe1@gmail.com Philippe Huneman: huneman@wanadoo.fr Thomas A.C. Reydon: reydon@ww.uni-hannover.de

More information about this series at https://link.springer.com/bookseries/8916

Luca Malatesti • John McMillan • Predrag Šustar
Editors

Psychopathy

Its Uses, Validity and Status

 Springer

Editors
Luca Malatesti
Department of Philosophy, Faculty
of Humanities and Social Sciences
University of Rijeka
Rijeka, Croatia

John McMillan
Bioethics Centre
University of Otago
Dunedin, New Zealand

Predrag Šustar
Department of Philosophy, Faculty
of Humanities and Social Sciences
University of Rijeka
Rijeka, Croatia

ISSN 2211-1948   ISSN 2211-1956   (electronic)
History, Philosophy and Theory of the Life Sciences
ISBN 978-3-030-82456-3   ISBN 978-3-030-82454-9   (eBook)
https://doi.org/10.1007/978-3-030-82454-9

This Springer imprint is published by the registered company Springer Nature Switzerland AG
The registered company address is: Gewerbestrasse 11, 6330 Cham, Switzerland

*In memory of Professor Scott Lilienfeld
(1960–2020)*

Acknowledgements

We would like to thank Marko Jurjako, who has considerably helped us in refereeing several contributions to this collection. Similarly, we are grateful to Inti A. Brazil for assisting in the evaluation of some contributions.

This book is an outcome of the project *Classification and explanations of antisocial personality disorder and moral and legal responsibility in the context of the Croatian mental health and care law* (CEASCRO) that was completely financed in the period 2014–2018 by the Croatian Science Foundation (project grant HRZZ-IP-2013-11-8071).

Contents

Contributors

Gwen Adshead is a Consultant forensic psychiatrist and psychotherapist. She has worked in high secure hospitals and prisons with high-risk offenders with personality disorder of both sexes. She is qualified as a group analyst and has held honorary Professor positions at Yale University and Gresham College.

Elvio Baccarini is Professor in the Department of Philosophy of the Faculty of Humanities and Social Sciences, University of Rijeka, Rijeka, Croatia.

Inti A. Brazil is Assistant Professor in the Department of Neuropsychology and Rehabilitation Psychology, Donders Institute for Brain, Cognition and Behaviour, Radboud University, Nijmegen, Netherlands.

Zdenka Brzović is a postdoctoral researcher in the Department Philosophy of the Faculty of Humanities and Social Sciences, University of Rijeka, Rijeka, Croatia.

Rachel Cooper is Professor in History and Philosophy of Science at Lancaster University, Lancaster, United Kingdom.

Johanna C. Glimmerveen is a doctoral student in the Department of Neuropsychology and Rehabilitation psychology, Donders Institute for Brain, Cognition and Behaviour, Radboud University, Nijmegen, Netherlands.

Stephanie Griffiths is College Professor in Psychology at Okanagan College in British Columbia and an adjunct faculty in the Werklund School of Education, University of Calgary, Penticton, Canada.

Velinka Grozdanić is Professor of Criminal Law at the Faculty of Law, University of Rijeka, Rijeka, Croatia.

Jelena Hodak is a doctoral student of Language and Cognitive Neuroscience at the Centre for Postgraduate Studies of the University of Zagreb, Croatia.

Jarkko Jalava is College Professor in Criminology at Okanagan College in British Columbia, Penticton, Canada.

Vesna Šendula Jengić is Associate Professor at the Faculty of Medicine, University of Rijeka, Rijeka, Croatia, a forensic psychiatrist, and the Director of the Rab Psychiatric Hospital, Rab, Croatia.

Marko Jurjako is Assistant Professor in the Department of Philosophy of the Faculty of Humanities and Social Sciences, University of Rijeka, Rijeka, Croatia.

Robert D. Latzman is Associate Professor and Associate Director of Clinical Training in the Department of Psychology at Georgia State University in Atlanta, USA.

Scott O. Lilienfeld was the Samuel Candler Dobbs Professor of Psychology at Emory University in Atlanta, USA.

Joseph H. R. Maes is Assistant Professor in the Department of Neuropsychology and Rehabilitation Psychology, Donders Institute for Brain, Cognition and Behaviour, Radboud University, Nijmegen, Netherlands.

Luca Malatesti is Associate Professor in the Department of Philosophy of the Faculty of Humanities and Social Sciences, University of Rijeka, Rijeka, Croatia.

Michael D. Maraun is Professor in the Department of Psychology, Simon Fraser University, Burnaby, Canada.

John McMillan is Professor at the Bioethics Centre, University of Otago, Dunedin, New Zealand.

Marga Reimer is Professor of Philosophy at the University of Arizona in Tucson, Arizona, USA.

Thomas A. C. Reydon is Professor of Philosophy of Science and Technology at the Institute of Philosophy and the Centre for Ethics and Law in the Life Sciences (CELLS), Leibniz Universität Hannover, Hannover, Germany.

Martin Sellbom is Professor in the Department of Psychology at the University of Otago in Dunedin, New Zealand.

Jeanne Snelling is Lecturer in the Bioethics Centre and Faculty of Law at the University of Otago, Dunedin, New Zealand.

Predrag Šustar is Associate Professor in the Department of Philosophy, Faculty of Humanities and Social Sciences, University of Rijeka, Rijeka, Croatia.

Armon J. Tamatea is a clinical psychologist and Senior Lecturer in the Division of Arts, Law, Psychology and Social Sciences, University of Waikato, Hamilton, New Zealand.

Dustin B. Wygant is Professor in the Department of Psychology at Eastern Kentucky University, Richmond, Kentucky, USA, as well as the Director of Clinical Training for the doctoral program in Clinical Psychology

Chapter 1
Introduction to the Volume

Luca Malatesti, John McMillan, and Predrag Šustar

Abstract This chapter introduces psychopathy and the interrelated issues concerning its use, validity, and status, as investigated in the other chapters in the collection. Psychopathy is a condition that is typically characterised as a personality disorder with low affectivity, a manipulative and deceptive interpersonal style, and a persistent pattern of antisocial and even criminal behaviour. Measures of psychopathy have been extensively applied in different countries, in forensic and other contexts, to support treatment, management, prediction of violent recidivism, general evaluation of risk and criminal accountability. These uses, that can have a serious impact on the lives of persons, raise different questions concerning the scientific robustness of these measures and the disorder status of psychopathy. We highlight the interdisciplinary nature of the investigations needed to address these questions. To delineate the general framework guiding the collection, we also delineate briefly how the interaction between the scientific study of psychopathy and philosophical insights, coming from philosophy of science and philosophy of psychiatry, can contribute to answering these questions.

Keywords Psychopathy · Ethics of diagnosis · Construct validity · Forensic applications · Interdisciplinary research

Psychopathy has become a focal point for scientific research over recent years. It is a condition that is typically characterised as a personality disorder with low affectivity, a manipulative and deceptive interpersonal style, and a persistent pattern of antisocial and even criminal behaviour. The Psychopathy Checklist Revised by Robert Hare (PCL-R) (Hare, 2003), that was elaborated in the seventies and further

L. Malatesti (✉) · P. Šustar
Department of Philosophy, Faculty of Humanities and Social Sciences, University of Rijeka, Rijeka, Croatia
e-mail: lmalatesti@ffri.uniri.hr; psustar@ffri.uniri.hr

J. McMillan
Bioethics Centre, University of Otago, Dunedin, New Zealand
e-mail: john.mcmillan@otago.ac.nz

© Springer Nature Switzerland AG 2022
L. Malatesti et al. (eds.), *Psychopathy*, History, Philosophy and Theory of the
Life Sciences 27, https://doi.org/10.1007/978-3-030-82454-9_1

refined into its current form in the following decades, has shaped the current paradigm for the scientific study of psychopathy. Thus, although many have offered and investigated alternative measures to PCL-R (Fowler & Lilienfeld, 2013), this measure informs a large number of the studies of behavioural and functional characteristics that are associated with psychopathy and their cognitive, neurological and genetic underpinnings (Patrick, 2018).

Being diagnosed as a psychopath can have a significant impact upon someone's life. In fact, PCL-R, and other measures of psychopathy, have been extensively applied in different countries, in forensic and other contexts, to support treatment, management, prediction of violent recidivism, general evaluation of risk and criminal accountability. These applications inform important decisions. In some countries being diagnosed as a psychopath might be a ground for denying parole, longer detainment, and be an aggravating factor in decisions concerning capital punishment (Edens et al., 2018). Similarly, psychopathy can be a relevant consideration in child custody cases (Lyon et al., 2016). Psychopathy, perhaps more than other psychiatric conditions, carries significant social stigma (Jurjako et al., 2019).

The impact of being diagnosed as psychopathic has motivated scientific study but also debates about the practical applications of this construct. Several researchers have thus far investigated how PCL/R and other measures of psychopathy fare, in terms of their reliability, when used in practical contexts (Boccaccini et al., 2012). Similarly, there are studies that investigate how judges, juries, or other decisional makers are affected by the information that a person is diagnosed as a psychopath (Edens et al., 2013). In any case, an important focus for research and debate has been the capacity of PCL-R or other measures of psychopathy to ground predictions of violent recidivism. Worries about this issue have even motivated a document by concerned experts in the uses of PCL-R for decisions about capital punishment (DeMatteo et al., 2020; cf., Olver et al., 2020). All these can be classified as worries internal to the paradigm of psychopathy studies.

To avert systematic misuses or extrapolation beyond the proper area of applicability of the construct of psychopathy, however, it is very important to frame the discussion by considering also more general issues. There are general concerns about the construct of psychopathy that can be addressed within philosophy of psychiatry (Fulford et al., 2013). There is, in fact, the problem of the overall plausibility of the construct of psychopathy. Answering this question requires also engaging with foundational issues in psychiatry that concern the nature and reliability of its classificatory practices. In addition, there are open related issues concerning the fact that certain values have a role in the formulation of the category of psychopathy. Also, the presence and significance of values in psychiatric theory and practice is a fundamental issue that has attracted the interest of the philosophers of psychiatry (Bolton, 2008; Sadler, 2005). Addressing these general issues renders explicit, clarifies, and probes general underlying assumptions that shape thinking and practices in research, clinical and forensic settings. We are strongly committed to the idea that this kind of explication is conducive to better practices in this specific case of translation of scientific knowledge.

The book documents and aims to promote research on these general issues about the construct of psychopathy. The volume includes chapters commissioned by different specialists who present leading and, whenever it is relevant, competing positions. The book is organized into three parts, each covering a range of important issues related to psychopathy. The first part, "Diagnosing psychopathy. Practices, case studies, and problematic areas", shows how psychopathy functions within the institutional and social practices of some countries. This overview delineates some major practical and ethical problems and perplexities that derive from these uses of the diagnosis of psychopathy.

The second part, "The plausibility and validity of psychopathy", offers, besides insights on the current diagnosis of psychopathy, functional and behaviour correlates, and certain explanatory hypotheses, more theoretical discussions of psychopathy. The plausibility of psychopathy as a construct and the validity of its measurements can be investigated within the boundaries of psychiatry and behavioural sciences. Psychometric research and studies about the neural, functional, and genetic correlates of psychopathy, as mentioned, abound and are likely to offer fundamental insights on these issues. However, theoretical, or philosophical considerations about the nature of personality disorders, psychiatric classification, and the reification of psychiatric conditions and disorders are relevant as well. In particular, the status of the construct of psychopathy is investigated in relation to the philosophical debate on the nature of scientific classifications.

Finally, the third part, "Psychopathy and values", discusses how different kinds of values are relevant to the construct of psychopathy and what consequences this has for the validity of the construct and its status as a medical disorder. Investigating the status of psychopathy as a disorder, the preferences and needs that govern the use of this construct and how they might be culturally specific, involves addressing general philosophical/foundational issues about the concept of mental disorder that are at the core of much current philosophy of psychiatry.

We would like to conclude this short introduction with a methodological consideration. The book is addressed to an interdisciplinary readership that, including graduate students, consists of behavioural scientists, different mental health practitioners, and philosophers. Our hope is that research on psychopathy and its practical significance could profit by the investigation of research hypotheses that focus on the issues covered in this book. But this will only be possible if these hypotheses can be formulated and investigated by interdisciplinary research groups that will be able to converse (or compromise) and agree on the relevant problems and how to address them. So far, with very few exceptions, it seems that there has been a lack of communication and collaboration amongst those who investigate psychopathy from these different perspectives. For instance, while several philosophical attempts at investigating the practical significance of psychopathy have relied on a too selective or inadequate reading of the neuropsychological literature (Jalava & Griffiths, 2017, this volume), the important conceptual and normative issues raised and debated in recent discussions in philosophy of psychiatry have had no significant impact on the scientific study of psychopathy.

References

Boccaccini, M. T., Turner, D. B., Murrie, D. C., & Rufino, K. A. (2012). Do PCL-R scores from state or defense experts best predict future misconduct among civilly committed sex offenders? *Law and Human Behavior, 36*(3), 159–169. https://doi.org/10.1037/h0093949

Bolton, D. (2008). *What is mental disorder? An essay in philosophy, science, and values.* Oxford University Press.

DeMatteo, D., Hart, S. D., Heilbrun, K., Boccaccini, M. T., Cunningham, M. D., Douglas, K. S., Dvoskin, J. A., Edens, J. F., Guy, L. S., Murrie, D. C., Otto, R. K., Packer, I. K., & Reidy, T. J. (2020). Statement of concerned experts on the use of the Hare psychopathy checklist—Revised in capital sentencing to assess risk for institutional violence. *Psychology, Public Policy, and Law, 26*(2), 133–144. https://doi.org/10.1037/law0000223

Edens, J. F., Magyar, M. S., & Cox, J. (2013). Taking psychopathy measure 'out of the lab' and into the legal system: Some practical concerns. In K. A. Kiehl & W. P. Sinnott-Armstrong (Eds.), *Handbook on psychopathy and law* (pp. 250–272). Oxford University Press.

Edens, J. F., Petrila, J., & Kelley, S. E. (2018). Legal and ethical issues in the assessment and treatment of psychopathy. In *Handbook of psychopathy* (2nd ed., pp. 732–751). The Guilford Press.

Fowler, K. A., & Lilienfeld, S. O. (2013). Alternatives to psychopathy checklist-revised. In K. A. Kiehl & W. P. Sinnott-Armstrong (Eds.), *Handbook on psychopathy and law* (pp. 34–57). Oxford University Press.

Fulford, K. W. M., Davies, M., Gipps, R. G. T., Graham, G., Sadler, J. Z., Stanghellini, G., & Thornton, T. (Eds.). (2013). *The Oxford handbook of philosophy and psychiatry.* (First published in paperback). Oxford University Press.

Hare, R. D. (2003). *The Hare psychopathy checklist revised* (2nd ed.). Multi-Health Systems.

Jalava, J., & Griffiths, S. (2017). Philosophers on psychopaths: A cautionary in interdisciplinarity. *Philosophy, Psychiatry, and Psychology, 24*(1), 1–12.

Jalava, J., & Griffiths, S. (this volume). Psychopathy: Neurohype and its consequences. In L. Malatesti, J. McMillan, & P. Šustar (Eds.), *Psychopathy. Its uses, validity, and status.* Springer.

Jurjako, M., Malatesti, L., & Brazil, I. A. (2019). Some ethical considerations about the use of biomarkers for the classification of adult antisocial individuals. *International Journal of Forensic Mental Health, 18*(3), 228–242. https://doi.org/10.1080/14999013.2018.1485188

Lyon, D. R., Ogloff, J. R. P., & Shepherd, S. M. (2016). Legal and ethical issues in the assessment of psychopathy. In C. Gacono (Ed.), *The clinical and forensic assessment of psychopathy: A practitioner's guide* (2nd ed., pp. 193–216). Routledge/Taylor & Francis Group.

Olver, M. E., Stockdale, K. C., Neumann, C. S., Hare, R. D., Mokros, A., Baskin-Sommers, A. R., Brand, E., Folino, J., Gacono, C., Gray, N. S., Kiehl, K. A., Knight, R. A., Leon-Mayer, E., Logan, M., Meloy, J. R., Roy, S., Salekin, R. T., Snowden, R., Thomson, N., … Yoon, D. (2020). Reliability and validity of the psychopathy checklist-revised in the assessment of risk for institutional violence: A cautionary note on DeMatteo et al. (2020). *Psychology, Public Policy, and Law, 26*(4), 490–510.

Patrick, C. J. (Ed.). (2018). *Handbook of psychopathy* (2nd ed.). The Guilford Press.

Sadler, J. Z. (2005). *Values and psychiatric diagnosis.* Oxford University Press.

Part I
Diagnosing Psychopathy. Practices, Case Studies and Practical Concerns

Chapter 2
Re-appraising Psychopathy

John McMillan

Abstract Psychopathy, as articulated in Hare's PCL-R, appears to reliably pick out a forensic category of troubled people. This chapter considers the use and utility of PCL-R by focussing upon two interrelated questions. Does philosophical investigation direct attention toward the issues that should interest us about psychopathy? Is being diagnosed as psychopathic or having ASPD clinically useful, as well as for judicial and sentencing purposes? While the research programmes that developed following the attention paid to psychopathy are warranted, more attention could be directed to the varied nature of psychopathy and the presentations of it that Cleckley described in the *Mask of Sanity*. It is important to understand psychopathy as it affects lives, as well as a forensic problem.

Keywords Psychopathy · Insight · Responsibility · Guilt · Personality Disorder · Ethics · Agency

2.1 Introduction

Psychopathy and sociopathy are conditions that are often portrayed in fiction and have folk or popular understandings. The idea that there are cold-blooded murderers in society, capable of violent offending without remorse is a trope that reappears in film and literature. Philosophy and law have studied psychopathy too, due in part to the sharp legal and philosophical questions it raises, but also because of the way in which violent offenders can create public concern and political action. This chapter begins by explaining why we tend to view the importance of psychopathy in the way we do. The reasons why we consider why a particular account of psychopathy of interest, and the issues that are subsequently thought worthy of further investigation, require scrutiny.

J. McMillan (✉)
Bioethics Centre, University of Otago, Dunedin, New Zealand
e-mail: john.mcmillan@otago.ac.nz

© Springer Nature Switzerland AG 2022
L. Malatesti et al. (eds.), *Psychopathy*, History, Philosophy and Theory of the
Life Sciences 27, https://doi.org/10.1007/978-3-030-82454-9_2

The second part of this chapter will discuss the concept of psychopathy that Cleckley developed in the *Mask of Sanity* (Cleckley, 1988). This work is justifiably well known for its influence upon Hare's Psychopathy Checklist Revised (PCL-R) (Hare & Neumann, 2012) and for the distinction between primary and secondary psychopaths, along with the claim that secondary psychopaths are often hiding in plain sight within society. However, when you read Cleckley's descriptions of the primary psychopaths he met clinically and tried to treat, a very different picture emerges than the folk conception of psychopaths as cold-blooded manipulators and killers. The people he presents do confusing, frustrating and ultimately self-destructive things, while at the same time appearing not to care that their life is by the standards of most people, well and truly off the rails. This chapter suggests that we need to keep this concept of psychopathy in view if we are to understand fully what it means for those who have it and are affected by it.

2.2   Psychopathy and Controlling Violence: Dangerous Severe Personality Disorder

One of the risks of allowing the discussion of those with personality disorders and who are troublesome in society to drive academic and medical inquiry is that moral panic can direct the research and policy response. In the late 1990s, Michael Stone became the poster boy for new public protection policy in the United Kingdom. In 1996, he murdered a mother and one of her daughters in a brutal, random, and unprovoked attack as they were walking through a park (Beck, 2010). It was unquestionably an appalling act and one that merited careful analysis and a response. The British media revealed that Stone was a diagnosed psychopath, well known by mental health services and shortly before the attack, he was denied admission to a psychiatric hospital because his condition was not treatable (Beck, 2010).

The Labour government's Home Office published a policy document in 1999 that defined a new category of personality disorder along with the creation of new facilities designed to safely house and treat those with personality disorders who were thought to present significant risk to the public (Home Office, 1999). The Report says:

> The phrase dangerous severely personality disordered (DSPD) is used in this paper to describe people who have an identifiable personality disorder to a severe degree, who pose a high risk to other people because of serious anti-social behaviour resulting from their disorder. p12

In this early iteration, DSPD required a causal link between a severe personality disorder and that person presenting a high risk to others in society. Funding for new institutions and research followed this policy statement and the concept of DSPD was refined further. Following the creation of a new legislative mechanism to detain those with DSPD and three hundred new institutional places, the Department of Health's DSPD Programme retained this causal link and gave the following benchmark for the level of risk

An individual will be considered to meet the criteria for treatment in DSPD high secure services if he is assessed as being more likely than not to reoffend, resulting in serious physical or psychological harm from which the victim would find it difficult or impossible to recover. (Ministry of Justice, 2005) p2

This is a therefore a balance of probabilities test, if it is judged more likely than not that a person will reoffend and that will result in harm that is hard or impossible to remediate, then the first arm of the DSPD test is met. That is also a test that clearly and perhaps predictably, Michael Stone would have met.

They describe DSPD in such a way that the severe personality disorder could draw upon the DSM-IV personality disorders (Association, 1994) and psychopathy as measured by Hare's Psychopathy Checklist Revised (PCL-R) (Hare & Neumann, 2012).

For the purpose of DSPD assessments, the criteria for severe personality
disorder includes:

- a PCL-(R) score of 30 or above (or the PCL-SV equivalent); or
- a PCL-(R) score of 25-29 (or the PCL-SV equivalent) plus at least one DSM-IV personality disorder diagnosis other than anti-social personality disorder; or
- Two or more DSM-IV personality disorder diagnoses. (Ministry of Justice, 2005) p15

One of the reasons why it was important for the DSPD construct to contain standard diagnostic categories was to avoid the human rights implications of the state preventatively detaining some of its citizens when they had not yet committed a crime. If the state can argue that preventative detention is in fact also about treatment and that it is because of a medical condition, then it is not simply a case of detention in the absence of a crime. Chapter three of this book, (Snelling & McMillan, this volume), discusses how 'medical' criteria are incorporated into similar initiatives in New Zealand for the same reason and it is something that has been repeated in many jurisdictions.

The plans for managing DSPD were criticised robustly at the time. Appelbaum described it as a case of the state experimenting with psychiatry as a means of "public protection" (Appelbaum, 2005). Mullen described the policy as "glaringly wrong and unethical" (Mullen, 1999). The well-known and influential anti-psychiatrist Thomas Szasz described the DSPD proposals as demonstrating the "apotropaic" nature of mental illness (Szasz, 2003, p. 227). Szasz meant that phrases such as "dangerousness to self and others" and "psychiatric treatment" function as incantations "to ward off dangers we fear". If ever there was an instance where psychiatry was asked to function as a means of social control in the way that antipsychiatry has always claimed, it is the creation of DSPD and similar developments in other jurisdictions (Szasz, 1962). Although DSPD does seem a clear cut case where an illness was created for the purposes of public protection, the way which Szasz and other anti-psychiatrists took this to vindicate what they claimed about mental illnesses in general is problematic. Antipsychiatry can be defined as the view that all mental illness are not genuine disease entities but are instead fictions created for controlling those deemed socially undesirable, and it is clear that this is not the case for many other mental illnesses (McMillan, 2003).

While a therapeutic orientation was supposed to be one of the functions of the DSPD programme, there are reasons for doubting how routine treatment is within these facilities (Burns et al., 2011). When discussing the clinical, legal and ethical implications of psychopathy it is important to be cognisant of the rhetorical force that public protection has played in the framing of policy. The research that is funded reflects an agenda that is often implicit. The issues taken to be central occur within a social context and this needs to be acknowledged. That is not to say that this is a reason for questioning the value or integrity of interdisciplinary research on psychopathy. Moreover, it is important to frame policy that protects the public from people like Michael Stone and to do what can be done therapeutically. It does mean that we should be mindful of why issues become worthy of debate and discussion and consider alternative ways of viewing the same phenomena.

2.3 Psychopathy and Responsibility

The flurry of interest that followed the English DSPD policies stimulated debate about whether or not psychopaths should be held morally or legally responsible for their actions. The connection between that issue and preventive detention is that if they often should not be held morally responsible, then that might strengthen the argument in favour of new institutions that have a therapeutic as well as public protection role. That research question led to the interdisciplinary work that was published in *Psychopathy and Responsibility: interfacing Law, Psychiatry and Philosophy* (Malatesti & McMillan, 2010). The reason why it was necessary to 'interface' these three disciplines was because the moral understanding and motivation of psychopaths is in part an empirical issue and research on the neurological and psychological abilities of psychopaths is the foundation of an account of responsibility. Legal tests of responsibility, such as the McNaghten Rule, converge to a limited extent with philosophical accounts of responsibility and there is therefore an "interfacing" requirement between legal and philosophical conceptions of responsibility. While this work did explore the DSPD policy response, it was a reaction to the question of whether that particular policy response could be justified.

2.4 PCL-R

Psychopathy and Responsibility defended Hare's concept of psychopathy as operationalized in the PCL-R. Since that time, the debate has broadened out and a range of related issues about PCL-R have been explored. Some are sceptical about whether it is a robust scientific construct, in much the same way that Szasz claimed mental illnesses are not genuine illnesses. However, when this objection is considered more carefully PCL-R shares many of the features of other scientific concepts such as reliably picking out phenomena and the role of disvalue is common in other scientific concepts that we consider robust (Malatesti & McMillan, 2014).

There are rival contemporary accounts of psychopathy, so we should not consider Hare's PCL-R the only or final word on the matter (Cooke & Michie, 2001). There is mounting pressure against PCL-R, especially from biocognitive approaches that recommend including more biological data in the categorisation (Brazil et al., 2018) (Jurjako et al., 2020). But, there are also those who continue to defend PCL-R as individuating a robust scientific category (Reydon, this volume) and it is still the most commonly used account of psychopathy (Malatesti & McMillan, 2014).

PCL-R has stimulated research into the biological and neurological foundations of psychopathy. Brain imaging studies have characterized the neurological correlates of PCL-R and such findings have informed legal debates (Umbach et al., 2015). The renewed interest in psychopathy and criminality has generated a number of legal issues that merit discussion (Kiehl & Hoffman, 2011). For example, if PCL-R is a robust scientific kind and correlates with diminished responsibility how and should that be reflected in sentencing? (Gonzalez-Tapia et al., 2017).

PCL-R was incorporated into the DSPD construct and it tends to be most commonly used in forensic psychiatry or criminal justice. It has utility when identifying individuals who have a psychological condition that underpins their repeat offending. That is despite the fact that PCL-R psychopathy seems less common and less strongly correlated with offending than antisocial personal disorder (ASPD). There is some consensus that psychopathy as measured by PCL-R is a less frequent diagnosis than the ASPD of DSV IV. Hare and Neumann report that ASPD has at three times the prevalence of psychopathy as measured by PCL-R (Hare & Neumann, 2010a, b) p102. The National Institute for Clinical Excellence (NICE) guidelines on the treatment of ASPD state that only about 10% of those diagnosed with ASPD meet the diagnostic criteria for psychopathy as measured by PCL-R (National Collaborating Centre for Mental, 2010) p20.

The extent to which PCL-R should incorporate criminality has been keenly debated following a paper by Skeem and Cooke in 2010 (Skeem & Cooke, 2010a). They objected that the way in which PCL-R is used for criminal risk prediction means that psychopathy can be equated with PCL-R, which is a tool for assessing it. Hare and Neumann responded that although PCL-R does include antisocial behaviour elements, that these are not the same thing as criminality and that it is a mistake to conflate the two (Hare & Neumann, 2010a, b). Skeem and Cooke were unconvinced and argued that the relationship between antisocial behavioural traits and the criminal behaviours themselves is too tight for the former to function as a test of the latter (Skeem & Cooke, 2010b). In doing so they drew upon points made in an earlier paper by Blackburn about the conceptual structure of the PCL-R (Blackburn, 2007).

Some research focusses upon a more clearly therapeutic approach to psychopathy and the prospect for helping those so diagnosed (Wilson & Tam-Ratea, 2013) (Tamatea Chapter two of this volume) (Ogloff & Wood, 2010) (Polaschek & Daly, 2013). Nonetheless it is striking that the primary usage of PCL psychopathy has drifted away from how and why it was articulated by Cleckley in the *Mask of Sanity* (Cleckley, 1988). As is often discussed, Hare constructed PCL-R as a diagnostic tool that operationalized Cleckley's account of psychopathy, so it is worth revisiting the way he describes psychopathy and why he seemed to be interested in it.

2.5   Understanding Psychopathy and the Mask of Sanity

Hare grounded PCL-R in the dysfunctions that Cleckley observed in the patients he saw and tried to help during his career as a psychiatrist. Given the importance of PCL-R, understanding Cleckley's psychopaths matters because the psychological traits he identified fed directly into the items of the PCL-R.

Another and more salient reason why we should continue to care about Cleckley's psychopaths is because he did. Not in the simplistic sense that he was a sympathetic and caring physician, what is striking about *The Mask of Sanity* is the richness, insight, and individuality of the descriptions. Oliver Sacks and Alexander Luria continue to be relevant and are influential partly due to their precise clinical observation of neurological conditions (Sacks, 1985; Luria, 1987). Hervey Cleckley applied a similar degree of rigour to his descriptions of psychopathy: he has a natural historian's attention to detail, but it is tempered by an implicit empathy for what the condition means for that person.

While ASPD and psychopathy are usually understood as related but distinct constructs, with some degree of overlap, Cleckley doesn't seem to think much hangs on the term used. He mentions antisocial personality, psychopathy and sociopathy and appears happy to use them interchangeably (Cleckley, 1988) p11. That doesn't mean that we should read him and PCL-R as articulating a construct that captures the contemporary senses of ASPD or sociopathy: it's clear that ASDP is a distinct construct while sociopathy seems to have a much more fluid meaning. That's another reason why grounding an understanding of PCL-R in Cleckley is important: Hare operationalized the construct so it could be assessed, Cleckley gives a rich account of what it means to be a psychopath, at least in the terms of the group of patients he met throughout his career. So, *The Mask of Sanity* is an indispensable guide to the meaning, presentation and perhaps even phenomenology of psychopathy.

The way in which Cleckley's presents his list of dysfunctions suggests an inductive approach in that he describes fifteen patients who have what he calls "full clinical manifestations": the category that is often called "primary psychopaths". *The Mask of Sanity* is known in popular culture because of its description of those who Cleckley described as incompletely manifesting or suggesting psychopathy: his so-called secondary psychopaths. These are the psychopaths who are said to be hiding in plain sight and often live out successful lives where their degree of pathology need not lead to them becoming institutionalized or imprisoned. Cleckley's descriptions of secondary psychopaths are interesting in that he presents them as possible people who might be occupying a particular role in society. For example, the subheadings for these sections are "The psychopath as businessman" p. 193 and "The psychopath as scientist". It may or may not be the case that he had a specific person in mind, but if he did perhaps this is a way of describing them that doesn't identify those people. The patients he describes, his examples of primary psychopaths, appear to be actual patients he either treated or knew and these sections are grouped under the first name of that patient. The patient or primary psychopath accounts are the most fine grained in their description and shed the most light on the nature of psychopathy and the way in which it fractures a life and affects those around that person.

Cleckley revised *The Mask of Sanity* four times after it was first published in 1941. In the preface, he notes that the construct is more fully described by the final edition and explains why this was necessary.

> It is not easy to convey this concept, that of a biologic organism outwardly intact, showing excellent peripheral function, but centrally deficient or disabled in such a way that abilities, excellent at the only levels where we can formally test them, cannot be utilized consistently for sane purposes or prevented from regularly working toward self-destructive and other seriously pathologic results. Vii

So, for Cleckley, part of the clinical and diagnostic difficulty with psychopathy is that patients present without any obvious deficiency of perception, cognition or affect. It is not until the more subtle underlying dysfunctions are observed that the destructive impact upon that person and those around them can be understood.

The book is structured so that the accounts of "primary" and "secondary" psychopaths appear first and are like observations or phenomena in need of explanation. He then derives 16 features that he thinks cluster and characterize psychopathy. Given that the book was revised a number of times and Cleckley mentions in the preface that over the years he encountered more patients and there were changes in the cohort he observed, including more female psychopaths, we shouldn't read this as clearly inductive. While Cleckley will have derived the following 16 dysfunctions from his clinical experience, the cases presented in the book are likely to have been selected to illustrate a range of presentations. His sixteen defining traits are as follows:

1. Superficial charm and good "intelligence"
2. Absence of delusions and other signs of irrational thinking
3. Absence of "nervousness" or psychoneurotic manifestations
4. Unreliability
5. Untruthfulness and insincerity
6. Lack of remorse or shame
7. Inadequately motivated antisocial behavior
8. Poor judgment and failure to learn by experience
9. Pathologic egocentricity and incapacity for love
10. General poverty in major affective reactions
11. Specific loss of insight
12. Unresponsiveness in general interpersonal relations
13. Fantastic and uninviting behavior with drink and sometimes without
14. Suicide rarely carried out
15. Sex life impersonal, trivial and poorly integrated
16. Failure to follow any life plan (Cleckley, 1988) p338

As noted, Hare credits Cleckley as illuminating psychopathy and says that the PCL-R is a way of measuring and classifying the underlying pathologies Cleckley observed. As was explained in the previous section, there is a dispute about the extent to which PCL-R is a test of Cleckley's construct or in fact a new construct that is slanted more toward antisocial behaviour and criminality. In any case, PCL-R is a tool that scores on the basis of 20 emotional and behavioural dysfunctions. Hare organized them into factors, which map onto the more narcissistic elements of psychopathy (Factor one) and the more antisocial elements (Factor two).

Interpersonal	Affective
1. Glibness/superficial charm	6. Lack of remorse
2. Grandiose self-worth	7. Shallow affect
4. Pathological lying	8. Lack of empathy
5. Conning/manipulative	16. Will not accept responsibility

Lifestyle	Antisocial
3. Need for stimulation	10. Poor behavioural controls
9. Parasitic lifestyle	12. Early behavioural problems
13.Lack of goals	18. Juvenile delinquency
14. Impulsivity	19. Revocation conditional release
15. Irresponsibility	20. Criminal versatility

Note that the item titles cannot be scored without reference to the formal criteria contained in the PCL-R Manual. Item 11, Promiscuous sexual behaviour, and Item 17, Many short-term marital relationships, contribute to the Total PCL-R score but do not load on any factors. (Hare & Neumann, 2012).

PCL-R does not score for all of the defining features that Cleckley observed. That is partly because Cleckley includes some negative features and PCL-R is a checklist for the features present. It is true that psychopathy tends not to be characterised by delusions or neurotic conditions, but that does not need to be operationalised in a checklist for psychopathy. We know that many psychopaths do have deficits in practical reason, and that's the case for many of the people Cleckley describes, so when he says "absence of irrational thinking" we might presume he has in mind, highly distorted thinking as we might expect if someone is experiencing a full blown mania or psychosis. He muddies the water somewhat later in the book when he elaborates on the lack of irrationality of psychopaths.

The psychopath, on the other hand, is free of all technical signs of this sort. There are no demonstrable defects in theoretical reasoning. At least he is free of them in the same sense that the general run of men and women are free. He carries out his activities in what is regarded as ordinary awareness of the consequences and without the distorting influences of any demonstrable system of delusions. His personality outline is apparently or superficially intact and not obviously distorted. (p247)

The claim that psychopaths have an unimpaired awareness of the consequences of their actions is one that we might qualify now. The means/ends reasoning of some psychopaths is impaired to a degree, particularly those who score highly on Hare's Factor 2 elements. In saying that, it is also clear that Cleckley's psychopaths do make very different choices from most people and perhaps his point is that there are other dysfunctions that drive this rather than a profound disruption of means/ends reasoning.

Some of the literature on psychopathy (Justman, 2020) and popular discussions of it have focussed upon the absence of guilt. That has resulted in a tendency to view it as primarily a condition whereby psychopaths are immune from the internal sanctions of morality and ruthlessly exploit that to their advantage, causing mayhem in the process.

It is important to consider the actual concepts used in Cleckley and how they combine to describe an agency that is damaged in a deep and pervasive sense. Cleckley tends to use the terms lack of "remorse" or "shame" and while these are similar to "guilt" there are some important differences. What is striking about the patients he describes is not so much they often don't experience guilt at moral wrongdoing, it's a more global phenomena where they don't react to events that would cause a profound reaction in most people. A good example of this is the way that they often do not feel shame at behaviour that is outrageous and would make the rest of us crimson with embarrassment. For example, Chester who was allowed to return home because of a death in the family.

> At the funeral the next day, having skilfully obtained more whiskey, Chester appeared at first poised and sober. Signs of drunkenness, however, soon appeared, and the patient proceeded to make a shocking and memorable impression. Vomiting and defecating while in the church contributed to this effect... The father sent him back to the hospital as soon as arrangements could be made. Although the patient could not have been unaware that his father has informed the hospital of his conduct... he said with the greatest assurance on arriving that he had voluntarily come back... He insisted that his conduct had been above reproach and demanded parole privileges. (p130)

A case *could* be made that Chester acted immorally and felt no moral guilt at what he did, but this is much better described as astonishingly and profoundly, shameless. That specific dysfunction is a feature of all of the primary psychopaths Cleckley describes. When explaining what he means by a lack of remorse or shame Cleckley says

> '...perhaps the true dignity of man is his ability to despise himself,' the psychopath is without a means to acquire true dignity. (p343)

So, this isn't merely the ability to feel the pull of moral disapprobation, it is a more general inability to regulate behaviour in the light of many norms. It fits with other items on the list such as 5, 12 and 15 where we rely upon what David Hume called "sympathy", or the ability to grasp or be moved by the feelings or beliefs of other people (Hume, 1739). We could also borrow Christine Korsgaard's terminology and describe these as deficiencies in the normativity required for constructing a stable and coherent sense of self (Korsgaard, 2009).

Several of Cleckley's patients are very capable and intelligent, yet incapable of using their talents to construct a coherent purpose or value system for their life. One of the more widely discussed cases is his patient Anna who despite performing to a very high level for short periods of time in education and other areas of life, would inevitably fall back on destructive and nihilistic behaviours. That was also a feature of her history of marriages and casual sexual encounters, all of which failed to have any deep emotional impact upon her. A good illustration of this is her engagement with literature and music.

> This patient spent a good deal of time reading. In contrast to many psychopaths who readily claim all sorts of entirely imaginary learning, she showed considerable familiarity with literature of many sorts. She seemed to read Shakespearean plays, the major Russian novels, pulp magazines, and comic books with about the same degree of interest. Her factual knowledge about what she had read seemed good, though it must be admitted she often falsified with assurance when questions led her into unknown areas.

> She played complicated music on the piano with fine technical skill and spent a good deal of time doing so. She had an accurate acquaintance with current scales of intellectual and aesthetic fashion and could probably have avoided offense even to the most snide of editors of the most avant-garde of little magazines. How she reacted to such matters in the innermost and final chamber of her being can only be surmised. My impression is that King Lear and Amazing Confessions elicited responses in no fundamental way different. (pp. 119–120)

The all-pervasive feature of Anna's psychopathy is not that she does not feel moral guilt nor is she aggressive or violent, instead she just seems to not be moved by any aesthetic, ethical, prudential or reputational considerations. In Harry Frankfurt's terms, she seems to completely lack any second order volitions about the kinds of considerations she wishes to act upon (Frankfurt, 1988). She is similar to Damasio's patient Elliot, who following a localised stroke in his prefrontal context, struggled to complete or care about tasks that he could routinely perform before (Damasio, 1994). That was even though on standard cognitive tests, he was not impaired. Elliot and Anna share Cleckley's sixteenth dysfunction, "Failure to follow any life plan."

The picture that emerges from Cleckley's psychopaths is of a group of people who potentially could be violent in a cold and guilt-free way, but who seemed unlikely to care enough about an issue to initiate such violence. Of course, we know that there are psychopaths such as Michael Stone, but many of those who present to Cleckley simply just don't seem to care enough about anything to become easily enraged. Dysfunctions 10–12 help explain why that could be the case. "General poverty in major affective reactions" is not just alerting us to an inability to feel or be moved by moral sentiments, this is a more global indifference to all major affective reactions, including rage. "Specific loss of insight" and "Unresponsiveness in general interpersonal relations" would seem to be closely related to why psychopaths just seem unmoved by events such as Chester's attendance at a funeral.

2.6   Conclusions

Returning to *The Mask of Sanity* might not lead us to reappraising psychopathy in the sense of discontinuing the legal, neurological and philosophical projects currently underway. Whether or not psychopaths know enough about the nature of their actions or are in sufficient control of them to counts as fully responsible agents continues to be an important policy question. That is also true of the need for societies to find ways of identifying those who might commit grave offences against other persons. But perhaps it gives us reason to pause and reflect upon what we want a concept of psychopathy for and whether that agenda is leading us to view this phenomenon in a way that misses its nuance and significance for those who have this affliction and those around them.

References

Appelbaum, P. S. (2005). Law & Psychiatry: Dangerous severe personality disorders: England's experiment in using psychiatry for public protection. *Psychiatric Services, 56*(4), 397–399.

Association, American Psychiatric. (1994). *Diagnostic and statistical manual of mental disorders* (4th ed.). American Psychiatric Publishing.

Beck, J. C. (2010). Dangerous severe personality disorder: The controversy continues. *Behavioral Sciences & the Law, 28*(2), 277–288.

Blackburn, R. (2007). Personality disorder and antisocial deviance: Comments on the debate on the structure of the psychopathy checklist-revised. *Journal of Personality Disorders, 21*(2), 142–159.

Brazil, I. A., et al. (2018). Classification and treatment of antisocial individuals: From behavior to biocognition. *Neuroscience & Biobehavioral Reviews, 91*, 259–277.

Burns, T., et al. (2011). Treatments for dangerous severe personality disorder (DSPD). *The Journal of Forensic Psychiatry & Psychology, 22*(3), 411–426.

Cleckley, H. (1988). *The mask of sanity: An attempt to reinterpret the so-called psychopathic personality* (5th ed.). Emily S Cleckley.

Cooke, D. J., & Michie, C. (2001). Refining the construct of psychopathy: Towards a hierarchical model. *Psychological Assessment, 13*, 171–188.

Damasio, A. R. (1994). *Descartes' error: Emotion, reason, and the human brain*. Putnam.

Frankfurt, H. G. (1988). Freedom of the will and the concept of a person. In M. F. Goodman (Ed.), *What is a person?* (pp. 127–144). Humana Press.

Gonzalez-Tapia, M., Obsuth, I., & Heeds, R. (2017). A new legal treatment for psychopaths? Perplexities for legal thinkers. *International Journal of Law and Psychiatry, 54*, 46–60.

Hare, R., & Neumann, C. (2010a). Psychopathy: Assessment and forensic implications. In L. Malatesti & J. McMillan (Eds.), *Responsibility and psychopathy: Interfacing law, psychiatry and philosophy* (pp. 93–123). Oxford Univrsity Press.

Hare, R., & Neumann, C. (2010b). The role of antisociality in the psychopathy construct: Comment on Skeem and Cooke (2010). *Psychological Assessment, 22*(2), 446–454.

Hare, R., & Neumann, C. (2012). Psychopathy and its measurement. In P. Corr & G. Matthews (Eds.), *The Cambridge handbook of personality psychology* (pp. 660–686). Cambridge University Press.

Home Office. (1999). *Managing Dangerous People with Severe Personality Disorder: Proposals for Development*. The Stationery Office.

Hume, David (1739), A treatise concerning human nature. (First published anonymously).

Jurjako, M., Malatesti, L., & Brazil, I. A. (2020). Biocognitive classification of antisocial individuals without explanatory reductionism. *Perspectives on Psychological Science, 15*(4), 957–972.

Justman, Stewart (2020), 'The guilt free psychopath', Philosophy, psychiatry and psychology. 28 (2):87-104

Kiehl, K., & Hoffman, M. (2011). The Criminal Psychopath: History, Neuroscience, Treatment and Economics. *Jurimetrics, 51*, 355–397.

Korsgaard, C. M. (2009). *Self-constitution: Agency, identity, and integrity*. Oxford University Press.

Luria, A. (1987). *The Man with a Shattered World: The History of a Brain Wound*. Harvard University Press.

Malatesti, L., & McMillan, J. (Eds.). (2010). *Responsibility and psychopathy: Interfacing law, psychiatry and philosophy*. Oxford University Press.

Malatesti, L., & McMillan, J. (2014). Defending psychopathy: An argument from values and moral responsibility. *Theoretical Medicine and Bioethics, 35*(1), 7–16.

McMillan, J. R. (2003). Dangerousness, mental disorder, and responsibility. *Journal of Medical Ethics, 29*(4), 232–235.

Ministry of Justice. (2005). *Dangerous and Severe Personality Disorder (DSPD) high secure services for men: Planning and delivery guide*. Ministry of Justice.

Mullen, P. E. (1999). Dangerous people with severe personality disorder. *British Proposals for Managing them are Glaringly Wrong—and Unethical, 319*(7218), 1146–1147.

National Collaborating Centre for Mental Health. (2010). Antisocial personality disorder: The NICE guideline on treatment, management and prevention. In *Antisocial personality disorder: Treatment, management and prevention* (Leicester (UK): British Psychological Society. Copyright © 2010, The British Psychological Society & The Royal College of Psychiatrists.).

Ogloff, J., & Wood, M. (2010). The treatment of psychopathy: Clinical nihilism or steps in the right direction? In L. Malatesti & J. McMillan (Eds.), *Responsibility and psychopathy: Interfacing law, psychiatry and philosophy* (pp. 155–184). Oxford University Press.

Polaschek, D., & Daly, T. (2013). Treatment and psychopathy in forensic settings. *Aggression and Violent Behaviour, 18*, 592–603.

Reydon, T. (this volume). Psychopathy as a scientific kind: On usefulness and underpinnings. In J. M. M. L. Malatesti & P. Šusta (Eds.), *Psychopathy. Its uses, validity, and status*. Springer.

Sacks, O. (1985). *The man who mistook his wife for a hat, and other clinical tales*. Summit Books.

Skeem, J. L., & Cooke, D. J. (2010a). Is criminal behavior a central component of psychopathy? Conceptual directions for resolving the debate. *Psychological Assessment, 22*(2), 433–445.

Skeem, J. L., & Cooke, D. J. (2010b). One measure does not a construct make: Directions toward reinvigorating psychopathy research–reply to Hare and Neumann (2010). *Psychological Assessment, 22*(2), 455–459.

Snelling, J., & McMillan, J. (this volume). Antisocial personality disorders and public protection orders in New Zealand. In L. Malatesti, J. McMillan, & P. Šustar (Eds.), *Psychopathy. Its uses, validity, and status*. Springer.

Szasz, T. (1962). *The myth of mental illness*. Secker and Wartung.

Szasz, T. (2003). Psychiatry and the control of dangerousness: On the apotropaic function of the term "mental illness". *Journal of Medical Ethics, 29*(4), 227–230.

Umbach, R., Berryessa, C., & Raine, A. (2015). Brain imaging research on psychopathy: Implications for punishment, prediction, and treatment in youth and adults. *Journal of Criminal Justice, 43*, 295–306.

Wilson, N. J., & Tam-Ratea, A. (2013). Challenging the 'urban myth' of psychopathy untreatability: The high-risk personality programme. *Psychology, Crime & Law, 19*(5–6), 493–510.

Chapter 3
Humanising Psychopathy, or What It Means to Be Diagnosed as a Psychopath: Stigma, Disempowerment, and Scientifically-Sanctioned Alienation

Armon J. Tamatea

Abstract Psychopathy is a personality disorder that has been described in various ways over the last two centuries but is popularly characterized in modern times by a collection of traits including interpersonal-affective features (e.g., lack of empathy, lack of remorse, superficial charm) and antisocial behaviour (e.g., interpersonal violence). Since an overwhelming proportion of the research surrounding psychopathy has focused on criminal justice and forensic populations, the label 'psychopath' widely elicits associations with criminals who commit serious violence with minimal chance of rehabilitation. The fear of 'psychopaths' remains highly present in the general population which perpetuates stigmatization. Yet, little is known about the impact of this stigma on those so-labelled as psychopathic. This chapter sketches an outline of clinical and research issues and argues that psychopathy engenders specific forms of stigma that is a consequence of a research tradition that has inadvertently marginalised an already-marginalised group with implications for research, assessment and clinical practice, as well as service delivery to this group. Furthermore, it is suggested that conscientiousness on the part of researchers and clinicians to reframe psychopathy as a health issue (rather than merely a criminal one), challenge scientific stereotypes, and develop inclusive research relationships with those from the psychopathic community will open up new ethical and conceptual spaces in knowledge development and a deeper understanding of this most challenging of populations.

Keywords Psychopathy · Stigma · Diagnosis · Assessment · Etiology · Treatment · Rehabilitation · Ethics

A. J. Tamatea (✉)
University of Waikato, Hamilton, New Zealand
e-mail: armon.tamatea@waikato.ac.nz

© Springer Nature Switzerland AG 2022
L. Malatesti et al. (eds.), *Psychopathy*, History, Philosophy and Theory of the Life Sciences 27, https://doi.org/10.1007/978-3-030-82454-9_3

'For most of those labelled as psychopathic it is a real stroke of bad luck. More often than not... the label has the effect of denying the patient psychiatric services, on the grounds that everyone knows psychopaths can't be successfully treated.' (Michael Cavadino, *Death to the psychopath*, 1998, p.7)

'Part of tolerance is believing they exist. It is one thing to discriminate against people. It is far worse to refuse to believe they exist.'' (Shepherd Mead, *The carefully considered rape of the world*, 1965, p. 7)

Psychopathy researchers concern themselves, among other things, with the development of knowledge and praxis about individuals identified as 'psychopathic'; the etiological pathways, prognosis, treatability, and interface between these people and the law. Psychopathy is characterised by an abnormal lack of empathy combined with antisocial conduct that is masked by an ability to appear outwardly 'normal' (Cleckley, 1976; Hare, 2003). Persons who display psychopathic traits are typically perceived by researchers, criminal justice professionals, and the public as ordinarily dangerous and fundamentally different from the rest of 'us'. For instance, psychopathy has been described in the psychological and psychiatric literature as (1) a deviant developmental disturbance reflecting inordinate instinctual aggression and *"the absence of an object relational capacity to bond...a fundamental disidentification with humanity"* (Meloy, 1988, p.5); (2) a *"socially devastating disorder"* comprised of affective, interpersonal, and behavioral characteristics (Hare, 1998, p.188); and, (3) an emotional disorder that puts the person at risk of *"repeated displays of extreme antisocial behavior"* (Blair et al., 2005, p.17).

The academic and clinical interest in psychopathy has long been associated with criminality, in particular its status as one of the best predictors of future antisociality (Fernandez-Suarez et al., 2018; Fox & DeLisi, 2019; Monahan et al., 2001; Salekin et al., 1996). Offenders with identified psychopathic traits have been reported to reoffend more frequently, more seriously, and demonstrate a wider range of offences than other offenders (Blackburn & Coid, 1998; Porter & Woodworth, 2006), including interpersonal violence (Asscher et al., 2011; Blais et al., 2014; Porter et al., 2001) and sexual coercion against women (Hanson & Morton-Bourgon, 2005; Hawes et al., 2013; Knight & Guay, 2006). Furthermore, the sequelae of these behaviours extend beyond the individual victim and include costs related to imprisonment, lost property, police, courts, prosecutors, public defenders, and jurors (Kiehl & Hoffman, 2011). Needless to say, psychopathic individuals incur disproportionate costs on communities, and as such, do not elicit much by way of public sympathy. Despite these somewhat ominous descriptions, the behaviour of individuals described as 'psychopathic' is complex and invites a deeper examination of what this label means. This chapter canvasses a number of issues as they pertain to research and clinical practice with psychopathic populations, in particular the meaning of psychopathy from the perspective of those who have the power to administer the label, as well as the impact (actual and potential) for those so-labelled.

3.1 What It Means to Be a 'Psychopath' I: From the Top-Down

Psychopathy has become a central concern for criminal justice agencies to identify and manage the behaviour of individuals who have the history and potential to threaten the security and wellbeing of others. The need for communities to monitor and regulate the behaviour of psychopathy relies on (1) an adequate definition of psychopathy, (2) accurate assessment approaches, (3) knowledge of etiological pathways, and (4) effective intervention design. These concerns reflect a biomedical view of persons, behaviour, and dangerousness. As we discuss in this section, the biomedical perspective is a long-standing research and clinical tradition that has shaped the prevailing paradigm of how psychopathy is perceived and managed.

3.1.1 A (Very) Brief Conceptual History of Psychopathy

Psychopathy has been described in various ways over the last two Centuries. The nineteenth century saw increased recognition of troublesome behaviour patterns and personality traits that would inform what we would now consider to be psychopathy. Despite a conceptual unevenness in the early literature, historical accounts from this era largely framed psychopathy as a failure of character (Haack et al., 2010) or an illness of the will – in short, a health issue with moral overtones (see Table 3.1).

Conceptual tensions between psychopathy and chronic antisociality emerged in the 1900s–1940s (see Table 3.2). Historical overviews of this era (e.g., Karpman, 1929, 1948; Maughs, 1941; Partridge, 1930) noted that psychopathy was an over-inclusive construct largely indistinct from long-term criminality. Typologies also emerged during this period in an attempt to distinguish psychopathy from other populations marked by persistent antisociality (Schneider, 1923) as did a view,

Table 3.1 Nineteenth century descriptions, hypotheses, and core ideas about psychopathy

A psychopathic person is/has/reflects…	According to…
'Manie sans delire' ('mania without delirium')	Pinel (1806)
'Moral depravity'	Rush (1812)
'Monomania'	Esquirol (1827; see Trichet & Lacroix, 2016)
'Moral insanity'	Prichard (1837)
'An unnatural monster'	Darwin (2009)
'Psychopathic personalities'	Kræpelin (1887/1915)
'Psychopathic inferiorities'	Koch (1891; see Millon et al., 1998)
'Moral imbecility'	Maudsley (1898)
'Given to excesses'	Krafft-Ebing (1899)

Table 3.2 Early twentieth century descriptions, hypotheses, and core ideas about psychopathy

A psychopathic person is/has/reflects…	According to…
'Sociopathic'	Birnbaum (1909)
'Psychopathic personalities'	Schneider (1923)
'Alloplasticity true criminality'	Alexander & Staub (1931)
Semantic aphasia	Cleckley (1941)
Deficient role-playing ability	Gough (1948)

Table 3.3 Mid-twentieth century descriptions, hypotheses, and core ideas about psychopathy

A psychopathic person is/has/reflects…	According to…
Disrupted attachment	Bowlby (1953)
A conditioning deficiency	Eysenck (1957, 1967)
Low fearfulness	Lykken (1957)
'Sociopathic reactions'	Thorne (1959)
A non-viable construct based on circular logic	Wootton (1959)
Insane and therefore ineligible for punishment	Haksar (1965)
Stimulation-seeking	Quay (1965, 1977)
A loss of inhibitory mechanisms	Hare (1970)
Pathological narcissism	Kernberg (1970, 1974)
A 'manipulative personality'	Bursten (1972)
'Morally dead' – No rights as persons	Murphy (1972)
Of two types (i.e., Primary/secondary psychopathy)	Blackburn (1975)
Poor ego development	Draughon (1977)

albeit not one that was widely adopted, that society itself was a significant contributor that enabled conditions for persistent antisocial behaviour to occur (Birnbaum, 1909).

Since the mid-twentieth century, an overwhelming proportion of psychopathy research has been rooted in the criminal justice and forensic fields, with typical samples being comprised of prisoners and convicted offenders. Indeed, the label 'psychopath' was widely associated with criminals and murderers with minimal chance of rehabilitation (Camp et al., 2013; Kiehl & Hoffman, 2011). Debates about the validity of psychopathy as a construct emphasised responsibility and agency as core issues (e.g., Haksar, 1965; Wootton, 1959), even questioning whether so-called psychopaths were valid members of the moral community (Bursten, 1972; Murphy, 1972). Experimental and developmental research emerged during this period as a major tradition for the identification of psychopathy as well as exploring individual differences (see Table 3.3).

In the latter half of the twentieth century, personality research examined psychopathic traits in a functional (or rather, dysfunctional) way with an increased emphasis on measurement precision and methodological rigour. The Psychopathy Checklist, and later the Psychopathy Checklist-Revised (PCL-R; Hare, 1991)

Table 3.4 Late-twentieth century descriptions, hypotheses, and core ideas about psychopathy

A psychopathic person is/has/reflects...	According to...
Sadistic personality pattern	Millon (1969, 1987)
Grandiose self-structure	Meloy (1988)
Competitive needs in a hostile, dog-eat-dog world	Beck & Freeman (1990)
A socially devastating constellation of interpersonal, affective, lifestyle factors and antisocial behaviour	Hare (1991, 2003)
A philosophy of life centred around trivialising others	Levenson (1992)
Deficient somatic markers	Damasio (1994)
Information-processing deficiency	Newman (1998)
Social response modulation difficulties	Lynam (1997)
Free-rider (evolutionary adaptation to exploit others)	Cavadino (1998)
Callous-unemotional traits in youth	Frick (1998)

emerged as the most credible and widely-used measure of psychopathy that refined the notion of psychopathy from thematically similar clinical descriptions to a researchable construct that opened the way for comparative investigations. Hare (1998) conceptualised psychopathy as a broad spectrum personality prototype that was characterised by a configuration of affective, interpersonal, and behavioral characteristics, or more pointedly as "intraspecies predators" (p.196) who use charm, manipulation, intimidation, and violence to control others and gratify their own egocentric needs.[1] Although this measure, and the wave of empirical research that was to follow, would expand the horizons of what would be known about psychopathy, the eventual use of the tool in violence risk assessment strongly reinforced the psychopathy-crime perception and the criteria by which psychopathy would be more or less defined (see Table 3.4).

The conceptualisation of psychopathy as a personality disorder that is characterized by distinctive interpersonal-affective traits (e.g., lack of empathy, callousness, superficial charm) and antisocial-lifestyle features (e.g., impulsivity, criminality) (Hare, 2003) has persisted into current times (see Table 3.5). The emergence of built-for-purpose self-report measures expanded research possibilities by decoupling the construct of psychopathy from the PCL-R items.

Despite the range of descriptions, points of emphasis, and theoretical developments, three observations can be made here: descriptions of psychopathy have been (1) diagnostic in nature, (2) largely devoid of cultural context, and (3) devised by researchers and practitioners who likely do not possess psychopathic traits. Firstly, diagnostic labels serve clinical purposes such as demarcating critical pathological differences between a target group and others. However, distress is an important element of illness – physical or psychological – yet this notion is largely absent from psychometrics and rating scales that pertain to psychopathy. Secondly, while diagnosis offers advantages of facilitating decision-making (e.g., treatment) and

---

[1] Some philosophers have had a sympathetic view, albeit with a less delicate approach to terminology (viz. "son-of-a-bitch"; Slovenko, 1999; "bastard"; Cavadino, 1998)

Table 3.5 Twenty-first century descriptions, hypotheses, and core ideas about psychopathy

A psychopathic person is/has/reflects…	According to…
Evolutionary adaptation: 'Cheaters' and 'warrior hawks'	Book and Quinsey (2004)
Pathology across multiple domains	Cooke et al. (2004)
Neurological impairment: Amygdala	Blair et al. (2005)
Social-information processing	Serin and Brown (2005), Porter (1996)
Self-centred impulsivity, coldheartedness, and fearless dominance	Lilienfeld and Fowler (2006)
Diminished capacity for rational self-governance	Litton (2008)
Disinhibition, boldness, and meanness	Patrick et al. (2009)
An adaptation to a disadvantageous environment	Gillett and Huang (2013)

standardising concepts across settings and over time, these benefits also serve agencies and researchers in the first instance, and not the person themselves. Personality and personality disorders are reflections of the degree of fit between a person and contexts over time, which means that context is an important element to make sense of behaviour and its success or failure to adapt or modify to the environment. What is determined to be appropriate or inappropriate behaviour is largely a matter of cultural context. For instance, a psychopathic profile that is characterised by an abundance of grandiosity, superficial charm and unrealistic long-term planning may present an aversive attention-seeking interaction style in one setting, but may also be essential to acquire social capital in another (e.g., an engaging cocktail party guest or academic networker!). Even a willingness to engage in violence may be seen as a desirable attribute in a partner if a community values macho attributes in spouses, parents, and other social roles.[2] Thirdly, the behaviour of psychopathic persons is often bewildering and self-defeating. Their oft-cited poor treatment response in offender rehabilitation, speaks to the black box problems of psychopathy and the vast body of research that has been dedicated to unpacking and understanding the mechanisms that govern their behaviour. However, psychopathic persons rarely, if ever, contribute to core research areas like diagnosis, assessment, and treatment. This is odd given how much other psychological disorders and mental wellness issues like depression, anxiety, and even borderline personality disorder are informed by those who have lived experience of these conditions (e.g., Zolnierek, 2011). In some respects, this is understandable given that researchers and clinicians learn – either formally and/or by hard experience in the field – to not trust psychopathic persons, especially if deceit, irresponsibility, and non-reciprocity in relationships (not to speak of violence, callousness, and crime) are common attributes.

---

[2] Just so we are clear, this author does not endorse interpersonal violence.

3.1.2 Diagnosis and Assessment

The aim of a conceptual definition of a given pathology is to adequately *describe* a phenotypic situation. In this regard, specificity and representativeness of core features are vital in order to inform the assessment and treatment of appropriate and unambiguous diagnosis, which means that prevalence, incidence, course, prognosis, and risk factors for vulnerability can be determined with greater precision. Definition is often informed by conceptual emphasis (e.g., psychodynamic, learning, genetic). Despite the vast body of literature on psychopathy (Skeem et al., 2012), the prevailing discourse has been framed in the criminal justice and forensic domain. Grisolia (2001) noted that most large-scale studies of psychopathy are based on behaviours, such as criminal arrests and childhood aggression, with rare reference to the specific diagnosis of the individual participants. This over-emphasis on criminal behaviours and social deviance as indices of psychopathy risks conceptual confusion especially given that psychopathic individuals are a minority in prisons – even amongst violent offenders – yet the commission of crimes is an inclusion criteria for both antisocial personality disorder (American Psychiatric Association, 2013) and psychopathy (Hare, 2003).

The most well-known description we have of a psychopathic prototype is Hare's (2003) widely accepted – but by no means unanimous – criteria that includes traits such as superficial charm, grandiosity, irresponsibility, deceitfulness, lack of empathy and antisocial behaviour across the lifespan. Such definitional specificity is important because it assists with reducing clinical 'noise' that can result from overly-inclusive or opaque diagnostic criteria. Once a psychopathic phenotype is established, identifying the salient contributions from genes and the environment (and their interaction) becomes a less complex task and could assist in improving approaches to early detection of vulnerability to developing psychopathic traits, and even offsetting the risk of future harm.

Clinical assessment of a disorder informs an ability to *detect*, within acceptable limits of accuracy, the presence and extent of a psychopathic phenotype. Contemporary approaches to the assessment of psychopathy have largely been developed by psychologists for use in clinical and forensic settings and include psychometric measures that are based on (1) clinical rating, (2) self-report, or (3) use of informants. Each of these approaches is designed to inform diagnosis (formal or provisional), recidivism risk, or responsiveness to treatment.

Perhaps the most widely-researched exemplar of a *clinical rating* scale for psychopathy assessment is the Psychopathy Checklist–Revised (PCL-R; Hare, 2003) and its derivations (i.e., the PCL: Youth Version (PCL-YV); Forth et al., 2003, and PCL: Screening Version (PCL-SV); Hart et al., 1995). These instruments require raters to integrate information from interviews and collateral sources to make inferences about specific traits. Factor analyses of PCL measures have yielded a model for understanding the disorder in the form of two prominent factors that reflect interpersonal (e.g., manipulative) and affective (e.g., callousness) components of psychopathy (Factor 1), while Factor 2 captures antisocial lifestyle (e.g., early antisocial behavior) and behavioral (e.g., need for stimulation) components (Hare, 2003; Harpur et al., 1988). Alternative factor structures for psychopathy have also been found such as a three-factor solution comprised of an arrogant and deceitful interpersonal style, deficient affective experience, and impulsive and irresponsible behavioral style (Cooke & Michie, 2001) that has offered an ostensibly better fit for some psychopathic populations (e.g., sex offenders) than the two-factor model.

Self-report scales, such as the Psychopathic Personality Inventory-Revised (PPI-R; Lilienfeld & Widows, 2005) have been designed in recognition of the shortcomings of clinical rating measures and omnibus personality inventories as well as diagnostic and research utility beyond criminal justice and forensic settings, especially where collateral or file information is non-existent or difficult to access. They have the added advantage of offering a more cost-effective and standardised approach to assessing psychopathy and can be administered to both criminal and non-criminal populations (Lilienfeld, 1994) and, due to the low-inference nature of these measures, can be completed without observer bias or subjectivity in scoring (Lilienfeld & Fowler, 2006). However, it is the *participant's* subjectivity that is the main limitation of self-report scales, especially where dishonest responding is a high likelihood, but not necessarily in the service of social desirability (Ray et al., 2013).

Thirdly, *informant rating* scales have been designed to assess psychopathy among children and adolescents and allow researchers the ability to assess a child by means of regular interaction and observation. Examples of these tools include the Antisocial Process Screening Device (APSD; Frick & Hare, 2001) and the Childhood Psychopathy Scale (CPS; Lynam, 1997) and can be completed by a parent, guardian, or teacher. Informant ratings are designed to detect psychopathic traits at a prodromal stage of development. However, the notion of 'psychopathic children' is controversial, not least due to the risk of over-diagnosing this disorder with young people (Salekin & Lynam, 2010).

Overall, while it is recognised that psychopathy consists of a broad range of behaviours and attributes, any given measurement procedure necessarily focuses on a limited set of traits or proxies (e.g., revocation of conditional release). Consequently, contemporary measures of psychopathy will be limited by a narrow coverage of behavioural repertoires, conceptualize symptom severity in global terms, and permit relatively crude categorical diagnoses (Hart & Cook, 2012). The formal assessment of psychopathy is complicated by the stigma of an imposed identity as assessors can further demonise an already demonised group, especially in criminal justice contexts where offenders are already the subject of prejudice by way of ethnicity, socioeconomic status, and criminal involvement, not to mention that official sources (e.g., prison files) are typically negative due to a criminal and antisocial bias of contents. Also, little is known about cultural differences, with some studies that include 'culture' as a variable that is really only appealing to 'nationality' or 'race' rather than the deeper and complex area of cultural *identity*. Both of these issues have resonance for psychopathy research, not least because stigma contributes to ongoing social adversity, and culture draws upon rich environmental contributions that are not typically explored in the criminal justice literature. Further, biologically-informed mechanisms from the domains of genetics, epigenetics and the increasingly popular neurosciences would broaden the scope of assessment beyond that of social deviance and antisocial behaviour (see Table 3.6). The inclusion of biomarkers that have a known empirical relationship with psychopathic disorder would inform biopsychosocial avenues for intervention. Although epigenetics and crime is an under-developed area, for epigenetic research to be meaningful it would need to (1) demonstrate incremental validity to current measures of psychopathy – particularly in relation to risk assessment, and (2) narrow the range of possible hypotheses in an individual case and hence offer prognoses that are more precise and inform treatment responsiveness (Tamatea, 2015).

3.1.3 Etiology and Developmental Issues

The etiology of a disease or psychological disorder informs the likely contributing factors in the development of a disease or psychological disorder as well as the likely trajectory of symptoms and what intervention possibilities may be appropriate to prevent, eradicate or reduce its impact. Despite the increasing interest in psychopathy research, surprisingly little is known about the etiology with a single cause of psychopathy yet to be universally accepted (Perez, 2012). However, the increase of assessment approaches to psychopathy has facilitated a growth of research that has focused on life course development (e.g., Lynam et al., 2007), affective processing (e.g., Patrick, 2007), neuroscience (e.g., Blair et al., 2005), personality traits (e.g., Frick & White, 2008; Lynam et al., 2009), and genetics (Gunter et al., 2010; Tikkanen et al., 2011; Viding & McCrory, 2012). Contemporary research reflects a shift from psychological towards biologically-informed models and perspectives.

At present, no single research tradition has adequately resolved or accounted for the range of psychopathic symptoms, telling us more about the complexity of the subject matter rather than the efforts of researchers. For instance, the proposal of a so-called 'warrior gene' – a variant of the normal MAOA gene – was linked to aggression and impulsivity in young Maori and Pasifika males (Gibbons, 2004). Although MAOA had been identified as a promising candidate gene, Buckholtz and Meyer-Lindenberg (2008) found that the effects of MAOA contributed a small amount of variance in risk of impulsive aggression – so not a 'violence gene' per se, but one contributor amongst many, since the substrate for behaviour involves multiple genes acting in concert rather than merely the actions of a single gene. It was further argued by Pickersgill (2009) that if there was such a thing as a gene for psychopathy, that the ontology of this personality disorder would be more amenable to being 'fixed', or at the very least, there would be less interpretative flexibility. As it is, these diverse developmental pathways could result in different types of pathology that could still fit in accordance with the concept of psychopathy. In the absence of a clear unitary relationship, the role of genes and environment has been difficult to elucidate.

Table 3.6 Critical differences between actuarial, clinical, psychometric, and biological approaches to measuring psychopathy

Practice aspect	Actuarial	Clinical assessment	Psychometric	Neurobiology	Epi/Genetic
Scope and focus	Risk of future offending (in general or specific e.g., child sex offending)	Detection of presence of psychopathic traits	Detection, discrimination, and severity of psychopathic traits	Patterns of neurological structure and functioning	Biological correlates; biomarkers
Use in risk prediction	Best practice	Well-established	Nil	Experimental	Nil
Detection of psychopathy	By proxy – Partial at best	Gold standard (i.e., PCL-R)	Accepted	Emergent	Contested
Emphasis	Statistical variables	Behavioural indices and antisociality	Personality traits	Neurological activity	Genetic sequencing, epigenetic modifications
Format/exemplars	Formal scoring rules, automated (e.g., VRAG, SORAG)	Interview, file and other corroborative sources (e.g., PCL measures)	Self-report (e.g., TriPM, PPI-R)	Apparatus: fMRI, CT, PET, etc.	Genetic analysis, DNA methylation, histone modification
Output	Bandwidth description of probability (e.g., 'high', 'medium' or 'low' risk)	Full, factor, and facet scores; diagnostic cut-off	Trait-based profiles	Visual imaging	Visual imaging
Strengths	Empirical, systematic, brief to administer, minimal scope for human error	Empirical, context-friendly, participatory, structured	Empirical, usable across contexts (region, population), ease of administration	Empirical, individualized	Empirical, link biology and context
Limitations	Disaggregation of the person, dependence on individual-to-group comparisons, no recognition of individual factors, context, or treatment effects	High degree of inference for many items, over-representation of antisociality, risk of negative emotional reactions to individual	Singular response modality (e.g., Likert scale), limited to fixed items, culturally-biased	Empirical support largely correlational, technology dependent, expensive, logistically challenging to train and administer	Findings largely from animal-based research; definition of 'environment' unclear
Assumption of cause of crime	Recorded criminal acts	Prototypical personality traits	Not applicable	Biomarkers: Neurocognitive	Biomarkers: Cellular

Biologically informative?	No	In some respects	No	Yes	Yes
Social costs to individual	Decontextualisation of individual	Imposed identity; labelling	Imposed identity; labelling	Discrimination of subgroups	'Biologization' of social constructs (violence)
Role of culture	'Objective' approach assumes cultural-neutrality	Culturally-biased: Some items reflective of socially aversive qualities in Canada/Nth America (e.g., superficial charm, grandiosity)	Culturally-biased: Items derived from specific cultural outlook	'Objective' problems similar in some ways to actuarial, but some consideration of impact of social adversity	Promotes mechanistic view of persons
Implications for offender management	Prioritisation of risk for treatment and resource	Inform rehabilitative needs	Inform treatment responsiveness	Inform risk and rehabilitation	Medicalisation of antisociality; mitigation of sentences

Stress signals from the environment that impact on the genome at various stages of development can lead to persistent epigenetic modifications. Candidate environmental factors that appear to contribute to a range of stress-related psychopathological conditions such as post-traumatic stress, depression in later life, and chronic physical aggression include maternal care, prenatal stress, substance abuse, and environmental enrichment (Dudley et al., 2011; Tremblay, 2008). In addition, social adversity, such as parental neglect and abuse, constant relocating (and consequently failing to develop strong attachments), low socioeconomic status and low IQ, membership in gangs and maternal drinking during pregnancy have been identified as exerting an impact on the behaviour of genes (Pickersgill, 2009). While these variables make intuitive sense as environmental stressors, a challenge for biological research in this area is to inform functional significance. For instance, callous-unemotional traits in adolescents was not only associated with Methylation of the oxytocin receptor gene, but in turn is associated with functional impairment in interpersonal empathy – a core psychopathic trait (Dadds et al., 2013, 2014).

In summary, psychological tests and assessment approaches tend to mirror the cultures in which they are practiced (Rogers, 1995). Technology, practices, knowledges and values are all incorporated into the system to form an assessment procedure that serves social and political needs, namely *role assignment* within the collective. Negative stereotypes belie the fact that psychopathy assessments are undertaken as part of an agenda to preserve the security and wellbeing of the community. However, there is a difference between impairment and lack of redeeming qualities, which can be difficult to appreciate if the diagnostic criteria and assessment tools define psychopathy according to the latter – a subject we turn to next.

3.2   What It Means to Be a 'Psychopath' II: From the Bottom-Up

In the prior section, issues of definition, diagnosis, development, and dysfunction were discussed in relation to psychopathy. In this section, the implications of top-down practices are unpacked and addressed, in particular the areas of stigma, treatment and justice.

3.2.1   Stigma

While much research exists to explore the breadth and scope of stigma and prejudice against people with mental illness more generally, the issues and impact of stigma for individuals identified as psychopathic have yet to be discussed fully in the literature. Stigma is an exclusion from full social acceptance that threatens all aspects of social relationships for people imposed with this burden. Its negative consequences can include diminished employment opportunities, lower quality of health care provision, and an impoverished social life (Panier et al., 2014; Sheehan et al., 2016).

Stigmatizing attitudes and misunderstandings are especially salient in the case of psychopathy, which remains poorly-understood by the public not least because of a lack of public awareness and information regarding the disorder (Edens et al., 2005; Edens et al., 2006; Helfgott, 1997). Furthermore, trait models of psychopathy obscure the situational complexity of individual's lives and overestimate the impact of personality variables on individual choices. By avoiding situated complexity, and variability of personality traits – and identity – individuals become vulnerable to being stereotyped and forced to engage in stigma-management strategies to negotiate through their lives.

The production of psychopathy stigma is evident in popular media (Federman et al., 2009). Lipczynska (2015) notes that cinema has a history of using mental disorders flippantly as dramatic or comedic devices without any real attempt to understand the disorder or present it in a meaningful way, adding that psychopathy is prone to particularly negative consequences for the diagnosis, not least because the kinds of films which use 'psychopaths' as a dramatic device to facilitate the perception that those with mental disorders are dangerous and to be feared. Whose perception determines whether and in what way an individual is stigmatized often goes unaddressed in much research (Meisenbach, 2010). There is some debate as to whether (and how) psychopathic persons can experience stigma-relevant emotions, such as shame (Cleckley, 1941, 1976; Hare, 2003; Morrison & Gilbert, 2001; Prado et al., 2016), but extant research has targeted interpersonal situations that are associated with social status, but not necessarily to do with being identified as a 'psychopath'. The dearth of psychopathy stigma research means that one can be forgiven for assuming that psychopathic individuals do not experience stigma to warrant meaningful research and/or that researchers in the field do not experience sufficient compassion with this group.

Stigma can only be created by over-simplifying complex situations. Stigma in relation to psychopathy is not measured and hence there is little research to indicate what negative consequences (social, emotional, or health-related) it may have on individual's lives. Some considerations include emphasising difference and enacting rejection and discrimination. Firstly, the decision to diagnose psychopathy is highly contextual in terms of culture and community, but also within an individual's life trajectory (i.e., from youth to adulthood). Furthermore, the decision to assess for the presence of psychopathic traits does not take place in a vacuum, instead it is a result of a particular set of often quite complex circumstances (e.g., reports to justice decision-makers about release feasibility or treatment suitability). The relative invisibility of psychopathy may also have an impact on prevalence data in community-based contexts. A mutually reinforcing cycle of invisibility and non-random assessment settings (i.e., prisons) makes it challenging to know the true prevalence of psychopathy in a given community.

Secondly, Feldman and Crandall (2007) identified three main factors for social rejection: personal responsibility, dangerousness, and rarity of the illness. Previous investigations have concluded that laypeople believe that people with psychopathic traits are responsible for their actions (Smith et al., 2014), that psychopaths are dangerous (Wayland & O'Brien, 2013), and that psychopathy has a prevalence rate of less than 1% in the general population (Smith et al., 2014). Psychopathy therefore fulfils all criteria for inducing stigmatization and social rejection.

Thirdly, the effectiveness of efforts to address stigma rely on an ability to understand stigma processes, the factors that produce and sustain stigma processes, and the mechanisms that lead to harmful consequences (Link et al., 2004). The lack of overt physical indicators for psychopathy makes stigma potentially different than stigma associated with a chronic or disfiguring illnesses. With psychopathy there is no contagion, causative agent, diagnosable disease or visible marking. The indistinctness of many who have been identified as psychopathic in criminal justice settings allows some men to avoid self-identifying or adopting a tainted identity linked to the experience. Even if it may be the case that psychopathic people may not experience stigma, they certainly experience the *sequelae* of stigma. The manifestations of enacted stigma on an interpersonal level may include denial of verbal or physical abuse, loss of employment, public shaming, ostracism, or poor quality of services. Historically, the perceived untreatability of psychopathy has led many clinicians and corrections authorities to avoid attempts to treat psychopathic offenders (Wong & Hare, 2005). In these instances, the allocation of treatment access itself is a form of structural discrimination.

Stigmatizing attitudes of health care professionals towards people with psychopathy may negatively affect service provision and healthcare delivery and could result in treatment avoidance, interruption, poor communication, diminished therapeutic alliance, and diagnostic overshadowing (i.e., misattribution of problems to pathology), and compromises in the person's perceptions of recovery and self-efficacy.

3.2.2   Treatment: Change or Compliance?

Treatment approaches are typically designed to favourably alter the course of a disease state or psychological disorder for the person so afflicted (and/or others that may be affected by association with them). Successful treatments tell us something about the mutability or even the reversibility of a phenotype. However, the literature in regard to the use of therapy to change the antisocial behaviour or attitudes associated with high-risk psychopathic offenders has presented a bleak picture – a result of early pessimism that had passed into clinical lore (Dolan & Coid, 1993; Hare, 1998; Hemphill & Hart, 2002; Lösel, 1998; Salekin, 2002; Wong & Hare, 2005). Indeed, many clinicians had boycotted the idea of even attempting to treat high-risk psychopathic offenders, and a number of corrections authorities have taken the position of 'sanctioned untreatability', that it is cost-effective to exclude high risk offenders from their standard treatment programs (Gunn, 1998). However, the development of experimental purpose-built programmes such as the Dangerous and Severe Personality Disorder units in the United Kingdom (Maden & Tyrer, 2003) or experimental treatment initiatives such as the High-Risk Personality Programme in

New Zealand (Wilson & Tamatea, 2013) have attempted to challenge this notion via highly structured psychosocial interventions – with mixed results.

The psychopathy treatment literature has been steadily growing since the 1990s – but not without issue. For instance, Salekin (2002) and Harris and Rice (2006) independently reviewed the same 42 studies of published treatment programmes for psychopathic offenders, revealing problems over diagnostic criteria, variable understandings of etiology, and poorly identified treatment targets. Hemphill and Hart (2002) identified a range of methodological issues within this literature such as lack of adequate control groups, failure to control for heterogeneity within treatment groups, inconsistent concepts and measures of psychopathy, lack of attention to developmental factors, inadequate definition and implementation of treatment, severely restricted outcome criteria, and absence of randomised controlled trials. More recently, challenges to the presumed untreatability of psychopathy have emerged emphasising treatment focus on co-morbid conditions (Felthous, 2011), realistic and definitive treatment goals (Wilson & Tamatea, 2013), and recognition of little empirical support for contraindications for treatment such as general capacity for behaviour change, incidence of treatment-interfering behaviour, or use of treatment for antisocial purposes (Polaschek, 2014).

A recent review of the psychopathy treatment literature by Polaschek and Skeem (2018) emphasised, again, the dearth of robust research in this area. They commented that treatment programmes that observed the risk, need and responsivity principles of offender management revealed the best outcomes for psychopathic clients in terms of offence reduction and community safety, but that no methodologically sound research demonstrated change in the *symptoms* of psychopathy as a function of treatment. They further argue that high-PCL-scoring clients are also high-risk offenders and as such should be a prioritised group for treatment rather than one considered to be ineligible merely because they are seen as too difficult to treat. Picking up on this thread, Rosenberg Larsen (2019) suggests that it might be time for the field to *"stop and more profoundly reconsider research and practices regarding the psychopathy diagnosis"* (p.263).

It should be noted here that when a psychopathic person enters a process of therapeutic change, this will likely be court-ordered or a driver for release from prison. The aim of change is rarely negotiated and imposed by the state in an effort to reduce harm. The lack of input from the psychopathic person on what matters to them is likely to be a reason why treatment-interfering behaviours are common in offender programmes.

3.2.3 Justice

Arguably, the utility of identifying psychopathic traits and predicting harm is a matter of governing current and future populations. Important questions that concern the acceptance and ethical use of these new discoveries includes responsible care, reducing social disadvantage, balancing predictions of dangerousness with wellbeing, and attitudes to punishment as an expression of discrimination.

Current diagnostic systems and risk assessment approaches may inadvertently facilitate a routinization of marginalising a vulnerable population. For instance, potential biomarkers for psychopathy may provide a biological rationale to exacerbate an existing prejudice against 'psychopaths' as being innately irredeemable and 'hard-wired' to offend by adding a quasi-medico legitimacy to a concept that is easily misunderstood and prone to unhelpful stereotypes (Jurjako et al., 2019). Furthermore, while little direct evidence exists to substantiate the notion of a genetically criminal underclass, the promotion of punitive interventions to curb 'undesirable' and 'undeserving' members of the community from full participation may indeed perpetuate further social disadvantage (i.e., unemployment, poverty) and the emergence of an 'epigenetic underclass'.

The concept of recidivism risk may need to expand to include (social and mental) health. For instance, what may have previously been seen as a behavioural problem may now be conceptualised as an organically-informed/biological disease entity, shifting the emphasis from a correctional problem to a *health* concern.

Whether there is a biological basis to psychopathy remains controversial; the particular genetic factors, neural structures, or other mechanisms have yet to fully account for psychopathy. However, extant epigenetic findings and their interpretations raise the possibility of organic correlates of psychopathy and criminal behaviour. This has led to speculation that some criminals may be less responsible for their crime than others (Glannon, 2008; Nadelhoffer & Sinnott-Armstrong, 2012; Robinson et al., 2011). Or, as Gillett and Huang (2013) argue, that if psychopathy (or more specifically, the downstream consequences of psychopathy, such as crime), emerge from highly adverse human ecologies, then we owe these individuals some remediation because of our collective responsibility for actively or tacitly allowing such adversity to exist and persist.

Fine and Kennett (2004) observed that psychopathy diagnosis can result in harsher sentences (Cox et al., 2013; Zinger & Forth, 1998) or even the imposition of the death sentence rather than a life sentence (Edens et al., 2001, 2013). Skeem et al. (2004) asserted that: "Public perceptions of psychopathy matter". Their study of laypersons as the trier of fact in mock jury trials supported the view that increasingly fearful societies support punitive criminal justice policies and legislation. In addition, if associated with a minority demographic, such as African Americans, psychopathy can be used as a means to further exacerbate disadvantage and inequality via sentencing laws.

3.3 Concluding Comment

This chapter has sketched a rough outline of two competing narratives, where academic, clinical, and criminal justice experts (from the top-down) have shared a contestable conceptual and ethical space with psychopathic persons (from the bottom-up) in a struggle to define identity (imposed vs. self-determined) and meaning (criminal vs. health). Psychopathy presents with a wide range of clinical, legal and social issues across a number of spheres that are complex and do not permit easy answers or rewarding experiences. This chapter has attempted to make three points: (1) approach psychopathy for what it is – a health issue, (2) challenge unhelpful and inaccurate stereotypes, and (3) develop inclusive research attitudes and modes of inquiry.

Firstly, *psychopathy as a health issue.* A reframe of psychopathy as a subject of clinical and scholarly inquiry beyond criminality and antisocial behaviour is in order. This is not to dismiss or ignore important issues, such as criminal activity and violence, but rather to bring a therapeutic lens to the fore. For instance, the standard clinical and research agenda has been to ask clients/participants: "Why/how are you different?" – which is an important diagnostic (but also alienating) question, and instead ask: "How did you come to be this way?" – which is more inclusive, dignifying, and acknowledges adversity. Developing an understanding of psychopathy – where criminality is acknowledged but not person-defining – has the potential to make a therapeutic contribution to those challenging individuals that are encountered in the field, but also an empathic stance when considering children with psychopathic potential.

Secondly, *interrogate received wisdoms and scientific stereotypes* to include and be informed by insider perspectives. Researchers need to develop more balanced accounts of psychopathy. For members of the psychopathy community to have a hand in research that would benefit them: What are the priorities for psychopathic persons? What research would benefit this group? The development of programmes and interventions in criminal justice, forensic, and mental health settings to reduce psychopathy stigma are needed. Interventions should consider lessons learned about stigma in other areas, especially decreased life opportunities and increased social inequities. Participatory research that has the potential for helping individuals to articulate the steps to dismantling stigma should be a central part of this approach. No known intervention exists that has been designed to measure and evaluate the experience of stigma with psychopathic participants. Indeed, most interventions were not designed with the objective of reducing stigma so causality may be challenging to attribute.

Thirdly, *democratize the research agenda.* As noted, most, if not all psychopathy research is conducted and disseminated by those who are not psychopathic (or are, perhaps, in denial!). As such, this body of research can be considered to be top-down *outsider* research where issues like self-determination and co-design by those who *experience* psychopathy are regarded as passive participants of experimental researchers or silent subjects for clinical researchers. A common practice has been to conduct research *on* psychopathic persons rather than *with* them (or even *for* them). There are methodological reasons as to why this is the case (dishonesty, lack of insight, etc.), but there are also reasons why persons with psychopathic traits should be able to contribute meaningfully in research that affects them. Firstly, to not include the wisdom of lived-experience is to perpetuate a low regard for this population by acting/researching 'in their best interests'. Secondly, given that psychopathy is an imposed identity with little room for the person so-labelled to negotiate, researchers can then focus on ethically and socially just practices with what is really a vulnerable group... albeit one that makes others vulnerable. Meaningful research about a given population is difficult to do without their input, their voice.

Research should be developed in conjunction with psychopathic persons – obviate the need for the roles of stigmatizer and stigmatized – and explore why these have been treated like incompatible activities.

Critically challenging the narrative of psychopathy from perpetuating scientifically-sanctioned stereotypes creates ethical and conceptual space to open up new ways of looking at long-held research, clinical, and forensic problems that have typically reflected one-way power relationships between researcher and researched, whilst denying self-determination and personal expertise on experience. It is critical to understand the social impact of psychopathy and stigma better, measure it empirically, deconstruct it in contexts of mental illness, dangerousness, and social rejection, and reconstruct its contours at a broader health and wellbeing-focused level.

References

Alexander, F., & Staub, H. (1931/1956). *The criminal, the judge, and the public; a psychological analysis* (rev. ed.) (G. Zilboorg trans.). Free press.

American Psychiatric Association. (2013). *Diagnostic and statistical manual of mental disorders: DSM-5*. American Psychiatric Association.

Asscher, J. J., van Vugt, E. S., Stams, G. J. J. M., Deković, M., Eichelsheim, V. I., & Yousfi, S. (2011). The relationship between juvenile psychopathic traits, delinquency and (violent) recidivism: A meta-analysis. *Journal of Child Psychology and Psychiatry, 52*(11), 1134–1143.

Beck, A. T., & Freeman, A. M. (1990). *Cognitive therapy of personality disorders*. Guilford Press.

Birnbaum, K. (1909). *Über psychopathische Persönlichkeiten. Eine psychopathologische Studie*. Bergmann.

Blackburn, R. (1975). An empirical classification of psychopathic personality. *British Journal of Psychiatry, 127*, 456–460.

Blackburn, R., & Coid, J. W. (1998). Psychopathy and the dimensions of personality disorders in violent offenders. *Personality and Individual Differences, 25*, 129–145.

Blair, R. J. R., Mitchell, D., & Blair, K. (2005). *The psychopath: Emotion and the brain*. Wiley-Blackwell.

Blais, J., Solodukhin, E., & Forth, A. E. (2014). A meta-analysis exploring the relationship between psychopathy and instrumental versus reactive violence. *Criminal Justice and Behavior, 41*(7), 797–821.

Book, A. S., & Quinsey, V. L. (2004). Psychopaths: Cheaters or warrior-hawks? *Personality and Individual Differences, 36*, 33–45.

Bowlby, J. (1953). *Child care and the growth of love* (2nd ed.). Pelican.

Buckholtz, J. W., & Meyer-Lindenberg, A. (2008). MAOA and the neurogenetic architecture of human aggression. *Trends in Neurosciences, 31*(3), 120–129.

Bursten, B. (1972). The manipulative personality. *Archives of General Psychiatry, 26*(4), 318–321.

Camp, J. P., Skeem, J. L., Barchard, K., Lilienfeld, S. O., & Poythress, N. G. (2013). Psychopathic predators? Getting specific about the relation between psychopathy and violence. *Journal of Consulting and Clinical Psychology, 81*(3), 467–480.

Cavadino, M. (1998). Death to the psychopath. *Journal of Forensic Psychiatry, 9*(1), 5–8.

Cleckley, H. (1941). *The mask of sanity: An attempt to reinterpret the so-called psychopathic personality*. Mosby.

Cleckley, H. (1976). *The mask of sanity: An attempt to reinterpret the so-called psychopathic personality* (5th ed.). Mosby.

Cooke, D. J., & Michie, C. (2001). Refining the construct of psychopathy: Towards a hierarchical model. *Psychological Assessment, 13*, 171–188.

Cooke, D. J., Hart, S. D., Logan, C., & Michie, C. (2004). *Comprehensive assessment of psychopathic personality – Institutional rating scale (CAPP-IRS)*. Unpublished manuscript.

Cox, J., Clark, J., Edens, J., Smith, S., & Magyar, M. (2013). Jury panel member perceptions of interpersonal-affective traits of psychopathy predict support for execution in a capital murder trial simulation. *Behavioral Sciences & the Law, 31*(4), 411–428.

Dadds, M. R., Moul, C., Cauchi, A., Dobson-Stone, C., Hawes, D. J., Brennan, J., & Ebstein, R. E. (2013). Polymorphisms in the oxytocin receptor gene are associated with the development of psychopathy. *Development and Psychopathology, 26*, 21–31.

Dadds, M. R., Moul, C., Cauchi, A., Dobson-Stone, C., Hawes, D. J., Brennan, J., & Ebstein, R. E. (2014). Methylation of the oxytocin receptor gene and oxytocin blood levels in the development of psychopathy. *Development and Psychopathology, 26*(1), 33–40.

Damasio, A. R. (1994). *Descarte's error: Emotion, reason, and the human brain*. Penguin.

Darwin, C. (2009). *The expression of the emotions in man and animals*. Penguin. (Original work published 1872).

Dolan, B., & Coid, J. (1993). *Psychopathic and antisocial personality disorders*. Gaskell.

Draughon, M. (1977). Ego-building: An aspect of the treatment of psychopaths. *Psychological Reports, 40*(2), 615–626.

Dudley, K. J., Li, X., Kobor, M. S., Kippin, T. E., & Bredy, T. W. (2011). Epigenetic mechanisms mediating vulnerability and resilience to psychiatric disorders. *Neuroscience and Biobehavioral Reviews, 35*, 1544–1551.

Edens, J. F., Petrila, J., & Buffington-Vollum, J. K. (2001). Psychopathy and the death penalty: Can the psychopathy checklist–revised identify offenders who represent "a continuing threat to society?". *Journal of Psychiatry and Law, 29*, 433–481.

Edens, J. F., Colwell, L. H., Desforges, D. M., & Fernandez, K. (2005). The impact of mental health evidence on support for capital punishment: Are defendants labeled psychopathic considered more deserving of death? *Behavioral Sciences & the Law, 23*, 603–625.

Edens, J. F., Marcus, D. K., Lilienfeld, S. O., & Poythress, N. G., Jr. (2006). Psychopathic, not psychopath: Taxometric evidence for the dimensional structure of psychopathy. *Journal of Abnormal Psychology, 115*, 131–144.

Edens, J. F., Davis, K. M., Fernandez Smith, K., & Guy, L. S. (2013). No sympathy for the devil: Attributing psychopathic traits to capital murders also predicts support for executing them. *Personality Disorders: Theory, Research, and Treatment, 4*(2), 175–181.

Eysenck, H. (1957). *The dynamics of anxiety and hysteria*. Routledge.

Eysenck, H. (1967). *The biological basis of personality*. Thomas.

Federman, C., Holmes, D., & Jacob, J. (2009). Deconstructing the psychopath: A critical discursive analysis. *Cultural Critique, 72*, 36–65.

Feldman, D., & Crandall, C. (2007). Dimensions of mental illness stigma: What about mental illness causes social rejection? *Journal of Social and Clinical Psychology, 26*(2), 137–154.

Felthous, A. R. (2011). The "untreatability" of psychopathy and hospital commitment in the USA. *International Journal of Law and Psychiatry, 34*, 400–405.

Fernandez-Suarez, A., Perez, B., Herrero, J., & Juarros-Basterretxea, J. (2018). The role of psychopathic traits among intimate partner-violent men: A systematic review. *Revista Iberoamericana De Psicología Y Salud, 9*(2), 84–114.

Fine, C., & Kennett, J. (2004). Mental impairment, moral understanding and criminal responsibility: Psychopathy and the purposes of punishment. *International Journal of Law and Psychiatry, 27*(5), 425–443.

Forth, A. E., Kosson, D. S., & Hare, R. D. (2003). *The hare psychopathy checklist: Youth version*. Multi-Health Systems.

Fox, B., & DeLisi, M. (2019). Psychopathic killers: A meta-analytic review of the psychopathy-homicide nexus. *Aggression and Violent Behavior, 44*, 67–79.

Frick, P. J. (1998). Callous-unemotional traits and conduct problems: Applying the two-factor model of psychopathy to children. In D. J. Cooke, A. E. Forth, & R. D. Hare (Eds.), *Psychopathy: Theory, research and implications for society (NATO ASI series)*. Springer.

Frick, P. J., & Hare, R. D. (2001). *Antisocial process screening device*. Multi-Health Systems.

Frick, P. J., & White, S. F. (2008). Research review: The importance of callous-unemotional traits for developmental models of aggressive and antisocial behavior. *Journal of Child Psychology and Psychiatry, 49*(4), 359–375.

Gibbons. (2004). Tracking the evolutionary history of a "warrior" gene. *Science, 304*(5672), 818.

Gillett, G., & Huang, J. (2013). What we owe the psychopath: A neuroethical analysis. *American Journal of Bioethics: Neuroscience, 4*(2), 3–9.

Glannon, W. (2008). Moral responsibility and the psychopath. *Neuroethics, 1*, 158–166.

Gough, H. G. (1948). A sociological theory of psychopathy. *American Journal of Sociology, 53*, 359–366.

Grisolia, J. S. (2001). Neurobiology of the psychopath. In A. Raine & J. Sanmartin (Eds.), *Violence and psychopathy* (pp. 79–87). Kluwer Academic.

Gunn, J. (1998). Psychopathy: An elusive concept with moral overtones. In T. Millon, E. Simonsen, M. Birket-Smith, & R. D. Davis (Eds.), *Psychopathy: Antisocial, criminal, and violent behavior* (pp. 32–39). Guilford.

Gunter, T. D., Vaughn, M. G., & Philibert, R. A. (2010). Behavioral genetics in antisocial spectrum disorders and psychopathy: A review of the recent literature. *Behavioral Sciences & the Law, 28*, 148–173.

Haack, K., Kumbier, E., & Herpertz, S. (2010). Illnesses of the will in 'pre-psychiatric' times. *History of Psychiatry, 21*(3), 261–277.

Haksar, V. (1965). The responsibility of psychopaths. *The Philosophical Quarterly, 15*, 135–145.

Hanson, R. K., & Morton-Bourgon, K. (2005). The characteristics of persistent sexual offenders: A meta-analysis of recidivism studies. *Journal of Consulting and Clinical Psychology, 73*, 1154–1163.

Hare, R. D. (1970). *Psychopathy: Theory and research (Wiley approaches to behavior pathology series)*. Wiley.

Hare, R. (1991). *The Hare psychopathy checklist–revised*. Multi-Health Systems.

Hare, R. D. (1998). Psychopaths and their nature: Implications for the mental health and criminal justice systems. In T. Millon, E. Simonsen, M. Birket-Smith, & R. D. Davis (Eds.), *Psychopathy: Antisocial, criminal, and violent behavior* (pp. 188–212). Guilford.

Hare, R. D. (2003). *The Hare psychopathy checklist-revised (PCL-R)* (2nd ed.). Multi-Health Systems.

Harpur, T. J., Hakstian, A. R., & Hare, R. D. (1988). Factor structure of the psychopathy checklist. *Journal of Consulting and Clinical Psychology, 56*, 741–747.

Harris, G. T., & Rice, M. E. (2006). Treatment of psychopathy: A review of empirical findings. In C. Patrick (Ed.), *Handbook of psychopathy* (pp. 555–572). Guilford.

Hart, S. D., & Cook, A. N. (2012). Current issues in the assessment and diagnosis of psychopathy (psychopathic personality disorder). *Neuropsychiatry, 2*(6), 497–508.

Hart, S. D., Cox, D. N., & Hare, R. D. (1995). *The Hare psychopathy checklist: Screening version*. Multi-Health Systems.

Hawes, S. W., Boccaccini, M. T., & Murrie, D. C. (2013). Psychopathy and the combination of psychopathy and sexual deviance as predictors of sexual recidivism: Meta-analytic findings using the psychopathy checklist--revised. *Psychological Assessment, 25*(1), 233–243.

Helfgott, J. (1997). The relationship between unconscious defensive process and conscious cognitive style in psychopaths. *Criminal Justice and Behavior, 24*(2), 278–293.

Hemphill, J. F., & Hart, S. D. (2002). Motivating the unmotivated: Psychopathy, treatment, and change. In M. McMurran (Ed.), *Motivating offenders to change* (pp. 193–219). Wiley.

Jurjako, M., Malatesti, L., & Brazil, I. A. (2019). Some ethical considerations about the use of biomarkers for the classification of adult antisocial individuals. *International Journal of Forensic Mental Health, 18*(3), 228–242.

Karpman, B. (1929). The problem of psychopathies. *Psychiatric Quarterly, 3*, 495–525.

Karpman, B. (1948). The myth of the psychopathic personality. *American Journal of Psychiatry, 104*, 523–534.

Kiehl, K. A., & Hoffman, M. B. (2011). The criminal psychopath: History, neuroscience, treatment, and economics. *Jurimetrics, 51*, 355–397.

Knight, R. A., & Guay, J.-P. (2006). The role of psychopathy in sexual coercion against women. In C. Patrick (Ed.), *Handbook of psychopathy* (pp. 512–532). Guilford.

Kræpelin, E. (1915). *Clinical psychiatry for students and physicians* (A.R. Diefendorf, Trans.). MacMillan. (Original work published 1915).

Krafft-Ebing, R. (1899). *Psychopathia sexualis* (10th ed.) (F.J. Rebman, Trans.). Rebman.

Levenson, M. R. (1992). Rethinking psychopathy. *Theory & Psychology, 2*(1), 51–71.

Lilienfeld, S. O. (1994). Conceptual problems in the assessment of psychopathy. *Clinical Psychology Review, 14*(1), 17–38.

Lilienfeld, S. O., & Fowler, K. A. (2006). The self-report assessment of psychopathy: Pitfalls, problems, and promises. In C. Patrick (Ed.), *Handbook of psychopathy* (pp. 107–132). Guilford.

Lilienfeld, S. O., & Widows, M. R. (2005). *Psychopathic personality inventory-revised: Professional manual*. Psychological Assessment Resources.

Link, B., Yang, L., Phelan, J., & Collins, P. (2004). Measuring mental illness stigma. *Schizophrenia Bulletin, 30*(3), 511–541.

Lipczynska, S. (2015). "We all go a little mad sometimes": The problematic depiction of psychotic and psychopathic disorders in cinema. *Journal of Mental Health, 24*(2), 61–62.

Litton, P. (2008). Responsibility status of the psychopath: On moral reasoning and rational self-governance. *Rutgers Law Journal, 39*(2), 349–392.

Litton, P. (2010). Psychopathy and responsibility theory. *Philosophy Compass, 5*(8), 676–688.

Loi, M., Del Savio, L., & Stupka, E. (2013). Social epigenetics and equality of opportunity. *Public Health Ethics, 6*(2), 142–153.

Lösel, F. (1998). Treatment and management of psychopaths. In D. J. Cooke, A. E. Forth, & R. D. Hare (Eds.), *Psychopathy: Theory research and implications for society* (pp. 303–354). Kluwer.

Lykken, D. T. (1957). A study of anxiety in the sociopathic personality. *Journal of Abnormal and Social Psychology, 55*, 6–10.

Lynam, D. R. (1997). Pursuing the psychopath: Capturing the fledgling psychopath in a nomological net. *Journal of Abnormal Psychology, 106*, 425–438.

Lynam, D. R., Caspi, A., Moffitt, T. E., Loeber, R., & Stouthamer-Loeber, M. (2007). Longitudinal evidence that psychopathy scores in early adolescence predict adult psychopathy. *Journal of Abnormal Psychology, 116*(1), 155–165.

Lynam, D. R., Charnigo, R., Moffitt, T. E., Raine, A., Loeber, R., & Stouthamer-Loeber, M. (2009). The stability of psychopathy across adolescence. *Development and Psychopathology, 21*(4), 1133–1153.

Maden, T., & Tyrer, P. (2003). Dangerous and severe personality disorders: A new personality concept from the United Kingdom. *Journal of Personality Disorders, 17*(6), 489–496.

Maudsley, H. (1898). *Responsibility in mental disease*. Appleton.

Maughs, S. (1941). A conception of psychopathy and psychopathic personality: Its evolution and historical development. *Journal of Criminal Psychopathology, 2*, 329-56 and 465-399.

Mead, S. (1965). *The carefully considered rape of the world*. Penguin.

Meisenbach, R. J. (2010). Stigma management communication: A theory and agenda for applied research on how individuals manage moments of stigmatized identity. *Journal of Applied Communication Research, 38*(3), 268–292.

Meloy, J. R. (1988). *The psychopathic mind: Origins, dynamics, and treatment*. Jason Aronson.

Millon, T. (1969). *Modern psychopathology: A biosocial approach to maladaptive learning and functioning*. Saunders.

Millon, T. (1987). *Manual for the MCMI–II* (2nd ed.). National Computer Systems.

Millon, T., Simonsen, E., & Birket-Smith, M. (1998). Historical conceptions of psychopathy in the United States and Europe. In T. Millon, E. Simonsen, M. Birket-Smith, & R. D. Davis (Eds.), *Psychopathy: Antisocial, criminal, and violent behavior* (pp. 3–31). Guilford.

Monahan, J., & MacArthur Violence Risk Assessment Study. (2001). *Rethinking risk assessment: The MacArthur study of mental disorder and violence*. Oxford University Press.

Morrison, D., & Gilbert, P. (2001). Social rank, shame and anger in primary and secondary psychopaths. *Journal of Forensic Psychiatry, 12*(2), 330–356.

Murphy, J. (1972). Moral death: A Kantian essay on psychopathy. *Ethics, 82*(4), 284–298.

Nadelhoffer, T., & Sinnott-Armstrong, W. (2012). Neurolaw and neuroprediction: Potential promises and perils. *Philosophy Compass, 7*(9), 631–642.

Newman, J. P. (1998). Psychopathic behaviour: An information processing perspective. In D. J. Cooke, A. E. Forth, & R. D. Hare (Eds.), *Psychopathy: Theory, research and implications for society* (pp. 81–104). Kluwer.

Panier, S., Van Remoortere, A., Van den Bogaert, A., & Uzieblo, K. (2014). Fearing the unknown? The relationship between familiarity and attitudes towards psychopathy. In *The treatment of psychopathy : Making the impossible possible? Abstracts* (Presented at the the treatment of psychopathy: Making the impossible possible?) (pp. 16–16). Thomas More.

Partridge, G. E. (1930). Current conceptions of psychopathic personality. *American Journal of Orthopsychiatry, 10*, 53–79.

Patrick, C. J. (2007). Getting to the heart of psychopathy. In H. F. Hervé & J. C. Yuille (Eds.), *Psychopathy: Theory, research, and social implications* (pp. 207–252). Lawrence Erlbaum.

Patrick, C. J., Fowles, D. C., & Krueger, R. F. (2009). Triarchic conceptualization of psychopathy: Developmental origins of disinhibition, boldness, and meanness. *Development and Psychopathology, 21*(3), 913–938.

Perez, P. R. (2012). The etiology of psychopathy: A neuropsychological perspective. *Aggression and Violent Behavior, 17*(6), 519–522.

Pickersgill, M. (2009). Between soma and society: Neuroscience and the ontology of psychopathy. *BioSocieties, 4*, 45–60.

Pinel, P. (1806). *A treatise on insanity* (D.D. Davis, Trans.). Cadell & Davies. (Original work published 1801).

Polaschek, D. L. L. (2014). Adult criminals with psychopathy: Common beliefs about treatability and change have little empirical support. *Current Directions in Psychological Science, 23*(4), 296–301.

Polaschek, D., & Skeem, J. (2018). Treatment of adults and juveniles with psychopathy. In C. Patrick (Ed.), *Handbook of psychopathy* (2nd ed., pp. 710–731). Guilford.

Porter, S. (1996). Without conscience or without active conscience? The etiology of psychopathy revisited. *Aggression and Violent Behavior, 1*(2), 179–189.

Porter, S., & Woodworth, M. (2006). Psychopathy and aggression. In C. Patrick (Ed.), *Handbook of psychopathy* (pp. 481–494). Guilford.

Porter, S., Birt, A. R., & Boer, D. P. (2001). Investigation of the criminal and conditional release profiles of Canadian federal offenders as a function of psychopathy and age. *Law and Human Behavior, 25*(6), 647–661.

Prado, C. E., Treeby, M. S., & Crowe, S. F. (2016). Examining the relationships between subclinical psychopathic traits with shame, guilt and externalisation response tendencies to everyday transgressions. *Journal of Forensic Psychiatry & Psychology, 27*(4), 569–585.

Prichard, J. C. (1837). *A treatise on insanity and other disorders affecting the mind*. Carey & Hart.

Quay, H. C. (1965). Psychopathic personality as pathological stimulation seeking. *American Journal of Psychiatry, 122*, 180–183.

Quay, H. C. (1977). Psychopathic behaviour: Reflections on its nature, origins, and treatment. In I. Č. Užgiris & F. Weizmann (Eds.), *The structuring of experience* (pp. 371–383). Plenum.

Ray, J. V., Hall, J., Rivera-Hudson, N., Poythress, N. G., Lilienfeld, S. O., & Morano, M. (2013). The relation between self-reported psychopathic traits and distorted response styles: A meta-analytic review. *Personality Disorders: Theory, Research, and Treatment, 4*, 1–14.

Robinson, L., Sprooten, E., & Lawrie, S. M. (2011). Brain imaging in psychosis and psychopathy – Ethical considerations. *Cortex, 47*(10), 1236–1239.

Rogers, T. B. (1995). *The psychological testing enterprise: An introduction*. Brooks/Cole.

Rosenberg Larsen, R. (2019). Psychopathy treatment and the stigma of yesterday's research. *Kennedy Institute of Ethics Journal, 29*(3), 243–272.

Rush, B. (1812). *Medical inquiries and observations*. J. Conrad.

Salekin, R. T. (2002). Psychopathy and therapeutic pessimism: Clinical lore or clinical reality? *Clinical Psychology Review, 22,* 79–112.

Salekin, R. T., & Lynam, D. R. (2010). Child and adolescent psychopathy: An introduction. In R. T. Salekin & D. R. Lynam (Eds.), *Handbook of child and adolescent psychopathy* (pp. 1–12). Guilford.

Salekin, R. T., Rogers, R., & Sewell, K. W. (1996). A review and meta-analysis of the psychopathy checklist-revised: Predictive validity of dangerousness. *Clinical Psychology: Science and Practice, 3,* 203–215.

Schneider, K. (1923). *Die psychopathischen Persönlichkeiten*. Deuticke.

Serin, R. C., & Brown, S. L. (2005). Social cognition in psychopaths: Implications for offender assessment and treatment. In M. McMurran & J. McGuire (Eds.), *Social problem solving and offending: Evidence, evaluation and evolution* (pp. 249–264). Wiley.

Sheehan, L., Nieweglowski, K., & Corrigan, P. (2016). The stigma of personality disorders. *Current Psychiatry Reports, 18,* 1–7.

Skeem, J. L., Edens, J. F., Camp, J., & Colwell, L. H. (2004). Are there ethnic differences in levels of psychopathy? A meta-analysis. *Law and Human Behavior, 28*(5), 505–527.

Skeem, J. L., Polaschek, D. L. L., Patrick, C. J., & Lilienfeld, S. O. (2012). Psychopathic personality: Bridging the gap between scientific evidence and public policy. *Psychological Science in the Public Interest, 12*(3), 95–162.

Slovenko, R. (1999). Responsibility of the psychopath. *Philosophy, Psychiatry, & Psychology, 6*(1), 53–55.

Smith, S., Edens, J., Clark, J., & Rulseh, A. (2014). "So, what is a psychopath?" Venireperson perceptions, beliefs, and attitudes about psychopathic personality. *Law and Human Behavior, 38*(5), 490–500.

Tamatea, A. J. (2015). 'Biologising' psychopathy: Ethical, legal, and research implications at the interface of epigenetics and chronic antisocial conduct. *Behavioral Sciences & the Law, 33*(5), 629–643.

Tikkanen, R., Auvinen-Lintunen, L., Ducci, F., Sjöberg, R. L., Goldman, D., Tiihonen, J., Ojansuu, I., & Virkkunen, M. (2011). Psychopathy, PCL-R, and MAOA genotype as predictors of violent reconvictions. *Psychiatry Research, 185*(3), 382–386.

Tremblay, R. E. (2008). Understanding development and prevention of chronic physical aggression: Towards experimental epigenetic studies. *Philosophical Transactions of the Royal Society, 363,* 2613–2622.

Trichet, Y., & Lacroix, A. (2016). Esquirol's change of view towards Pinel's mania without delusion. *History of Psychiatry, 27*(4), 443–457.

Van Draanen, J., Jeyaratnam, J., O'Campo, P., Hwang, S., Harriott, D., Koo, M., & Stergiopoulos, V. (2013). Meaningful inclusion of consumers in research and service delivery. *Psychiatric Rehabilitation Journal, 36*(3), 180–186.

Viding, E., & McCrory, E. J. (2012). Genetic and neurocognitive contributions to the development of psychopathy. *Development and Psychopathology, 24,* 969–983.

Wayland, K., & O'Brien, S. D. (2013). Deconstructing antisocial personality disorder and psychopathy: A guidelines-based approach to prejudicial psychiatric labels. *Hofstra Law Review, 42*(2), 519–588.

Wilson, N. J., & Tamatea, A. (2013). Challenging the 'urban myth' of psychopathy untreatability: The high-risk personality programme. *Psychology, Crime & Law, 19*(5–6), 1–18.

Wong, S., & Hare, R. D. (2005). *Guidelines for a psychopathy treatment program*. Multi-Health Systems.

Wootton, B. (1959). *Social science and social pathology*. Macmillan.

Zinger, I., & Forth, A. (1998). Psychopathy and Canadian criminal proceedings: The potential for human rights abuses. *Canadian Journal of Criminology, 40,* 237–276.

Zolnierek, C. D. (2011). Exploring lived experiences of persons with severe mental illness: A review of the literature. *Issues in Mental Health Nursing, 32*(1), 46–72.

Chapter 4
Antisocial Personality Disorders and Public Protection Orders in New Zealand

Jeanne Snelling and John McMillan

Abstract Over the last two decades, there has been a trend in New Zealand's criminal justice system toward longer, harsher sentences for serious crimes, and a greater emphasis on preventive justice. The introduction of post-sentence civil detention in 2014 is a high-water mark in this regard. Imposition of Preventive Detention, Extended Supervision Order's, and post-sentence Public Protection Orders (PPOs) are based on statutory criteria and informed by expert evidence. PPO's in particular, are premised on evidence that the circumstances of the offender fall within specified psychological and social criteria. This chapter explores the use of such criteria to justify the imposition of PPOs in New Zealand.

Keywords Justice · Harm · Public protection · Personality Disorder · Offending

4.1   Introduction

In New Zealand, as in many other countries, 'law and order' is a political platform that can shape election manifestos and determine their result. Protecting the public from those who are considered at risk of serious sexual and/or violent offending is an issue that crosses the political divide, with the rise of penal populism an ongoing concern in New Zealand.[1]

In 2011, New Zealand's (centre right) National Party proposed, as part of its election campaign, new 'civil detention orders' as a centrepiece of its law and

---

[1] J Pratt *A Punitive Society: Falling crime and Rising Imprisonment in New Zealand* (Bridget Williams Books, 2013).

J. Snelling (✉)
Bioethics Centre and Faculty of Law at the University of Otago, Dunedin, New Zealand
e-mail: jeanne.snelling@otago.ac.nz

J. McMillan
Bioethics Centre, University of Otago, Dunedin, New Zealand

© Springer Nature Switzerland AG 2022
L. Malatesti et al. (eds.), *Psychopathy*, History, Philosophy and Theory of the Life Sciences 27, https://doi.org/10.1007/978-3-030-82454-9_4

order policy.[2] The primary objective of the subsequently introduced the Public Safety (Public Protection Orders) Act 2014 is to "protect members of the public from the *almost certain harm* that would be inflicted by the commission of serious sexual or violent offences. (emphasis added)".[3]

Significantly, the PPO regime permits the ongoing and indefinite detention of an eligible person *after* that individual has already completed the sentence originally imposed by the sentencing court,[4] subject to specified risk predictions and behavioural criteria.[5] These orders, which significantly extend New Zealand's turn to 'preventive justice', are controversial from a human rights perspective.

4.2 The New PPO Regime and the Context of 'Preventive Justice'

New Zealand has witnessed an increasing trend towards what has been labelled "preventive" justice in recent years.[6] The "preventive" justice concept is distinguishable from traditional retributive account of criminal law, whereby the law derives its legitimacy from imposing punishment that is responsive and proportionate to the offender's culpability for acts already carried out.[7] Preventive justice, in contrast, is forward-looking and risk-oriented, concerned less with what the offender has already done, than with what he is thought likely to do in future.

Preventive detention in the form of indeterminate sentences is a feature of many countries.[8] The European Court of Human Rights has held that given the States obligation to protect the public, if at the time of sentencing an offender is sentenced to an indeterminate sentence for a serious crime such as murder, no human rights issues arise as long as that sentence is proportionate to the gravity of the crime and it is subject to a guaranteed right to review (i.e. is reducible).[9] Similarly, the United Nations Human Rights Committee has found that a sentence of preventive detention does not breach the right against arbitrary detention, *provided* that there is

---

[2] Derek Cheng "Collins Talks Tough on Detaining Sex Offenders" *The New Zealand Herald* (New Zealand, online ed., Auckland, 8 November 2011).

[3] Public Safety (Public Protection Orders) Act (2014), s 4. http://www.legislation.govt.nz/bill/government/2012/0068/latest/DLM4751015.html

[4] PPO Act, s 7.

[5] PPO Act, s 13.

[6] For example, see A Ashworth, A Lee and L Zedner "Oxford Preventive Justice Project" University of Oxford, Faculty of Law <www.law.ox.ac.uk>.

[7] SJ Morse "Blame and Danger: An Essay on Preventive Detention" (1996) 76 Boston University Law Review 113 at 121 citing P Robinson "Foreword to the Criminal-Civil Distinction and Dangerous Blameless Offenders" (1993) 83 Criminal Law & Criminology 693 at 706–708.

[8] (17 September 2013) 693 NZPD 13445.

[9] *Vinter v United Kingdom* (2013) 57 EHRR. Section 9 of the New Zealand Bill of Rights Act affirms "the right not to be subjected to torture or to cruel, degrading, or disproportionately severe treatment or punishment".

a compelling justification and it is reviewable.[10] In New Zealand, indeterminate sentences imposed at *sentencing* generally have a specified parole period, after which the Parole Board must assess whether the individual still poses a risk to the community. However, detaining a person indefinitely after he has served a finite sentence is a very different endeavour.[11]

4.3   New Zealand's PPO Regime

When the PPO Bill was first introduced in Parliament, then Minister of Justice Judith Collins claimed that PPOs are necessary to fill an apparent gap in the current penal framework. Such orders would 'improve public safety and save potential victims from almost certain serious harm, or worse'.[12] PPOs would supplement existing penal mechanisms (i.e. parole conditions, extended supervision orders, and preventive detention) and would respond to 'situations where an offender presents an unacceptable risk that cannot be managed through these existing measures.'[13] It was claimed that there were no alternative legal options for managing those persons who have finished serving their sentence, but pose a 'very high level of *imminent* risk' (emphasis added).[14] Despite the fact that PPOs are only applicable to prisoners who have served sentences for serious criminal offences, it was claimed that PPOs would not seek to punish for past crimes:

> Although these people have offended in the past, they will not be detained for their previous crimes. They will be detained because of their imminent risk of serious sexual or violent offending, at the time of the application. The test for the risk of imminent future offending will be difficult to meet.

The Parliamentary debates during the enactment of the regime indicated cross-party support for the PPO regime, although it was generally acknowledged that it was in tension with civil and political rights domestically and internationally. The Opposition emphasised that the challenge lay in ensuring that the proposed law is "justified on the basis of an intense risk from a very small number of people".[15] When reporting on the consistency of the PPO Bill with the New Zealand Bill of

---

[10] *Rameka v NZ* (2003) 7 HRNZ 663. See also United Nations Human Rights Committee Views: Communication No 2502/2014 121 CCPR/ C/121/D/2502/2014 (7 November 2017) [*Miller v New Zealand*].

[11] Because PPOs are not "criminal" orders as such, parole periods are not imposed, although the Act requires review of the justification for a PPO by the review panel within 1 year of the Order, and then annually s 15. The court must review the continuing justification of an Order within 5 years of the Order being made, and again 5 years later. Thereafter, a PPO must be reviewed at five yearly intervals unless the court directs a review at intervals of not more than 10 years s 16(1),(2).

[12] NZPD Vol: 693, 17 September 2013 at 13441.

[13] Ibid.

[14] Ibid.

[15] NZPD Vol: 693, 17 September 2013 at 13441.

Rights Act 1990 (NZBoRA), New Zealand's Attorney General noted that the powers contained in the legislation "are new and far-reaching … it is possible that some detainees might never be released. Even those ultimately released would have been detained beyond, and possibly well beyond, their original sentences".[16]

Despite containing the hallmarks of a penal regime, the Act states that it is not its objective to 'punish persons against whom orders are made under this Act.'[17] Further, it sets out specific principles that underpin which includes the principle that a PPO "should only be imposed if the magnitude of the risk posed by the respondent justifies the imposition of the order".[18]

Notwithstanding the claim that PPOs are not imposed to punish persons,[19] the requirement of prior conviction as a "triggering event" is strongly indicative of their penal character. Further, the Court of Appeal has previously held that restrictive measures in response to criminal convictions "amount to punishment."[20] Significantly, while the Act permits the detention of an individual subject to a PPO in a 'residence', the lived experience of a person subject to a PPO is comparable to that of a convicted prisoner. Persons detained under a PPO are in the custody of the Chief Executive of the Department of Corrections, and their movements, visitors and day-to-day management are subject to restriction. The manager of a residence has power to control a detainees' correspondence and phone calls, to undertake drug and alcohol tests, to conduct searches and to put detainees in seclusion and/or to use forcible restraint. While detainees may work, such work may only be undertaken within the residence or in a prison.[21]

If PPOs are properly seen as punitive measures, it has significant implications for human rights. Section 26 (2) of the New Zealand Bill of Rights Act 1990 (NZBORA) provides that "[no] one who has been finally acquitted or convicted of, or pardoned for, an offence shall be tried or punished for it again."[22] The designation of PPOs as criminal may also be relevant to the question of whether they infringe against the

---

[16] See NZBORA, s 7. Office of the Attorney-General, *Public Safety (Public Protection Orders) Bill – Consistency with the New Zealand Bill of Rights Act 1990* (14 October 2012) at [3] Ministry of Justice <www.justice.govt.nz>.

[17] PPO Act, s 4(2).

[18] PPO Act, s 5(b).

[19] Section 4(2) states: "It is not an objective of this Act to punish persons against whom orders are made under this Act." Section 5(a) states: "orders under this Act are not imposed to punish persons and the previous commission of an offence is only 1 of several factors that are relevant to assessing whether there is a very high risk of imminent serious sexual or violent offending by a person."

[20] *R v Peta* [2007] NZCA 28. para 13. See also *Belcher v Chief Executive of the Department of Corrections* [2007] 1 NZLR 507, para 26. Both cases concerned the imposition of extended supervision orders under the Parole Act 2002 (as amended), but it is evident that analogous considerations apply to PPOs.

[21] Office of the Attorney-General, *Public Safety (Public Protection Orders) Bill – Consistency with the New Zealand Bill of Rights Act 1990* (14 October 2012) at [13] Ministry of Justice <www.justice.govt.nz>.

[22] *Police v Gilchrest* (1998) 16 CRNZ 55, at [60]. The judge noted in the District Court that the "defendant is entitled to the certainty that, after the passing of sentence and any time for appeal, her case is over and she can get on with life. There must be finality — an end to proceedings".

right not to be arbitrarily arrested or detained, a right protected in section 22 NZBORA and reflected in Article 9 of the International Covenant on Civil and Political Rights (ICCPR). Arguably the right may be triggered when PPOs are applicable only to people convicted of a serious criminal offence, but not those who have not been convicted but who pose the same level of risk.

4.4   The PPO Act: Eligibility and Threshold Test

It is a central tenet of liberal democracies that when limiting a fundamental human right and freedom such as the right to liberty, the objective of the limitation must be of such importance as to warrant overriding the right, and the means used must be reasonable and demonstrably justified.[23] Sections 7 and 13 of the PPO Act attempt to provide criteria that is commensurate with this principle.

Section 7 sets out the 'threshold' for the imposition of a PPO. A person is 'eligible' if they are over the age of 18 and are either: detained in a prison under a determinate sentence for a serious sexual or violent offence and must be released within 6 months; or is subject to an ESO and is, or has been, subject to specified conditions under the Parole Act 2002; or is subject to a protective supervision order; or the person has arrived in NZ less than 6 months after ceasing to be subject to any sentence, supervision conditions, or order imposed for a serious sexual or violent offence by an overseas court and intends to reside in NZ. Section 13(1) authorizes the court to make a PPO if the respondent meets the threshold for a PPO set out in s 7 and, after "considering all of the evidence offered", and "in particular, the evidence given by 2 or more health assessors[24] including at least 1 registered psychologist" it is *satisfied*:

(b) there is a very high risk[25] of imminent[26] serious sexual or violent offending by the respondent if,—

---

[23] *R v Hansen* [2007] 3 NZLR 1 (SC).

[24] A health assessor is defined as a health practitioner who is deemed to be, or is registered with the MCNZ as a practicing psychiatrist or a registered psychologist (pursuant to the Health Practitioners Competence Assurance Act 2003) (s 3).

[25] The RIS recommended the following definitions: "very high risk" means that the offending is considered extremely likely, "serious" means that the predicted offending would cause serious physical and/or psychological harm to one or more other persons, 'imminent' means that the offending is expected to occur when, provided with a suitable opportunity, the offender would immediately inflict serious harm on a vulnerable victim" [22]. Department of Corrections, Regulatory Impact Statement Management of High Risk Sexual and Violent Offenders at end of sentence (March 2012).

[26] See comment in Department of Corrections, *Regulatory Impact Statement Management of High Risk Sexual and Violent Offenders at End of Sentence* (March 2012) at [31] regarding the importance of retaining "imminence" so that the scope of the legislation is not too wide. It states: "In general, high risk violent offenders typically do not meet the imminence test in respect of future violent offending". Widening it (ie removing the criterion of imminence) would mean many persons would be detained who would not go on to re-offend in a "seriously violent manner and some who would never re-offend in a violent manner at all". [30]

(i) where the respondent is detained in a prison, the respondent is released from prison into the community; or

(ii) in any other case, the respondent is left unsupervised.

The Act defines "imminent" as meaning that a "person is expected" to commit a serious sexual or violent offence "as soon as he or she has a suitable opportunity to do so".[27] Clearly this meaning is not the mainstream definition of "imminent"— i.e. about to happen at any moment. Instead the Act places a gloss on the meaning of "imminent", making it sufficiently broad to encompass opportunistic sex offenders.

Before a court may make a finding that the person poses a high risk of imminent serious sexual or violent offending, the court must be satisfied that the person exhibits specific behavioural deficits set out in s 13(2). Hence, a PPO may only be made if the person "exhibits a severe disturbance in behavioural functioning", which is established by evidence to a "high level"[28] of four specified characteristics. These are:

(a) an intense drive or urge to commit a particular form of offending;

(b) limited self-regulatory capacity, evidenced by general impulsiveness, high emotional reactivity, and inability to cope with, or manage, stress and difficulties;

(c) absence of understanding or concern for the impact of offending on actual or potential victims…;

(d) poor interpersonal relationships or social isolation or both.

Significantly, the Court of Appeal has suggested that a broad interpretation of "exhibits" should be adopted in the context of establishing whether a person has "an intense drive or urge to commit a particular form of offending".[29] The Court has suggested that a "latent" drive is sufficient to satisfy s 13(2)(a), meaning that if a person has exhibited a pattern of offending in the past, and there is nothing to suggest that the relevant traits and behavioural characteristics no longer subsist, it may be inferred that they continue to be present.[30]

The characteristics specified in section 13(2) are premised on a lack of "adaptive functioning" in terms of empathy, social skills, impulse control and judgment. While it requires that "volition", or "control" must be impaired in conjunction with an absence of "understanding" or concern for victims, it differs from the definition of "mental disorder" adopted in the civil commitment regime provided under the Mental Health (Compulsory Treatment and Assessment 1992) Act 1992 (MHA) to justify compulsory assessment and treatment.[31]

The MHA permits compulsory assessment for "mental disorder", defined as an "abnormal state of mind, characterised by delusions, or by disorders of mood or

---

[27] PPO Act, s 3.

[28] 'High level' has been interpreted by the courts as requiring evidence showing that the characteristic is present to a high level.

[29] R v Alinizi [2016] NZCA 3184.

[30] R v Alinizi [2016] NZCA 3184 at [26]. While a trait or behavioural characteristic must be present, they "need not be externally manifest" at the time an application is made.

[31] Mental Health (Compulsory Assessment and Treatment) Act 1992 http://www.legislation.govt.nz/act/public/1992/0046/latest/whole.html

perception or volition or cognition, of such a degree that it seriously diminishes the person's ability to care of themselves, or that it "poses a serious danger to their, or others, health or safety."[32] The MHA deliberately eschews diagnostic categories, instead describing symptoms of an abnormal state of mind that poses a risk of harm to self or others, or seriously diminishes an individual's capacity for self-care.

Offenders with Anti-Social Personal Disorder (ASPD) or psychopathy simpliciter are not ordinarily considered as falling within the definition of the MHA. Rather, the criminal justice system has been the legal route whereby offences committed by such people are managed. Significantly, the PPO Act states that a PPO should not be imposed on a person who is eligible to be detained under the MHA or the Intellectual Disability (Compulsory Care and Rehabilitation) Act 2003.[33]

The PPO criteria set out in section 13 (2) differs from the MHA definition of "mental disorder" largely in that the deficits are more easily categorised as character or behavioural traits, rather than symptoms of mental illness. Indeed, it seems closer to describing the causation of criminal behaviour itself, rather than defining any mental abnormality.

There are clearly significant challenges in determining if someone is at high risk of imminent serious sexual or violent offending. The Department of Corrections, who were largely responsible for drafting section 13(2), stated that it would require psychological assessment including the use of psychometric and actuarial risk assessment procedures.[34] It provided examples of evidence that would suggest an "intense drive or urge to commit a particular form of offending" for the purposes of 13(2)(a). This included: "recurrent and intense deviant fantasy; compulsivity in relation to deviant urges; a pattern of repetitive and opportunistic offending; rapid re-offending following previous releases from custody". The Department claims[35]:

> [o]ffenders of this type display few gains from rehabilitation or are unwilling to participate satisfactorily, usually as a result of low intelligence or other cognitive deficits. Most of these offenders would be child sex offenders, although adult sex offenders may also fall within this group. A very small number of violent offenders may also have the identified characteristics and may meet the imminence test.

As noted above, the s 13(2) criteria list specific behavioural deficits that must be exhibited by a person before a court may make a PPO. Given the Act's rationale of public protection, one might wonder why PPOs could not be made in relation to anyone who presents "a very high risk of imminent serious sexual or violent offending". Arguably, introduction of "clinical" criteria seeks to enable the Act's categorisation as a "civil" measure more plausible, closer to mental health legislation than penal measures, side stepping human rights-based objections to the Act. There is a

---

[32] MH(CAT) Act, s 2.

[33] PPO Act, s 5(c).

[34] Department of Corrections, Regulatory Impact Statement Management of High Risk Sexual and Violent Offenders at End of Sentence (March 2012) at [26].

[35] Department of Corrections, Regulatory Impact Statement Management of High Risk Sexual and Violent Offenders at End of Sentence (March 2012) at [24] and [25].

genuine connection between these characteristics and human rights arguments that, to justify civil detention, those detained must be disordered in some way that results in a diminished/absent ability to control their actions. This argument requires deeper analysis. Of significance for this book is that section 13(2) appears to characterize many of the attributes associated with psychopathy.

4.4.1 Section 13(2) and the Psychopathic Individual

The most utilized assessment tool to identify psychopathy is the psychopathy checklist (PCL-R) developed by Dr. Robert Hare.[36] It assesses 20 emotional and behavioural features that define psychopathy.[37] Of those 20 items, two distinct factors are apparent (with further subcategories in each factor): the first factor (F1) relates to emotional and interpersonal characteristics, the second (F2) to impulsive and antisocial behaviour. They are outlined in the following table:

Interpersonal	Affective
1. Glibness/superficial charm	6. Lack of remorse
2. Grandiose self-worth	7. Shallow affect
4. Pathological lying	8. Lack of empathy
5. Conning/manipulative	16. Will not accept responsibility

Lifestyle	Antisocial
3. Need for stimulation	10. Poor behavioural controls
9. Parasitic lifestyle	12. Early behavioural problems
13. Lack of goals	18. Juvenile delinquency
14. Impulsivity	19. Revocation conditional release
15. Irresponsibility	20. Criminal versatility

Note that the item titles cannot be scored without reference to the formal criteria contained in the PCL-R Manual. Item 11, Promiscuous sexual behaviour, and Item 17, Many short-term marital relationships, contribute to the Total PCL-R score but do not load on any factors.[38]

---

[36] R. Hare, The Hare Psychopathy Checklist-Revised (Multi-Health Systems, Toronto 1991).

[37] Ibid.

[38] Robert Hare and Craig Neumann, "Psychopathy and Its Measurment," in *The Cambridge Handbook of Personality Psychology*, ed. Philip Corr and Gerald Matthews (Cambridge: Cambridge University Press, 2012) at 62.

The characteristics listed in section 13 (2) are arguably similar to the attributes set out in the Lifestyle/Antisocial (F2) items of the PCL-R which, together, are closely associated with Anti-Social Personality Disorder (ASPD).[39] Given that the PPO Act seems to target individuals with high levels of psychopathic traits, it is worth considering its implications for this particular category of offenders.

4.5 Psychopathy and the Law

Psychopathy presents a category of individuals that pose challenges for the law. Although a person may score highly on the PCL-R, and their capacity for self-control may be reduced due to moral and cognitive deficits, such individuals still possess a degree of self-control. Fox and colleagues state[40]:

> Psychopaths lack a number of attributes that are ascribed to an ordinary moral agent: (1) they lack the ability to empathize with the aversive conditions of others, (2) they do not understand the difference between conventional and moral rules, and (3) they do not learn from error in a way that nonpsychopathic persons do. These attributes appear to dispose a psychopath toward being able to commit antisocial acts remorselessly, without regard for needs beyond his or her own, in a manner that is otherwise consistent with an ordinary agent, but nonetheless in the presence of a substantially diminished capacity for ordinary moral reasoning.

The New Zealand Court of Appeal has characterized psychopathy as:[41]

> ... a severe form of personality disorder with distinctive emotional, inter-personal and anti-social features. Highly psychopathic offenders are characterised by emotional deficits such as a lack of empathy or remorse, a manipulative and exploitative interpersonal style, and a blatant disregard for the rights of others. Research has consistently found psychopathy to have a strong relationship to a variety of negative criminal justice outcomes. These include poor response to available treatment interventions, increased involvement in institutional misconduct while incarcerated and high levels of violent and sexual re-offending as compared to less psychopathic offenders.

Consequently, the question of whether, and to what extent, psychopaths should be excused from criminal responsibility is both complex and contested. One problem relates to psychopathy's contested status as a mental illness. While DSM-5 lists "antisocial personality disorder", which contains significant features in common

---

[39] Typically, psychopathic individuals are generally diagnosed with Anti-Social Personality Disorder. A Fox, T Kvaran, and RG Fontaine "Psychopathy and Culpability: How Responsible Is the Psychopath for Criminal Wrongdoing?" (2013) 38 *Law & Social Inquiry* 1–26 at 3.

[40] A Fox, T Kvaran, and RG Fontaine "Psychopathy and Culpability: How Responsible Is the Psychopath for Criminal Wrongdoing?" (2013) 38 *Law & Social Inquiry* 1.

[41] *R v Peta* [2007] NZCA 28 at [39].

with psychopathy,[42] as a personality disorder, it contains no explicit reference to psychopathy itself.[43]

Some empirical studies have indicated that psychopathic individuals have deficits in moral reasoning.[44] It's important to note there is an ongoing debate about whether such studies have significant implications for criminal responsibility.[45] But, it is reasonable to suppose that some psychopathic individuals may not possess the cognitive attributes sufficient for being attributed with *full* responsibility for their actions. The relevance of impaired control to the attribution of criminal responsibility is, however, unclear and controversial. To be able to be found criminally nonresponsible, the legal presumption of sanity must first be rebutted.[46] The defence must essentially *prove* insanity, albeit to the lower legal standard of "balance of probabilities", and more specifically, must demonstrate that, at the time of the offence, they were labouring under "natural imbecility" or "disease of the mind" to such an extent as to render them unable to either: understand either the nature and quality of their actions, or know that they were morally wrong.[47] If it can be established in the circumstances that there was indeed an complete *absence* of understanding, or an inability to know whether the nature of an action is morally wrong, an individual may be relieved of criminal and moral responsibility for that act. This is premised on the view that it is contrary to ethical and legal principles to punish an individual for a criminal act that was not subject to the perpetrator's voluntary control.[48]

In New Zealand, the defence of insanity relies on the presence of a "disease of the mind",[49] and whether a particular condition meets that description is ultimately

---

[42] Coid and Ullrich, for example, have written of 'considerable symptom overlap' between psychopathy and ASPD; "Antisocial personality disorder is on a continuum with psychopathy" *Comprehensive Psychiatry* (2010); 51: 426–433, at 432.

[43] Despite this, in evidence provided to the Court of Appeal, a high PCL-R score was equated with ASPD on the basis that the PCL-R score satisfied diagnostic criteria of ASPD in DSM IV. See *Duvell Chaz Antonio v R* CA New Zealand Court of Appeal (unreported, 198/03 16 October, 5 November 2003 McGrath J, Goddard J, Laurenson).

[44] A Fox, T Kvaran, and RG Fontaine "Psychopathy and Culpability: How Responsible Is the Psychopath for Criminal Wrongdoing?" (2013) 38 *Law & Social Inquiry* 1.

[45] Jefferson, A., & Sifferd, K. (2018). Are Psychopaths Legally Insane? *European Journal of Analytic Philosophy*, 14(1), 79–96. https://doi.org/10.31820/ejap.14.1.5; Jurjako, M., & Malatesti, L. (2018). Neuropsychology and the Criminal Responsibility of Psychopaths: Reconsidering the Evidence. *Erkenntnis*, 83(5), 1003–1025. https://doi.org/10.1007/s10670-017-9924-0; Borg, J. S., & Sinnott-Armstrong, W. (2013). Do psychopaths make moral judgments? In K. A. Kiehl & W. Sinnott-Armstong (Eds.), *Handbook on psychopathy and law* (pp. 107–128). Oxford University Press.

[46] The rule in *R v M'Naghten* (1854) 10 Cl & Fin 200; 8 ER 718 provides that every person is presumed sane and to possess sufficient reason to be responsible for their crime unless the contrary is proved. The *M'Naghten* rules are codified in section 23 of the Crimes Act 1961.

[47] Crimes Act 19611, s 23(2). See *R v Dixon* (2007) 23 CRNZ 911 (CA).

[48] "By definition, an agent who is not morally responsible for behavior does not deserve moral blame and punishment for it." SJ Morse, "Psychopathy and Criminal Responsibility" *Neuroethics* (2008) 1:205–212, at 208.

[49] Crimes Act 1961, s 23(2).

a question of law for the judge.[50] Yet attempting to establish an insanity defence without expert psychiatric support is likely to be, to say the least, a difficult task. Describing the present legal situation in New Zealand, authors of a leading criminal law text conclude that psychopathy could be regarded as a "disease of the mind" for the purposes of an insanity defence, "if there is medical evidence that the condition is regarded as a mental illness".[51]

Even if this were established, however, a defence based upon a diagnosis of the disease of the mind may well fail if the further elements of insanity are not met.[52] New Zealand law defines insanity in cognitive terms; it requires that a criminal defendant does not understand the nature of his action (including its moral nature) as a result of disease of the mind. It does not, however, provide a defence for what might be termed *volitional* insanity. Thus, it has been said that:[53]

> Provided a person's cognitive processes are functioning at a level sufficient to enable the accused to grasp the nature and wrongfulness of his act, the fact that his emotional and volitional capacities are abnormal will not detract from the judgment that he was legally sane.

Insofar as psychopathy or ASPD impact upon an individual's volitional capacities – i.e. his ability to control his behaviour – he will not be excused culpability for criminal acts. What, though, of the cognitive effects? Psychopathy is not typically accompanied by delusions of the type that would deprive the affected person of the ability to understand the nature of their act. Whether it would prevent them from "knowing that the act or omission was morally wrong" is a considerably more contested question, which continues to divide opinion among prominent legal theorists and philosophers.[54] Furthermore, not only may psychopathy be rejected as an excusing or mitigating condition, it may even serve as an aggravating factor in sentencing.[55]

The extent to which psychopathy, or indeed antisocial personality disorder, provide valid grounds for civil detention under New Zealand's mental health legislation

---

[50] *Adams on Criminal Law*, at CA23.05

[51] *Adams on Criminal Law*, at CA23.06

[52] *Adams on Criminal Law*, at CA23.06

[53] AP Simester AP and WJ Brookbanks. *Principles of Criminal Law*. 4th ed. Wellington: Thomson Reuters, 2012, at 356.

[54] N Levy "Psychopaths and blame: The argument from content", *Philosophical Psychology* (2014); 27(3): 351–367 with E Aharoni, W Sinnott-Armstrong, KA Kiehl. "What's wrong? Moral understanding in psychopathic offenders" *Journal of Research in Personality* (2014); 53: 175–181. In Australia too, there is uncertainty whether psychopathy would provide a basis for an insanity defence; see D Lanham, D Wood, B Bartal, R Evans. *Criminal Laws in Australia*. The Federation Press, 2006, at 13. Though the decision of the Australian High Court in *Willgloss v R* [1960] HCA 5 is often thought to suggest that it would not, the judgment – which is now in any event over a century old - did not entirely preclude this possibility.

[55] "To the extent psychopathy is considered at all in sentencing, it will virtually always be considered an aggravating factor, such as using it as a risk factor for dangerousness in capital sentencing."SJ Morse, "Psychopathy and Criminal Responsibility" *Neuroethics* (2008) 1:205–212, at 206. See also A Fox, T Kvaran, and RG Fontaine "Psychopathy and Culpability: How Responsible Is the Psychopath for Criminal Wrongdoing?" (2013) 38 *Law & Social Inquiry* 1 at 2.

is another contested question. As already noted, the MHA defines "mental disorder" in terms of abnormal states of mind characterized by "delusions, or by disorders of mood or perception or volition or cognition." It is known that psychopaths are volatile and struggle to control impulses, and that many have diminished cognition.[56] However, it is also seems clear that the legislative intent and clinical interpretation of the "mood" and "cognition" requirements in the MHA tend to be interpreted as symptoms of mental illness, such as full blown clinical depression or schizophrenia.

4.6 The Ethics of Psychopathy and Treating the Psychopathic Individual

Classic theoretical accounts of when we might restrict liberty so as to prevent harm to others weigh the harm to be prevented, against the ability of an agent to act voluntarily.[57] In cases where someone has a seriously reduced capacity for voluntary action, perhaps due to an acute psychotic episode, intervening to curtail actions that might lead to others, or themselves, being harmed would be justifiable. However, in the context of mental illness, there is also an obligation to act therapeutically for that person, in addition to preventing harm. Consequently depriving someone of their liberty because they pose a risk of harm to themselves or others, because they are unable to understand or control their actions due to mental illness, is justified insofar as we are working toward their recovery. This is why New Zealand's MHA is claimed to be justified in a liberal democracy.

Conversely, those who have only a diminished ability to understand or control their actions (such as psychopathic individuals) may not be covered by Mental Health legislation, because they would often not meet the statutory threshold for "mental disorder". But insofar as the proposed PPO system rests on analogous reasons – diminished voluntariness combined with a risk of harm to others – it is at least arguable that it should give rise to similar obligations.

If a person who scores highly on the PCL-R and is deemed to be a psychopath does not have the same propensity to observe social norms as most people, and given that their liberty has been restricted in a way that would not ordinarily be warranted, the state arguably incurs an ethical obligation to do what can reasonably be done to assist them moderate or control their violent or sexual impulses to the greatest extent possible and to enable them to achieve liberty. Indeed, New Zealand's Attorney-General expressly noted that detention justified on the basis of "apprehended risk" must occur in a "distinct clinical and presumptively

---

[56] C Harenski, R Hare, K Kiehl "Neurodevelopmental bases of psychopathy: a review of brain imagining studies" in Malatesti and McMillan eds *Psychopathy and Responsibility: interfacing law, psychiatry and philosophy*. Oxford University Press 2010.

[57] J Feinberg *Harm to Self: The Moral Limits of the Criminal Law* (Oxford University Press, New York 1986).

therapeutic context" if it is to comply with the sorts of standards required by the European Court of Human Rights.[58]

If it is possible to justify civil commitment post-sentence for the protection of the public, this justification is premised on the correlative rights of others—i.e. the rights of society not to exposed to a heightened risk of harm.[59] However, this correlative rights justification depends upon establishing such a *risk* of harm. While risk is always difficult to quantify, there is some evidence that psychopathy, particularly if it co-exists with sexual deviancy,[60] provides a strong indicator of future sexual re-offending.[61] While there may be elevated risk when an individual has high levels of psychopathic traits, this risk level is not necessarily fixed or wholly resistant to therapeutic interventions.

New Zealand clinical psychologists Nick Wilson and Armon Tamatea note that the treatment or management of "psychopathic behaviour" is yet to receive rigorous study.[62] However there is some evidence that focusing on dynamic-risk factors associated with offending i.e. the factors that may alter with intervention/treatment, rather than on focusing on the psychopathic individual's basic personality, may reduce sexual or violent offending. So while psychopathy poses barriers to treatment, addressing the factors that tend to produce criminality in general may present a promising path and the development of "innovative programmes," for individuals with high level of psychopathic traits.[63] Wilson and Tamatea's experimental treatment programme run by the Department of Corrections, the High-Risk Personality Programme, targeted reducing violence in a psychopathic group. After completing

---

[58] Office of Attorney General, 'Public Safety (Public Protection Orders) Bill – Consistency with the New Zealand Bill of Rights Act 1990' (14 October 2012), para 23.2.

[59] The correlative rights principle claims that because others are potentially placed at risk by "dangerous/psychopathic" persons, this justifies protective mechanisms to affect the rights of other(s) to be safe/protected from harm. A Fox, T Kvaran and RG Fontaine "Psychopathy and Culpability: How Responsible Is the Psychopath for Criminal Wrongdoing?" (2013) 38 *Law & Social Inquiry* 1–26.

[60] In this context sexual deviancy constitutes sexual activity with children or coercive sex with non-consenting adults. *R v Peta* [2007] NZCA 28 at [41]. See also *Kerr v CEDC* [2017] NZHC 2366.

[61] *R v Peta* [2007] NZCA 28 at [39] and [42] citing Hildebrand, de Ruiter and de Vogel "Psychopathy and Sexual Deviance in Treated Rapists: Association with Sexual and Nonsexual Recidivism" (2004) 16 Sexual Abuse: A Journal of Research and Treatment 1. The study found a sexual reconviction rate of 82% over an average follow-up of 11.8 years for offenders who were both psychopathic and sexually deviant, in comparison to 18% for offenders who were both non-psychopathic and non-deviant. See also Rice and Harris "Cross-validation and Extension of the Violence Risk Appraisal Guide for Child Molesters and Rapists" (1997) 21 Law and Human Behaviour 231.

[62] N Wilson and A Tamatea "Challenging the 'Urban Myth' of Psychopathy Untreatability: the High-Risk Personality Programme" (2013) 19 Psychology, Crime & Law 493. See also D Polaschek and Tadhg Daly "Treatment and Psychopathy in Forensic Settings" (2013) 18 Aggression and Violent Behavior 592.

[63] N Wilson and A Tamatea "Challenging the 'Urban Myth' of Psychopathy Untreatability: the High-Risk Personality Programme" (2013) 19 Psychology, Crime & Law 493 at 495, citing S Wong and R Hare *Guidelines for Psychopathy Treatment Program* (Toronto, Ontario, Multi-Health Systems, 2006).

the programme most of the 12 participants were able to have their "high" security classification reduced following the programme.[64] The authors state:[65]

> Despite mixed offending results and the limitations of one small sample, the HRPP outcomes appear promising, suggesting that not only do psychopathic offenders – as a group – appear to benefit from correctional programmes (Polaschek et al, 2005), but they may further benefit from purpose-built interventions designed to target specific features.

While the limitations of the study should be acknowledged and care taken not to over-generalise the outcome of the programme to high-risk psychopathic populations, the study demonstrates measurable and positive changes for participants.[66] They conclude[67]:

> The more we understand these offenders in terms of their functional differences the greater our ability to assist those able to change and to identify those who, at this stage, remain a significant risk to others if released ... The HRPP outcomes while positive should be interpreted as small steps towards understanding this population as a group defined by challenging behaviours.

4.7   Conclusion

The Public Safety (Public Protection) Orders Act introduced far-reaching powers for post sentence civil detention in New Zealand. The PPO regime is claimed to constitute a civil commitment regime, and targets offenders deemed high risk of imminent serious sexual or violent offending. At its core, it seeks to prevent those considered to be at high risk of committing a serious sexual or violent offence if given a suitable opportunity, from being availed of such an opportunity. When imposing a PPO, the High Court must be satisfied that the offender meets the threshold requirements for imposition of a PPO, the offender exhibits specific maladaptive behavioural characteristics, and that specified social criteria are met. In addition, the court must be satisfied that no lesser restraint is adequate to meet the perceived risk to public safety. New Zealand's Supreme Court has stated that the PPO Act 'is to be

---

[64] This involved violence as measured by the Violence Risk Scale (VRS) which was developed by Wong and Gordon and comprises a structured clinical judgment involving both static and dynamic variables. See N Wilson and A Tamatea "Challenging the 'Urban Myth' of Psychopathy Untreatability: the High-Risk Personality Programme" (2013) 19 Psychology, Crime & Law 493 at 499.

[65] N Wilson and A Tamatea "Challenging the 'Urban Myth' of Psychopathy Untreatability: the High-Risk Personality Programme" (2013) 19 Psychology, Crime & Law 493 at 505.

[66] N Wilson and A Tamatea "Challenging the 'Urban Myth' of Psychopathy Untreatability: the High-Risk Personality Programme" (2013) 19 Psychology, Crime & Law 493 at 505.

[67] N Wilson and A Tamatea "Challenging the 'Urban Myth' of Psychopathy Untreatability: the High-Risk Personality Programme" (2013) 19 Psychology, Crime & Law 493 at 507.

interpreted and applied in the context of human rights obligations protective of liberty and suspicious of retrospective penalty.'[68]

Despite the pervasive pessimistic view regarding the ability to provide treatment, rehabilitate and reintegrate high-risk offenders, several innovative and intensive programmes run in New Zealand prisons have reported that participants have made modest, but nevertheless significant, improvements in relation to diverse treatment goals.[69] If this indeed is a category of individuals targeted by the Public Safety (Public Protection Orders) Act 2014, there is a strong argument that early treatment and intervention programmes to reduce the likelihood of those individuals becoming subject to a PPO at the end of their sentence should be provided. Additionally, the right to rehabilitation contained in section 36 of the Act is highly relevant in this respect: it states that a 'resident' is entitled to receive rehabilitative treatment' *if* that the treatment has a 'reasonable prospect of reducing the risk to public safety posed by the resident" (emphasis added). Given the emerging research regarding treatment of high-risk offenders and the extreme limitations placed on an individual's liberty by a PPO, the conditional nature of this 'right' to rehabilitative treatment is problematic.

The need to ensure any on-going preventive detention is reasonable, necessary, subject to review (and is therefore not arbitary), *and* is aimed at facilitating a detainee's rehabilitation and reintegration, is reinforced by the UN General Comment No. 35 on Article 9 of the ICCPR.[70] That this is a pressing issue in New Zealand has been made apparent in a recent Communication from the UN Human Rights Committe.[71]

---

[68] *Chisnall v Chief Executive of the Department of Corrections* [2017] NZSC 114.

[69] D Polaschek, J Yesberg and P Chauhan "A Year Without Conviction: an Integrated Examination of Potential Mechanisms for Successful Reentry in High-Risk Violent Prisoners" (2018) 45 Criminal Justice and Behavior 425; D Polaschek "How to Train Your Dragon: An Introduction to the Special Issue on Treatment Programmes for High-Risk Offenders" (2013) 19 Psychology, Crime & Law 409; D Polaschek and T Kilgour "New Zealand's Special Treatment Units: the Development and Implementation of Intensive Treatment for High-Risk Male Prisoners" (2013) 19 Psychology, Crime & Law 511; N Wilson, G Kilgour and D Polaschek "Treating High-Risk Rapists in a New Zealand Intensive Prison Programme" (2013) 19 Psychology, Crime & Law 527.

[70] UN General Comment No. 35 on Article 9 of the ICCPR, at [21].

[71] United Nations Human Rights Committe Views: Communication No 2502/2014 121 CCPR/C/121/D/2502/2014 (7 Novermber 2017) [*Miller v New cealand*].

Chapter 5
Psychopaths – A "Tough Nut" of Forensic Psychiatry Practice in the Republic of Croatia

Vesna Šendula Jengić, Velinka Grozdanić, and Jelena Hodak

Abstract Currently in the Republic of Croatia, in the clinical context, the term "psychopathy" is used according to the International Classification of Diseases 10th revision. In the legal context, it falls under the Criminal Code category of "some other serious mental disturbance".

The chapter describes the historical and the current legal context related to psychopathic and other mentally disturbed perpetrators in the Republic of Croatia as well as the position and the difficult task of the psychiatric expert witness in legal proceedings. It also presents a case-law research on the Supreme Court final judgments in criminal proceedings dealing with psychopathic perpetrators of serious criminal offences, namely murder and aggravated murder, and describes the cooperation of legal authorities and psychiatrists in such proceedings, and give examples of reasonings of final judgements.

In conclusion, the authors emphasize the need to appreciate the fact that our knowledge about psychopathy is still wanting. Awareness of these limitations should render us more careful when it comes to diagnostic assessment of these persons, as well as their penal and/or medical management.

Keywords Psychopathy · Antisocial personality disorder · Mental capacity · Criminal responsibility · Expert witness · Criminal code · Forensic psychiatry

V. Šendula Jengić (✉)
Rab Psychiatric Hospital, Rab, Croatia

Faculty of Medicine, University of Rijeka, Rijeka, Croatia

V. Grozdanić
Faculty of Law, University of Rijeka, Rijeka, Croatia

J. Hodak
Rab Psychiatric Hospital, Rab, Croatia

Faculty of Health Studies, University of Rijeka, Rijeka, Croatia

© Springer Nature Switzerland AG 2022
L. Malatesti et al. (eds.), *Psychopathy*, History, Philosophy and Theory of the Life Sciences 27, https://doi.org/10.1007/978-3-030-82454-9_5

'I can't,' he said, 'my dear man, I can't get well, because I'm not ill. I am as I am, and one can't be cured of oneself.'
Ivo Andrić "Devil's Yard"
1961 Nobel Prize winner

5.1 Introduction

The collaboration of lawyers and doctors dates back for centuries. It resulted from the need for medical knowledge in solving legal problems. In this context, a special place belongs to forensic psychiatry which deals with persons who come into conflict with legal norms due to their mental disorder, temporary mental disturbance, insufficient mental development or some other serious mental disturbance, i.e. with persons who require a special legal status because of their mental health condition.[1] The legal issues related to this category of persons are numerous and very different: forced hospitalization, criminal responsibility of persons with mental disorders, their legal capacity, capacity to enter contracts, marry and be parents, capacity to drive, own firearms, etc. Among all these issues and legal provisions regulating them, which belong to different branches of law (criminal, civil, family, labour, administrative), the most important one in forensic psychiatry is the study and interpretation of those behaviours of persons with mental disorders that violate the norms of criminal law, i.e. that perform criminal acts. This is understandable because the serious consequences of some criminal offences cause discomfort, fear, and sometimes disturb the public to a great extent. However, when the perpetrator of a criminal offence is a person with mental disturbances, the problem acquires a new dimension, and it cannot be resolved without the cooperation of lawyers and psychiatrists.

The closest cooperation of law and psychiatry is found in the field of medical expertise for the purposes of court proceedings: criminal, civil and non-litigation. Indeed, the study of the mental state of perpetrators of serious criminal offences for the purpose of determining their criminal responsibility is what gave birth to forensic psychiatry. Psychiatrists are faced with requests to explain the various, sometimes very bizarre, and/or monstrous crimes in "terms of the mind". Expert witnesses as "attorneys of the unconscious" *in foro* had to use their best knowledge and

---

[1] The development of forensic psychiatry is directly linked to the historical development of psychiatry and legal science and has taken place in accordance with the laws of the historical development of civilization and society in general. This development was not perpendicular, on the contrary, there were numerous oscillations, ups and downs, advances, and setbacks. Thus, in periods of advanced civilization, culture, science, especially medicine (end of Ancient Age and end of the Middle Ages), people with mental illness were treated relatively humanely, as opposed to the period of scientific darkness, Christian dogmatism (5th – 15th ct.) when treatment of the mentally ill was cruel, inhumane, pervaded by physical torture and led to their physical destruction. For more on the historical development of the legal status of mentally incapable persons, see Kozarić-Kovačić et al. (2005) pp.12–20.

experience while at the same remaining very aware of their role in the court process, especially the possible counter-transference reactions. Today however, the role of forensic psychiatry in the field of criminal law is much wider. Psychiatric expertise, apart from its primary task of assessing the mental capacity of mentally disturbed perpetrators, also includes assessment of their current and future danger to the environment. The expert opinions and recommendations for further treatment of such delinquents, as well as planning of the treatment-related activities, are of great value to judges.

And thus, two completely different scientific disciplines – psychiatry and law – which belong to different fields of science, have different systems of thinking and knowledge, apply different methodologies and use different terminology, come together on a common task of understanding human beings and their behaviour, all for the sake of a fairer trial. There is no doubt that the task at hand is extremely demanding. In fact, one can justifiably ask whether there is anything more demanding than understanding a human being and their behaviour. Is it possible for us to understand others when often we do not understand ourselves? How are we to understand those sudden, completely unexpected behaviours that inflict mental suffering, grievous bodily harm or even death? Can we offer the right solutions in absence of proper understanding? There are many difficult and complex questions, and very few clear and unambiguous answers. Nevertheless, in forensic psychiatry for each specific case of a mentally disturbed offender both law and psychiatry strive to offer interpretation, reasoning, and solution within the realm of the scientific reach of their respective disciplines. Sometimes these explanations and solutions seem logical, justified and fully socially acceptable. However, there are also instances when, despite the efforts of both professions, the cases remain unclear, incomprehensible, intractable, leaving an impression of insufficiency of psychiatry as well as the law to adequately deal with them. Such a "hard nut" for both psychiatry and the law is encountered in cases of psychopathic offenders.

5.2 Psychiatric Approaches

Since the first edition of Cleckley's *Mask of Sanity* in 1941 (Cleckley, 1988), which initiated more profound research into psychopathy, medicine has come a long way, and psychopathy has been the focus of research from many different medical fields: genetics, psychophysiology, neurophysiology, endocrinology, neurobiology: all in the attempt to find the biological correlates of psychopathic traits reflected in morality, emotions or antisocial and violent behaviour (Glenn, 2011, Waldman et al., 2018; Glenn et al., 2011; Dadds et al., 2014; Flórez et al. 2017; Perez, 2012). Better understanding or psychopathy aims to enhance the diagnosis, prevention, and treatment options of the affected population. In recent decades we have witnessed a fascinating development in neuroscience and various brain imaging methods (PET, SPECT, fMRI, DTI...), and so far abnormalities have been observed at a functional or structural level across a range of brain regions, including: prefrontal cortex, amygdala, temporal cortex, hippocampus, striatum, corpus callosum, and others (for a literature review see Cummings, 2015; Yang & Raine, 2018). However, despite numerous discoveries, the questions of causation, origin, predisposing, precipitating and environmental factors, and many others related to psychopathy and antisocial behaviour, remain open. In a legal context however, cause and developmental dynamics of the disorder are less significant. Instead, it is necessary to link an impairment, symptom, or disorder, or more particularly its direct impact, to a legal norm.

Psychopathy is a term that has historically been used for a variety of psychopathologies. Until the 1990s and Hare's PCL, a reliable diagnostic instrument did not exist, but diagnosis was based on unstructured assessments and the clinician's personal impression and experience. Over time, and especially with the emergence of the PCL-R diagnostic tool (Hare, 1991), the meaning narrowed down to a specific construct, but the term is still often inaccurately and inconsistently used or confused with sociopathy, antisocial and dissocial personality disorder. The two main internationally recognized classifications of mental illnesses: the Diagnostic and Statistical Manual of Mental Disorders 5th Edition – DSM-5 (American Psychiatric Association, 2013) and the International Classification of Diseases 10th Edition – ICD-10 (World Health Organization, 1992) – still do not list psychopathy as a diagnosis but as a subtype of Antisocial or Dissocial Personality Disorder. For example, in the first edition of DSM "sociopathic personality disorder" was further clarified: "the term includes cases previously classified as 'constitutional psychopathic state' and 'psychopathic personality'" (for a review of the history of the term in the DSM see Crego & Widiger, 2014). In the current version of the DSM classification of mental disorders, DSM-5 (American Psychiatric Association, 2013), although the categorical approach of the previous version of this manual is retained in the section on personality disorders, in an attempt to better identify the different variations in psychopathology, the Alternative DSM-5 Model for Personality Disorders has been introduced. It offers a dimensional approach to the diagnosis of these disorders based on the domains of personality features or facets within the domains, namely: Negative Affectivity, Detachment, Antagonism, Disinhibition and Psychoticism. In this section, the term psychopathy is offered as an additional feature in the diagnosis of antisocial personality disorder, manifested by the "lack of anxiety or fear and by a bold interpersonal style that conceals maladaptive behaviours (e.g. fraudulence)." (American Psychiatric Association, 2013).

Similarly, while ICD-10 lists psychopathy as a subtype of 'Dissocial Personality Disorder' along with amoral, antisocial, asocial and sociopathic, the most recent revision of this classification, the ICD-11, introduces a change from categorical towards dimensional diagnostic approach. It focuses on the severity of a personality disorder (Mild, Moderate, Severe, Unspecified) and trait domains which are further specified within each of the following categories: Negative Affectivity, Detachment, Disinhibition, Dissociality, and Anankastia (World Health Organization, 2018). In this, it largely coincides with the Alternative DSM-5 Model for Personality Disorders, with the exception of the absence of the domain of psychoticism and the addition of anankastia.

In clinical practice, the terms *psychopathy* and *antisocial personality* are often used as synonyms, thus the probability of misapplication of these terms is considerable since antisocial behaviour by itself is a nonspecific symptom present in various psychiatric conditions. However, when it comes to clinical diagnostics, psychiatrists strictly apply the nomenclature of the current International Classification of Diseases ICD-10 and adhere to the criteria of Dissocial Personality Disorder of psychopathic type. The degree of psychopathic deviation is determined by using different clinical instruments, such as Emotions Profile Index – PIE (Plutchik & Kellerman, 1974), Eysenck Personality Questionnaire – EPQ (Eysenck & Eysenck, 1991), Personality Assessment Inventory – PAI (Morey, 2007), Minnesota Multiphasic Personality Inventory – MMPI-2 (Butcher et al., 2001), Freiburg Personality Inventory Revised – FPI-R (Fahrenberg et al., 2001), Structured Clinical Interview for DSM-IV Axis II Personality Disorders – SKID-II (First et al., 1997).

Although the *International Classification of Diseases* (ICD-10) is the official classification for mental health disorders in the Republic of Croatia, it is not uncommon to use the DSM classification for research and for a better understanding of mental disorders. The above-named instruments may contribute to the description of the personality type and may be useful for forensic assessment since they can describe the presence and intensity of a trait. However, they do not illustrate all aspects of the emotional, interpersonal, and behavioural spectrum, nor their inter-relationships, which would be decisive in forming a forensic psychiatric evaluation. The question of the capacity for insight and responsibility can be addressed only after reaching a conclusion about the possible psychopathological motivation for a specific behaviour.

The concept of psychopathy is linked to negative behaviour and is highly stigmatized. In the forensic legal context, judgment is formed with respect to past behaviour and acts, and the diagnosis itself has to be subject to legal norms in such a way as to quantify the legally relevant severity of the disorder and to indicate its impact on a specific illegal behaviour. Psychiatrists in the Republic of Croatia have different stances regarding personality disorders, depending on the psychotherapy type or school to which they belong. From a psychiatric perspective, the operationalization of this term would include a personality disorder along with the relationship of the diagnosis to the degree of danger as well as to other diagnoses such as substance abuse and others. It should be emphasized that the entirety of clinical psychiatry is based on a kind of philosophy of continuity in which the individual is viewed in their complexity, and through their relationship with their environment. Personality is formed during childhood and continues to be moderated throughout the course of life. It is affected by many factors including biological predispositions, prenatal, perinatal, and postnatal factors, age and developmental stage, influence of the family while growing up, broader cultural influences, life events and more. Personality and behaviour patterns in adulthood are usually stable, but they may change under different circumstances. Therefore, personality should be viewed as a certain continuum – from healthy to disturbed – and the intensity of the problem can be assessed directly or indirectly. When evaluating a personality, it is important to look at a person's identity, but also object relations and tolerance of affect.

Healthy personalities see themselves and others in a stable and secure way. They are able to maintain satisfactory relationships with others, to perceive and tolerate other people's affect and to regulate their own impulses and affect by flexible use of defence mechanisms and strategies. They are morally responsive and consistent in terms of mature integration of the superego, the ideal self-concept, and the ego ideal. They test reality adequately and demonstrate satisfactory resilience (resistance) to stress as well as rapid recovery from painful events.

On the other hand, in a neurotic personality most of the above-mentioned abilities are maintained, although there are some problems in functioning that these individuals are able to describe realistically. In case of borderline personality organization (which is different from borderline personality disorder), a person is often not able to describe problems, but they need to be inferred through observation or testing. This type of personality organization more often has problems and limitations in the domains of identity, object relations, affect tolerance and regulation, integration of superego, and in case of narcissistic and antisocial personalities also ego resilience. Persons with neurotic disorders can have good work results and other achievements outside the spectrum of their difficulties. They can maintain quality relationships with others and tolerate dysphoric affect without somatization. Unlike neurotic disorders, in the case of borderline personality organization there are significant disabilities in overall psychosocial functioning.

Considering the above, in any individual case of a possible behavioural disorder the psychiatrist must invest a great deal of effort, knowledge and experience to provide an accurate diagnosis. If the individual with a personality disorder is an offender, the challenge is even greater. The psychiatrist then accepts the additional forensic role, which means that they are confronted with numerous legal texts, rules, and norms, which they must be familiar with in order to fulfil the very demanding role of a psychiatric expert witness in legal proceedings. In the process of expert witness evaluation of the defendant, the psychiatrist observes the subject's emotions, behaviour, attitude toward the act and life values. The interviewing and the report should be exclusively of observatory nature. However, a lot of experience is needed, as well as continuous work on oneself, for the expert witness to be able to shut out the experiential ego which might direct the identification with the subject or the penal system (Sattar et al., 2004). Indeed, it is in the very nature of every human being to observe irrational behaviours of other persons more easily than one's own. That is why a continuous education of psychiatric expert witnesses is imperative in order for them to better understand their own defence mechanisms, so that they could preserve their moral integrity, understand the dynamics of court proceedings, and partake in it in an utmost objective manner.

In the forensic legal sense, the evaluation of the severity of a mental disturbance is oriented toward its clinical effect on behaviour rather than toward a structural or functional disorder. In other words, a serious mental disturbance would be a more or less permanent readiness for abnormal behaviour resulting from abnormal personality predispositions and environmental stressors. The roots of the disorder are of secondary importance, and the disorder must be severe to the extent that it has its psychopathological expression in the narrow sense, which in its wider sense comprises a sociopathological expression and calls into question the issue of loss of control (Šendula Jengić, 2008).

5.3 Legal Perspective

5.3.1 Mental Health Legislation in the Republic of Croatia

In modern societies, the legal status of persons with mental disturbances is determined by laws that, by defining their rights and regulating the conditions of application of medical procedures on them, represent extremely important legal mechanisms functioning as protection of this category of persons. The sensitivity of the position of the mentally ill stems not only from their inability or difficulty to exercise their rights, but more often from their inability to sometimes judge what is in their own interest, and occasionally from behaviours that endanger their lives and the health, life, or safety of others. It is for these reasons, and under certain conditions prescribed by law, that involuntary placement and treatment of these persons may be justified. Since from the perspective of the person to whom it applies involuntary placement and treatment mean the deprivation or restriction of their fundamental human rights, such as the right to liberty, the right to move, the right to undergo medical procedures only with consent, the right to autonomy and self-determination, it must be based on the constitution, determined by law, and subject to judicial decision and control. For this reason, contemporary states pay special attention to this in the legislative field and define normative content seeking to provide mechanisms for the protection of persons with mental disorders through legal definition of the behaviour of all the participants involved in the treatment of persons with mental disorders (criminal, civil, non-litigation, etc.).

The standards aiming to achieve a high level of protection for persons with mental disturbances have been set by resolutions adopted by the World Health Organization (WHO), the Council of Europe, the Council of the European Union, and the General Assembly of the United Nations. Of particular importance is the Resolution WHA66.8 of 27 May 2013, as it adopted a Comprehensive Mental Health Action Plan for the period 2013 to 2020.[2]

In the Republic of Croatia, the regulations governing the treatment of persons with mental disorders are numerous, such as the Health Care Act,[3] the Criminal

---

[2]Amongst the most important are: WHO Resolution EB130.R8 on Global Burden of Mental Disorders and the Need for a Comprehensive, Coordinated Response from Health and Social Sectors at the Country Level of 20th Jan. 2012; V. COE Resolution 1946 (2013) on Equal access to health care of 26th Jun., 2013.; V. COEU Resolution (2000/C 218/03) on Action on Health Determinants of 29th Jun., 2000.; UN General Assembly Resolution 65/238 on Scope, Modalities, Format and Organization of the High-Level Meeting of the General Assembly on the Prevention and Control of Noncommunicable Diseases of 24th Dec., 2010.

[3] Health Care Act, Official Gazette 100/2018.

Code,[4] the Criminal Procedure Act,[5] the Family Act,[6] the Social Welfare Act,[7] The Execution of Prison Sentence Act,[8] the Labour Act[9] and, most importantly, the Act on the Protection of Persons with Mental Disorders.[10]

The first Act on the Protection of Persons with Mental Disorders in the Republic of Croatia was adopted in 1997.[11] It was an important legal mechanism with the function of protecting persons with mental disorders. It incorporated the already well-known and generally recognized international standards that had, decades earlier, been the foundation of the legal status of these people in most Western European and American states (see Kallert & Torres-Gonzales, 2006). However, it can also be said that this law presented a significant problem for all those, both lawyers and psychiatrists, who had to apply its provisions. Specifically, the discrepancy between the high standards prescribed by the law itself and the existing practice of treatment of persons with mental disorders posed an insurmountable obstacle to the implementation of the law (see Grozdanić & Tripalo, 2013). That is why in 1999, just over a year after its adoption, the Act on the Protection of Persons with Mental Disorders was substantially amended, abandoning the prescribed high standards, and seeking to mitigate the gap between the legal provisions and the existing level of practice in the treatment of persons with mental disorders.[12] A significant change to this law was made again in 2002 in the provisions regulating involuntary hospitalization of persons lacking mental capacity.[13]

The current Act on the Protection of Persons with Mental Disorders was passed in 2014 and came into force on 1 January 2015. The reasons for its adoption result from the need to align its provisions with the new Criminal Code and the new Criminal Procedure Code. In addition, the relevant legally binding conventions have in the past 15 years set new, far higher standards in the protection of people with

---

[4] Criminal Code, Official Gazette 125/2011, 144/2012, 56/2015, 61/2015, 101/2017 and 118/2018.

[5] Criminal Procedure Act, Official Gazette 152/2008, 76/2009., 80/2011, 121/2011 cleared text, Official Gazette 91/2012 Decision of the Constitutional Court of the Republic of Croatia, 143/2012, 56/2013, 145/2013, 152/2014 and 70/2017.

[6] Family Act, Official Gazette 103/2015.

[7] Social Welfare Act, Official Gazette 157/2013, 152/2014, 99/2015., 52/2016, 16/2017 and 130/2017.

[8] Execution of Prison Sentence Act, Official Gazette 128/1999, 55/2000., 59/2000, 129/2000, 59/2001, 67/2001, 11/2002, 190/2003 – cleared text, 76/2007, 27/2008, 83/2009, 18/2011, 48/2011, 125/2011, 56/2013 and 150/2013.

[9] Labour Act, Official Gazette 93/2014 and 127/2017.

[10] Act on the Protection of Persons with Mental Disorders, Official Gazette 76/2014.

[11] Act on the Protection of Persons with Mental Disorders, Official Gazette 111/1997, 27/1998, 128/1999, 79/2002.

[12] Act on Amendments to the Act on the Protection of Persons with Mental Disorders, Official Gazette 128/99.

[13] Act on Amendments to the Act on the Protection of Persons with Mental Disorders, Official Gazette 79/02.

intellectual disabilities.[14] Finally, another significant reason to pass a new Act on the Protection of Persons with Mental Disorders is the existence of the already rich case law of the European Court of Human Rights concerning the protection of persons with mental disorders, including specific judgments of that court against the Republic of Croatia, which arose due to inadequate legal provisions and treatment of psychiatric patients.[15]

In addition to the fact that the Act on the Protection of Persons with Mental Disorders is the most important legal mechanism in the protection of psychiatric patients in the Republic of Croatia, unlike all other regulations which affect the legal status of persons with mental disabilities, this Act is *lex specialis*. Thus, in case of different regulations for the same content, the provisions of this Act are always applied.

5.3.2   Psychopaths as Perpetrators of Criminal Offences

5.3.2.1   Criminal Code Regulations

When it comes to psychopathic criminal offenders the Criminal Law and Criminal Procedure Code occupy a special place. Psychopaths are common among perpetrators of offences against life and limb, property, and sexual offences, but also among recidivists (Grozdanić et al., 2013). This statement is based on research conducted in the Republic of Croatia. For example, a survey conducted at the Centre for Forensic Psychiatry in Vrapče on a sample of 150 persons diagnosed with personality disorder as their first diagnosis, produced the following results: 40% were perpetrators of criminal offences against life and limb, 26% against property, and 22% were sexual delinquents. Recidivism rate among the subjects in this sample was as high as 74% (Kozarić-Kovačić et al., 2005). Although in most cases they are of sound mind *tempore criminis*, sometimes their mental capacity may be diminished (the cumulative effect of either specific affective states or toxic effects of alcohol or drugs), or very rarely it can be absent. Therefore, the provisions of the Criminal Code governing the institutes of diminished responsibility, insanity (mentally

---

[14] For example, Convention for the Protection of Human Rights and Dignity of the Human Being with regard to the Application of Biology and Medicine: Convention on Human Rights and Biomedicine, Official Gazette – International Contracts (OG-IC) 13/2003, 18/2003, Universal Declaration on Bioethics and Human Rights of 19 October 2005, Convention on the Rights of Persons with Disabilities and the Optional Protocol to the Convention on the Rights of Persons with Disabilities, OG-IC 6/2007, 3/2008, 5/2008.

[15] For specific cases against the Republic of Croatia v. Đurđević, Z., Ivičević – Karas, E. (Eds.): Judgments of the European Court of Human Rights against the Republic of Croatia in Criminal Matters, Faculty of Law, University of Zagreb, Zagreb, 2013. On the rights of persons with mental health problems in recent case law The Court has discussed the cases ECtHR, Djordjevic v. Croatia, no. 41526/10, dated 24 July 2012; ECtHR, A.K. and L. v. Croatia, no. 37956/11, of 8 January 2013; ECtHR, M.S. v. Croatia, no. 36337/10, dated 25 April 2013.

incapable persons) and self-induced mental incapacity (voluntary intoxication) are particularly relevant to this category of persons.

According to the provisions of the Criminal Code, psychopaths can fall into the category of persons with *a serious mental disturbance*. In addition to mental disorder, temporary mental disturbance and insufficient mental development, the legal category *some other serious mental disturbance* represents the biological basis of mental incapacity (insanity) or substantially diminished responsibility. Although a psychiatric definition of this category does not exist, it is arbitrarily understood to cover personality disorders including psychopathy, neurotic and impulse disorders. The criminal law applies only to those mental disorders that can be classified as serious. Since the Criminal Code does not define *serious mental disturbances*, the relevant definition may be that of the Act on the Protection of Persons with Mental Disorders, which states that a serious mental disturbance refers to a disorder as defined by valid internationally recognized classifications of mental disorders, and which by its nature and intensity restricts or impairs mental functions of a person to the extent that they need psychiatric help (Art. 3; P.1; T17).

It should be emphasized that even when it comes to psychopathy falling into the category of a serious mental disturbance, this still does not impact the perpetrator's accountability for the crime. Namely, according to the biological-psychological method, prevalent in most European criminal laws and adopted in the Criminal Code of the Republic of Croatia, it is not enough to determine the existence of psychopathy as a serious mental disturbance, but evaluation must be given as to whether and to what extent has such a person's condition *tempore criminis* influenced their ability to grasp the meaning of their actions (cognitive component) and the ability to govern their own will (volitional component) (Jurjako & Malatesti, 2018). In other words, the question of mental capacity/incapacity never remains at the level of the established diagnosis. Namely, the diagnosis itself, including those in the field of personality disorders, does not imply a specific forensic-psychiatric or legal conclusion, i.e. the diagnosis itself does not imply either the causes or the possible consequences of the impairment. Therefore, as a rule each case is evaluated in terms of severity and its association with some forensically relevant event (Goreta et al., 2004).

Establishing a diagnosis is only the first step. It is followed by an assessment of the impact of the diagnosed condition on the defendant's ability to reason and make decisions. Therefore, forensic psychiatrists characterize personality disorders as severe mental disorders only when the intensity of the personality disorder compromises the psychosocial functioning of the person to such an extent that it may affect their accountability. There is also the question of how much it is possible to distinguish whether the abnormality stems from the disease or not, as well as whether exact quantification is possible and of which psychopathological phenomena and variables, with respect to their extent and intensity, that could produce a "serious mental disturbance" (Šendula Jengić, 2008). Finding the existence of psychopathic personality traits in a particular offender and determining their impact on their ability to reason and to control their actions at the moment of perpetration of the crime constitutes a significant factor in evaluating a person's mental capacity and the risk they may pose to the environment.

In analysing the psychiatric meaning of the legal notion of *other serious mental disturbance* in the context of mental capacity, psychiatrists emphasize the existence of an intermediate area between permanent or temporary mental illness, delayed mental development and the diversity of human behaviour with consequent unlawful acts. From a psychiatric perspective, these persons are not mentally ill in the nosological sense, but at the same time they are significantly mentally damaged (and very difficult to treat). Although researchers have different views of a more accurate description of *serious mental disturbances*, most agree that a psychiatric diagnosis outside the spectrum of psychotic disorders is not in and of itself sufficient for assessing a person's accountability. Thus, any forensic psychiatric assessment should be based on establishing the existence of the disorder and then on the evaluation of its intensity. The produced effect of *other serious mental disturbance* on the inference of mental incapacity should have its quantitative equivalent of psychotic in terms of both understanding and managing one's own actions. Inadequate and unprofessional forensic-psychiatric judgment can have significant detrimental effects on a person with a serious mental disturbance, but also on society in general (Šendula Jengić & Bošković, 2001).

It is in the field of forensic psychiatry, through psychiatric expert witnesses, that we find the closest cooperation between a legal (judge, public prosecutor, lawyer) and a medical professional (psychiatrist). The roles of lawyers and psychiatrists as well as their mutual relationship and cooperation are clearly set out in the Criminal Procedure Act. The cooperation starts when there is a possibility of reduced or lack of accountability of the perpetrator for a criminal offence, when it may have been caused by substance abuse and when the person is unable to stand trial due to mental illness. The defendant's accountability may be questioned on the basis of different information, such as: information on previous medical treatment in a psychiatric institution, cruel and unmotivated manner of committing a crime, bizarreness in the perpetration of a crime, defendant's behaviour deviating considerably from the behaviour of a mentally healthy person, etc. It should be emphasized that a psychiatric examination of the defendant and expert witness testimony is mandatory whenever there is even the slightest suspicion of one's mental capacity *tempore criminis*. In practice, as a rule, expert witnesses are optional, i.e., the court assesses whether certain expert knowledge is required to establish a fact from outside the legal field. If the court estimates that the defendant's mental state had no effect on the perpetration of a criminal offence, this opens the possibility that a number of mentally disturbed perpetrators, especially in the domain of personality disorders, are left unevaluated, which calls into question the purposefulness of the type of the sentence.

In assessing the defendant's mental capacity, the expert witness is expected to determine whether *tempore criminis* they had any: mental disorder, temporary mental disturbance, insufficient mental development, or some other serious mental disturbance. They also need to determine the nature, type, degree, and permanence of a mental disturbance and give their opinion as to the effect such a mental state had on the defendant's understanding of the meaning of their conduct and the management of their own will. Based on their findings, the expert witness must also give their opinion on the defendant's future behaviour, i.e., if it presents a threat in the future. Thus, if the expert witness estimates that at the time of committing the criminal offence the defendant's intellect or free will (or both) could not be accounted for, or that the defendant's capacity to understand the meaning of his/her acts was considerably diminished, the expert must then give their opinion on the degree of likelihood that the defendant could repeat a serious crime due to their mental disorder, but also whether psychiatric treatment could eliminate this danger.

Based on the findings and opinion of the expert witness, the court decides on the: sanity (intact mental capacity), diminished responsibility, substantially diminished responsibility or mental incapacity (insanity) of the defendant. Each of these decisions will have different legal consequences. Since mental capacity is the first component of culpability, thus will mental incapacity exclude culpability, and consequently, based on the generally accepted principle *nulla poena sine culpa,* so will the possibility of punishment be excluded. However, because of the potential risk of repeating a criminal offence, the court will decide on the defendant's inpatient or outpatient psychiatric treatment which will be carried out in accordance with the provisions of the Act on the Protection of Persons with Mental Disorders. Substantially diminished responsibility establishes culpability but may be a basis for a mitigating punishment along with imposing a precautionary measure of mandatory psychiatric treatment in the case of a perpetrator who committed a crime for which the prescribed punishment is one or more years' imprisonment and there is a danger of repeating a serious criminal offence. Sanity and diminished responsibility (which in some cases may be judged as a mitigating circumstance) do not question the issue of culpability nor punishment of the perpetrator, nor do they allow a possibility of psychiatric interventions through the criminal sanction of compulsory psychiatric treatment.

The findings and opinions of a psychiatrist expert witness are considered evidence in the court proceedings. Thus, based on the principle of free evaluation of the evidence, just as any other evidence, the court critically examines them for their acceptance if they are understandable, logical and clear or rejection if they are vague, contradictory and incomplete. In the latter case, the Criminal Procedure Act prescribes a number of mechanisms (repeated questioning of expert witnesses, repeated expert witnessing with the same or another expert witness, entrusting expert witnessing to expert institutions) in order to obtain a correct, accurate and convincing opinion. Although the issue of mental capacity is a purely criminal legal concept and decided on exclusively by the court, the court decision is based on expert opinion. Therefore, it is extremely important that the findings be clear and accurate. If this is not the case, i.e., if the court is not convinced of the correctness of the expert witness's findings and opinions, after having taken all the actions to which the court is authorized under the Criminal Procedure Act (additional expert evaluation, repeated expert evaluation), the court will consider the expert examination unsuccessful and the question of mental capacity/incapacity will be declared undetermined. The court will then, as in any other case where a fact remains doubtful, apply the principle *in dubio pro reo* and evaluate the fact in favour of the defendant. What would be more favourable for the defendant in a particular case is *questio facti* and cannot be answered in advance. However, since mental incapacity eliminates culpability and consequently punishment, and substantially diminished mental capacity is an optional basis for lenient punishment, it is reasonable to assume that among these categories, mental incapacity (insanity) would be the most favourable for the defendant, and substantially diminished mental capacity more favourable than sanity (intact mental capacity).

5.3.2.2 Psychopaths in Court Practice

For the purpose of this chapter, a case law research was conducted in order to obtain a useful impression of the cooperation of lawyers and psychiatrists through the prism of psychiatric expertise of psychopathic persons in criminal proceedings. The survey included final judgments of the Supreme Court of the Republic of Croatia in the period range of more than 10 years (from 1 January 2009 to 1 September 2019). The sample is limited to final judgments relating to the offences of murder and aggravated murder. The reason for this is that previous research has shown that psychopaths most often commit this type of crime, but also the fact that psychiatric expert witnesses, as explained earlier, conduct their expertise according to explicit legal regulations only if the perpetrator's mental capacity is questioned. Serious, violent offences that cause fatal consequences almost always raise the issue of the mental state of the offender at the time the crime was committed, thus psychiatric expert assessments are also almost always conducted in such cases. It is quite clear that statistical monitoring of psychopathic persons among the perpetrators of criminal offences is very difficult. Namely, it is reasonable to assume that psychopaths, like everyone else, commit very different types of crimes. However, in the case of other crimes, where there is no apparent physical violence (e.g., criminal offences against economy, official duty, human health, the environment, general security, employment and social security, human rights and fundamental freedoms, privacy, honour and reputation, etc.), the perpetrator's mental capacity is much less likely to be questioned. In this way, the personality disorders of some offenders remain undetected. Therefore, any numerical indicators on the participation of psychopaths in crime are highly questionable.

Keeping in mind the mentioned ambiguity of statistical indicators from the case law research conducted for the purposes of this paper, the following results are reported:

Among all the perpetrators of the criminal offences of murder and aggravated murder, over a period of just over 10 years, 36 were identified by psychiatric expert witnesses as persons with personality disorders (30 committed ordinary and 6 aggravated murder),[16] which was upheld after the appeals proceedings by decisions of the Supreme Court of the Republic of Croatia. Therefore, these are 36 indisputable perpetrators of these crimes with a personality disorder. Their actual number among homicide offenders could not be determined from this research. Namely, many cases have been returned for reconsideration and have not yet become final. In addition, although psychiatric expertise is a rule of thumb in cases of such a serious crime, there are exceptions where the clarity of the situation, especially the motivation of the perpetrator, does not cast any doubt on his or her mental capacity. Finally, reaching a judgment based on the agreement between the prosecution and the defence limits the court's ability to problematize the issue of a perpetrator's accountability.

The term *psychopath* (or psychopathic structure or psychopathy) was not used in any of the judgments. The term used was *personality disorder*, which is in line with the current international classifications of mental disorders. Accepting the established diagnosis of personality disorder, in explanations of their decisions on the perpetrator's culpability, the courts took verbatim sentences or excerpts from psychiatric expert witness reports. To illustrate this, below we give you some examples of the reasoning sections of several Supreme Court (SC) judgements:

(i) "…his personality characteristics show that he is prone to spontaneous and disorganized behaviour, alcohol and drug abuse, along with personality traits of emotional instability, immaturity and cold disposition, impulsiveness, low threshold of tolerance for frustration, which are all indicative of a personality disorder." (SC Number: I Kž 90/15-11);

(ii) "It has been established that the accused is a person diagnosed with narcissistic personality disorder. A characteristic of these persons is that even less serious situations, some irrelevant provocations, can provoke narcissistic rage, anger and aggression resulting from their narcissistic personality. They are dominant, they like to control the situation, they feel superior, they are often without empathy towards the environment and are primarily oriented towards themselves, their sense of superiority over others as well as denial of their weaknesses (…), an established personality disorder characterized by aggressive outbursts disproportionate to the actual cause of this reaction, which can reach the dimensions of the so called narcissistic rage, and which resulted in death due to a fatal injury." (SC Number: I Kž 375/2017-8);

---

[16]The Criminal Code of the Republic of Croatia for the so-called ordinary murder ("who kills another") prescribes a sentence to imprisonment for a term between 5 and 20 years. The crime of aggravated murder is punishable by imprisonment of at least 10 years or long imprisonment (21–40 years). Aggravated murder consists of killing another person in a cruel or treacherous manner, killing a person who is especially vulnerable due to their age, severe physical or mental impairment or pregnancy, killing a close person whom they had previously abused, killing out of self-interest, unscrupulous revenge, hatred or other petty motives, killing for committing or concealing another crime or killing an official in connection with their performing their official duty (Articles 110 and 111 of the Criminal Code).

(iii) "The defendant is a person with dominant traits of dissociative personality disorder with a tendency to take various addictive substances. Because of these personality traits the defendant's mental capacity was substantially diminished *tempore criminis*." (SC Number: I Kž 386/06-3);

(iv) "His diminished mental capacity *tempore criminis*, according to the findings and opinion of a psychiatrist expert witness, is exclusively a consequence of the personality disorder, which includes impulsivity, lower threshold of tolerance for frustration, with an increased sense of security and power due to the effects of alcohol and cocaine."(SC Number: I Kž 401/09-6);

(v.) "A psychiatric expert identified a personality disorder in the defendant with characteristics of impulsive-aggressive response, egocentricity, emotional immaturity and difficulty in dealing with stress" (SC Number: I Kž 433/11-9);

(vi) "In the domain of personality traits, the defendant demonstrates a symptomatology of adjustment disorder with unrestrained characteristics and excessive alcohol consumption. It thus appears that the defendant shows traits of dissociative disorder, impulsiveness, recklessness and latent aggression that are characteristic of his personality." (SC Number: I Kž 889/08-7).

The results that we found while doing this research confirmed the finding mentioned earlier in this paper, that personality disorder does not, as a rule, exclude a person's ability to understand the meaning of his or her actions or control over his or her will *tempore criminis*, but it is reduced in combination with alcohol, drugs, or some strong affective states. Thus, in none of the cases was the person judged to lack mental capacity, 18 persons were with diminished capacity and 6 persons were judged as those with substantially diminished mental capacity. In all the cases the established diminished mental capacity of any degree was taken into account as a mitigating circumstance, and in a number of cases it led to the application of security measures: in 8 cases compulsory psychiatric treatment was applied, and in 9 cases compulsory addiction treatment.

Since this sample involved serious crimes, all the cases resulted in imprisonment and some even in the most severe punishment in the criminal justice system of the Republic of Croatia, the punishment of long-term imprisonment. The latter sentence was imposed on four defendants, one of whom was sentenced to 23 years in prison, while the other three were sentenced to 32, 38 and 40 years in prison for committing two homicide offences of aggravated murder.

In this sample, there were a total of 13 recidivists (36%) and the previously committed offences were predominantly with elements of violence.

With all the limitations of this research mentioned earlier, and considering the fact that the research covers a long period of adjudication of serious crimes against the perpetrators for whom during the criminal proceedings, and through psychiatric expert witnesses, a personality disorder was identified, the following can be concluded without doubt:

1. The diagnosis of personality disorder does not eliminate the perpetrator's culpability, and homicide offenders are always sentenced to imprisonment. In addition to imprisonment, in some cases where mental capacity is reduced to a significant

level (substantially diminished responsibility), security measures of compulsory psychiatric treatment or compulsory treatment for addiction are applied, and they are carried out within the prison system. The established fact of diminished responsibility is taken into account in sentencing, but we do not know to what extent, since it is not possible to determine that from the reasoning of the court judgments, because this circumstance is only listed together with other mitigating circumstances.

2. Recidivism of persons with personality disorders, seen in 36% of the cases in this sample, is an extremely worrying fact. Namely, these are the most serious crimes, those against life and limb, and earlier offences were also characterized by violence (personal injury, serious physical injury, domestic violence, robbery, homicide, causing a risk to life and limb). The question here is whether subsequent homicides could have been prevented had the criminal justice system responded adequately and identified the perpetrators' disorders from earlier crimes. In answering this question, one should not neglect the fact that the criminal justice system, when it comes to persons with personality disorders, is highly dependent on the current reaches of psychiatric science. In addition, it should always be borne in mind that criminal law is only the *ultima ratio societas* and that it "enters the scene" only when all other systems have failed. It is illusory to expect a solution to the problem from the most repressive part of the legal system. This is evidenced by the entire history of human society in which repression through criminal sanctions has never achieved satisfactory results in either preventing or reducing crime.

3. Exact answers to the question of the cooperation of psychiatrists and lawyers in judicial proceedings cannot be obtained from this research which is based on the study of court judgments. Although at first glance this cooperation appears to be very successful since the courts generally accept expert findings and opinions and quote them in the reasoning of their decisions on the defendant's mental capacity. However, the question is whether this is because the expert's findings are logical, clear, explained well and persuasive, or is it simply easier to accept the opinion of a psychiatrist because lawyers are not competent to assess the mental state of the offender? Regardless of which answer is correct, while examining the court decisions one cannot escape the impression of templates, enumeration, unnecessary repetition, and lack of consideration of the causal relationship of the personality disorder and behaviour within the framework of the committed crime that would make the conclusion clear and unambiguous. In addition, in some of the reasonings of the sentence, the distinction between the roles of the court and the expert witnesses is largely blurred, so it remains unclear who made the decision on the mental capacity of the perpetrator, and this is an exclusively legal and not a psychiatric term.

Considering all that has been said so far, it is crystal clear that both in legislation and the case law practice, the law relies on psychiatry to a great extent. Therefore, good cooperation of lawyers and psychiatrists in legal proceedings against persons with mental disorders is a *conditio sine qua non* of a successful and lawful trial. However, theorists and practitioners of both disciplines often point to their poor cooperation. Therefore, it should be emphasized that good cooperation is enabled by legislation, not only through the Criminal Procedure Act, but also the Act on the Protection of Persons with Mental Disorders. Namely, both laws regulate the roles of lawyers and psychiatrists in detail, set precise deadlines and define their relationship clearly. Therefore, consistent adherence to legal provisions eliminates, or at least reduces misunderstandings about rights and obligations, or the role of one or the other in criminal proceedings, including problems related to transferring one's own responsibility, assuming someone else's powers, or psychiatrization of criminal law and laicism in the field of criminal justice (Grozdanić, 1987). Of course, this is only the first step in eliminating misunderstandings. Mutual trust is also required. It cannot be imposed by law, but it can be established by persistent fair behaviour, while adhering to the code of ethics of one's own profession and by mutual respect. And what seems most important now is honesty and openness in considering the realistic possibilities of truly assessing one's mental state at the time of the crime. This last requirement is particularly prominent when it comes to psychopaths, who still represent a great unknown, and thus they are a major challenge within psychiatric science. Needless to say, any doubts, uncertainties, insecurities, and lack of argumentation in psychiatric expert witness reports made for the purpose of criminal proceedings can result in a myriad of negative repercussions, especially from the point of view of the rights of persons with mental disorders and the fairness of their trial.

5.4 Conclusion

The word *psychopathy* is well known not only in professional circles but also by the public. This word evokes different associations and a whole spectrum of diverse, predominantly negative emotions. Moreover, if the word psychopathy is associated with a specific name, face, or behaviour, it can lead to discomfort, anxiety and even fear among those who know the person. Without prejudice to the justification or unjustifiability of such reactions to psychopathic figures, they are completely understandable for several reasons. Namely, despite extensive research and new insights in the field of both natural and social sciences and humanities, psychopathy remains unclear from the psychiatric perspective. There is no unambiguous stance among experts as to what this is about. Several theories that seek to provide a framework within which to obtain acceptable answers to questions about the causes, development, and implications of psychopathy come to very different, sometimes conflicting conclusions. Different attitudes about psychopathy impressively span from seeing it as an incurable personality disorder to normal functioning of people in abnormal social circumstances. And as long as science, in conjunction with clinical practice, does not offer clear and precise answers about the aetiology and phenomenology of psychopathy, it will arouse the interest of the professional as well as the lay public and provoke very different reactions. Also, one should consider the mere human nature, which is prone to prejudice, stigmatization, condemnation, fear, and most often when it is based on lack of understanding. In fact, it seems that the easiest thing to do when we do not understand something is to consider it abnormal. In other words, the easiest way for us to deal with that which we do not understand, be it a condition or behaviour, is to simply pathologize it. The media, especially the well-known and very popular horror films, are also contributing to the dark perception of psychopathy by portraying a psychopath whose extremely negative characteristics will lead or have already led to cruel, unimaginable crimes. When all the above is considered, one should not be surprised to see psychopathy as an extremely dangerous phenomenon in society.

However, experts dealing with psychopathy are not entitled to one-sided and especially not to distorted perceptions. In the case of criminal proceedings for a crime committed by a person diagnosed with a personality disorder of psychopathic type the decisions of experts must be lawful, correct, and fair. In order to get closer to this ideal of a fair trial and to make lawful and fair court decisions, both disciplines, psychiatry, and law, seek to exclude as much as possible arbitrariness and subjectivity in decision making. They do this in a way that they systematize, categorize, and classify. Thus, psychiatry acts according to the already mentioned internationally recognized classifications of mental disorders, and criminal law defines criminal acts and prescribes criminal sanctions, and precisely and in detail determines the basic principles (legality, culpability), specific institutes (insanity, substantially diminished responsibility, self-induced mental incapacity) and their criminal legal effects (punishment or compulsory treatment). In this way, systems are established where consistent application in each individual case should guarantee the correct outcome expressed in a final court ruling.

It should be noted here that this whole system rests on a rather uncertain basis. Namely, it is a system based on previous intelligence, knowledge, experience, or based on the achievements of psychiatric and legal science. This can mean a lot, but also not enough, depending on the degree of development of both disciplines. However, it is clear that although we are addressing the problem of psychopathy to the best of our abilities and efforts, it is only within the limits of our knowledge, which may be far from the actual truth. In other words, we are "playing a game" that we have constructed ourselves, which can be significantly different from the true state of things that we are not yet able to comprehend. Awareness that our knowledge of psychopathy is scarce, and our abilities for comprehension are still quite limited, protects us from the impression that we have found good solutions and that we simply have to follow the rules we have set. Likewise, awareness of our limitations of comprehension and our ignorance makes us more cautious, humble, and responsible in the application of present solutions, but it also obliges and encourages us in further research into psychopathy.

Acknowledgement The research for this chapter was made possible with the financial support of the Croatian Science Foundation (HRZZ) based on the project *Classification and Explanations of Antisocial Personality Disorder and Moral and Legal Responsibility in the Context of the Croatian Mental Health and Care Law* (CEASCRO) (contract number HRZZ-2013-11-8071).

References

American Psychiatric Association. (2013). *Diagnostic and statistical manual of mental disorders* (5th ed.). Author.

Butcher, J. N., Graham, J. R., Ben-Porath, Y. S., Tellegen, A., Dahlstrom, W. G., & Kaemmer, B. (Eds.). (2001). *Minnesota multiphasic personality Inventory-2: Manual for administration and scoring* (2nd ed.). University of Minnesota Press.

Cleckley, H. (1988). *The mask of sanity, an attempt to clarify some issues about the co-called psychopathic personality* (5th ed.). Emily S. Cleckley.

Crego, C., & Widiger, T. A. (2014). Psychopathy and the DSM. *Journal of Personality, 83*(6), 665–677.

Cummings, M. A. (2015). The neurobiology of psychopathy: Recent developments and new directions in research and treatment. *CNS Spectrums, 20*(3), 200–206.

Dadds, M. R., Moul, C., Cauchi, A., Dobson-Stone, C., Hawes, D. J., Brennan, J., Urwin, R., & Ebstein, R. E. (2014). Polymorphisms in the oxytocin receptor gene are associated with the development of psychopathy. *Development and Psychopathology, 26*(1), 21–31.

Eysenck, H. J., & Eysenck, S. B. G. (1991). *Manual of the Eysenck Personality Scales*. Hodder & Stoughton.

Fahrenberg, J., Hampel, R., & Selg, H. (2001). *Das Freiburger Persöhnlichkeitsinventar (FPI) überarbeitete und neu normierte Auflage*, Manual. Hogrefe.

First, M. B., Spitzer, R. L., Gibbon, M., & Williams, J. B. W. (1997). *Structured clinical interview for DSM-IV® Axis II personality disorders SCID-II*. American Psychiatric Association.

Flórez, G., Vila, X. A., Lado, M. J., Cuesta, P., Ferrer, V., García, L. S., Crespo, M. R., & Pérez, M. (2017). Diagnosing psychopathy through emotional regulation tasks: Heart rate variability versus implicit association test. *Psychopathology, 50*(5), 334–341.

Glenn, A. L. (2011). The other allele: Exploring the long allele of the serotonin transporter gene as a potential risk factor for psychopathy: A review of the parallels in findings. *Neuroscience & Biobehavioral Reviews, 35*(3), 612–620.

Glenn, A. L., Raine, A., Schug, R. A., Gao, Y., & Granger, D. A. (2011). Increased testosterone-to-cortisol ratio in psychopathy. *Journal of Abnormal Psychology, 120*(2), 389–399.

Goreta, M., Peko-Čović, I., & Buzina, N. (Eds.). (2004). *Psihijatrijska vještačenja. Zbirka ekspertiza – Knjiga prva: Kazneno pravo*. Naklada Zadro.

Grozdanić, V., & Tripalo, D. (2013). Novosti u Zakonu o zaštiti osoba s duševnim smetnjama. *Hrvatski ljetopis za kazneno pravo i praksu, 2*(2), 795–800.

Grozdanić, V. (1987). Psihijatrijski nalaz i mišljenje kao osnova odluke o krivičnoj odgovornosti. *Zbornik Pravnog fakulteta Sveučilišta u Rijeci, 8*, 59–76.

Grozdanić, V., Škorić, M., & Martinović, I. (2013). *Kazneno pravo: opći dio*. Pravni fakultet Sveučilišta u Rijeci.

Hare, R. D. (1991). *Hare psychopathy checklist—Revised manual*. Multi-Health Systems.

Jurjako, M., & Malatesti, L. (2018). Neuropsychology and the criminal responsibility of psychopaths: Reconsidering the evidence. *Erkenntnis, 83*, 1003–1025.

Kallert, T., & Torres-González, F. (2006). *Legislation on coercive mental health care in Europe: Legal documents and comparative assessment of twelve European countries* (1st ed.). Peter Lang, International Academic Publishers.

Kozarić-Kovačić, D., Grubišić-Ilić, M., & Grozdanić, V. (2005). *Forenzička psihijatrija, 2. izdanje*. Medicinska naklada.

Morey, L. C. (2007). *Personality Assessment Inventory professional manual*. Psychological Assessment Resources.

Perez, P. R. (2012). The etiology of psychopathy: A neuropsychological perspective. *Aggression and Violent Behavior, 17*(6), 519–522.

Plutchik, R., & Kellerman, H. (1974). *Manual of the Emotions Profile Index*. Western Psychological Services.

Sattar, S. P., Pinals, D., & Gutheil, T. G. (2004). Countering countertransference, II: Beyond evaluation to cross-examination. *The Journal of the American Academy of Psychiatry and the Law, 32,* 148–154.

Šendula Jengić, V. (2008). *Kriminogene specifičnosti psihotičnih počinitelja kaznenih djela* (Doctoral dissertation). Medicinski fakultet Sveučilišta u Zagrebu.

Šendula Jengić, V., & Bošković, G. (2001). Forenzički značaj „drugih, težih duševnih smetnji". *Zbornik Pravnog fakulteta Sveučilišta u Rijeci, 22*(2), 635–654.

Waldman, I. D., Rhee, S. H., LoParo, D., & Park, Y. (2018). Genetic and environmental influences on psychopathy and antisocial behavior. In C. J. Patrick (Ed.), *Handbook of psychopathy* (2nd ed., pp. 335–353). The Guilford Press.

World Health Organization. (1992). *The ICD-10 classification of mental and behavioural disorders: Clinical descriptions and diagnostic guidelines.* World Health Organization.

World Health Organization. (2018). *International classification of diseases for mortality and morbidity statistics (11th Revision).* Version: 04/2019. Retrieved from https://icd.who.int/browse11/l-m/en#/http%3a%2f%2fid.who.int%2ficd%2fentity%2f941859884. Accessed on 2/7/2019.

Yang, Y., & Raine, A. (2018). The neuroanatomical bases of psychopathy. A review of brain imaging findings. In C. J. Patrick (Ed.), *Handbook of psychopathy* (2nd ed., pp. 380–400). The Guilford Press.

Chapter 6
Psychopathy: Neurohype and Its Consequences

Jarkko Jalava and Stephanie Griffiths

Abstract Many argue that psychopaths suffer from a stable pattern of neurobiological dysfunctions that should be taken into account in sentencing and treatment decisions. These arguments are compelling only if the neuroimaging data are consistent. It is possible that such consistency is created by reviewers who ignore contradictory findings. To evaluate this, we examined how accurately forensic literature reported neuroimaging findings on psychopaths in a theoretically central structure – the amygdala. We found that forensic commentators consistently under-reported null-findings, creating a misleading impression of the data's consistency. We discuss this misrepresentation from the perspectives of spin and neurohype, and examine their causes and consequences.

Keywords Psychopathy · Psychopathy Checklist – Revised · Neuroimaging · Spin · Neurohype · Bias · Forensic · MRI

Psychopathy is often described as a brain-based disorder. In the scientific literature psychopathy is frequently called a "neurodevelopmental" or "neuropsychiatric" disorder (Anderson & Kiehl, 2014, p. 103; Gao et al., 2009, p. 813; Sethi et al., 2018, p. 88) or some variant thereof, and titles such as *the neurobiology of psychopathy* (Cummings, 2015; Gao et al., 2009; Glenn & Raine, 2008; Herba et al., 2007; Stratton et al., 2015) are common. The first contemporary neuroimaging studies on psychopathy were conducted in the early 2000s; by 2002, one leading researcher told an interviewer that psychopathy is "definitely a biologically based condition in the sense that the amygdala is functioning poorly" (Blair, quoted in Purdie, 2002). "They're wired differently than other people" explained another (Raine, quoted in University of Southern California, 2004). "The consistency of their brain abnormalities" a prominent psychopathy researcher noted, "never ceases to amaze me" (Kiehl, 2014, p. 262).

J. Jalava (✉) · S. Griffiths
Okanagan College, Penticton, BC, Canada
e-mail: jjalava@okanagan.bc.ca

© Springer Nature Switzerland AG 2022　　　　　　　　　　　　　　　　　　79
L. Malatesti et al. (eds.), *Psychopathy*, History, Philosophy and Theory of the
Life Sciences 27, https://doi.org/10.1007/978-3-030-82454-9_6

This apparent consensus naturally raised the question of what to do with psychopaths. Many argued for reduced moral or criminal responsibility (e.g., Blair, 2008; Knabb et al., 2009). Some maintained that psychopaths should not be held criminally responsible at all (e.g., Glenn et al., 2011). Others offered treatment suggestions, ranging from positive reinforcement (e.g., Reidy et al., 2015) to neurosurgery (De Ridder et al., 2009). Most agreed that psychopaths should not be punished in the same way as those without neurobiological deficits. A philosopher explained that "psychopathy is a mental illness with bio-genetic etiology .... At this point, this is something everyone in the debate about the philosophy of punishment should simply accept" (Nadelhoffer, 2015, pp. 174–175). In 2009, a team of researchers imaged American serial killer Brian Dugan's brain, and presented the results as mitigating evidence in his sentencing hearing. Dugan, the defense argued, suffered from "an emotional developmental disorder (i.e., psychopathy)" (Gaudet et al., 2014, p. 44), and the court found that the testimony passed the *Frye* test for admissibility.[1] Judges have admitted neurobiological evidence of psychopathy in other cases as well (Denno, 2015).

But the consistency of the neurobiological data on psychopaths seemed to depend on who one asked and how closely one scrutinized the data. Popular media articles, which usually included interviews with leading psychopathy researchers, tended to treat the neurobiology of psychopathy as a scientific fact (e.g., Bhattacharjee, 2018; Hagerty, 2017; Ridley, 2018). "We have a fairly good idea," an article in *The Atlantic* explained "what an adult psychopathic brain looks like" (Hagerty, 2017). Researchers cited in these stories often had neurobiological theories of their own, theories the researchers argued were supported by the data (e.g., Anderson & Kiehl, 2012, 2014; Gao et al., 2009; Glenn & Raine, 2014; Kiehl, 2006; Umbach et al., 2015).

Reviews by authors without theoretical commitments painted a different, though not always cohesive, picture. Most reviews by non-theorists tended to find the data inconsistent (Brook et al., 2013; Griffiths & Jalava, 2017; Koenigs et al., 2011; Pujara & Koenigs, 2014; Santana, 2016; but also see also Del Casale et al., 2015). One meta-analysis, however, found consistent abnormalities in a number of brain regions [Poeppl et al., 2018; this finding contrasted with an earlier meta-analysis (Yang & Raine, 2009), which did not find abnormalities].

Regardless of theoretical inclinations, all writers who found the data consistent shared one thing in common: they included few or no null-findings. The meta-analysis above, for instance, reported on 753 findings from 155 studies, with no nulls (Poeppl et al., 2018). The most-cited review studies by the most-cited psychopathy researchers (Blair, 2003; Gao & Raine, 2010; Kiehl, 2006) included findings from a combined seventeen different sMRI, fMRI and SPECT studies. Together, they described no disconfirming null findings (the only nulls they reported were to distinguish between factor scores, successful and unsuccessful psychopaths, and

---

[1] The court did not allow actual fMRI images to be shown, however, but did allow general diagrams of the brain.

white and grey matter). The absence of null-findings from data-sets this large is statistically improbable: Even if a neural profile for psychopathy did exist, null findings should still occur by chance. For one, neither neuroimaging methodology nor personality measures, including psychopathy measures, are reliable enough to produce consistent and/or high correlations (see e.g., Vul et al., 2009). When we tabulated all findings in published neuroimaging studies between 1997 and 2017 using the most common definition of psychopathy (PCL-R, the same definition all but one study in the review above studies also used) we found that the most consistent findings were, in fact, nulls (Griffiths & Jalava, 2017). In other words, those writers who argued that the data showed consistent abnormalities had done something akin to evaluating baseball batters by counting the number of safe hits and where the ball had landed while, mostly, leaving out the times the batters had swung and missed.

If there is no neural profile of psychopathy, it is unclear whether psychopaths share anything in common aside from their diagnosis. This is because the evidence for theoretically relevant deficits in psychopaths, such as in emotion (Brook et al., 2013), moral judgment (Marshall et al., 2016), experience of fear (Hoppenbrouwers et al., 2016), and attention (Smith & Lilienfeld, 2015) is also relatively inconsistent, and in some cases possibly skewed by publication bias (Smith & Lilienfeld, 2015). These inconsistencies may be due to the way psychopathy is diagnosed: According to one calculation, the 30 item PCL-R (with 3 possible levels of severity for each item) allows for more than 15,000 different symptom combinations at or above the diagnostic cutoff score (Rogers, 1995).

If lack of clarity on psychopaths' neurobiology is a problem for basic science, it is an even larger problem for applied science. Incomplete or biased reviews of the data can lead readers to misconstrue the state of the science. Consequences are much worse if the data are used to determine how we should deal with actual people. It is for this reason that applied arguments require a higher burden of proof than those of pure science – consider for example the different levels of proof required for scientific claims in a single drug study and the acceptance of that drug for medical use.

How, then, does applied literature on psychopathy approach the data on psychopaths' neurobiology? Do philosophers, lawyers, and social scientists concerned with things such as criminal responsibility and punishment exercise caution traditionally expected of applied science or do they engage in what is commonly known as *neurohype* – the making of exaggerated claims about neuroscientific data? (see e.g., Lilienfeld et al., 2018). Exactly what kinds of conclusions do applied writers draw from neuroimaging data on psychopaths?

6.1   Reporting Accuracy

We reviewed applied arguments about how the criminal justice system should deal with psychopaths by searching PsycINFO, PubMed, and Google Scholar for published literature using the terms "psychopathy and responsibility" and "psychopathy and punishment." We also examined sources cited in these publications for anything our first search missed. Overall, we found 65 publications since 2000, the year the first MRI study on psychopaths was published (Raine et al., 2000). Of these, 85% argued for one kind of application or another, the most common of which concerned legal responsibility (reduced criminal responsibility, no criminal responsibility or, more generally, that the empirical evidence should somehow affect responsibility; we did not include publications that argued about moral responsibility only). Other proposed applications involved punishment, treatment, crime prediction, and preventive detention. The arguments showed varying levels of prescriptiveness; some called for immediate application while others only hinted at the possibility of application and offered caveats. Overall, 80% of the publications (52 out of 65) cited neuroimaging evidence for their conclusions (see Table 6.1). The vast majority cited individual studies, a few cited reviews by psychopathy theorists only, and a single paper cited a critical review study by non-theorists (Table 6.1).

Given that most articles focused on single-study findings, we evaluated the accuracy of data reporting in the applied literature. To do this, we compared the narrative description of neuroimaging data in the 52 applied articles to the empirical findings in the published studies. We focused on the amygdalae, one of the most widely studied and theoretically important brain regions in psychopathy (e.g. Blair, 2010). The amygdaloid bodies contain many specialized nuclei that are important in processes such as emotions (especially fear) and motivated behavior. These structures have obvious theoretical relevance to psychopathy as psychopathy includes symptoms such as callousness, shallow affect, and antisocial behavior. We focused on the most common diagnostic measure, the PCL-R (Hare, 2003), widely considered the 'gold standard' in the field (e.g., Glenn & Raine, 2008).

Of the 52 publications in our sample that argued for application based on neuroimaging evidence, most authors cited more than one primary neuroimaging study to support their arguments (see Table 6.2). None of the arguments rested exclusively on neuroimaging of the amygdala – other neurobiological data were cited as well. We only included publications that cited original research (i.e., not review studies) on adult psychopaths, and publications that clearly stated the direction of the findings (+ = increased structure or function, – = decreased structure or function, or 0 = null finding, bilaterally or in either right or left amygdala). This allowed us to evaluate whether the narrative description of the neuroimaging findings of the amygdalae were congruent with the original research findings themselves.

The original findings are summarized in Table 6.3 (modified from Griffiths & Jalava, 2017; also includes PCL-SV, Hart et al., 1995 screening version of the PCL-R). Roughly half of the original neuroimaging studies, both structural and functional, found no differences between psychopaths and non-psychopaths. The remaining studies showed either reduced structure and function or increased structure and function in psychopaths. Therefore, if the applied literature accurately described neuroimaging in psychopathy, null findings should feature prominently. However, across all of the specific neuroimaging effects described in the applied literature, only three were null results (3% of the total 108 effects described in Table 6.2). Since one of these effects was in reality *not* a null finding, the applied

Table 6.1 Summary of applied literature on psychopathy and neuroimaging

Authors	Primary conclusion[a]	Neuroimaging Y/N
N articles included = 65	*N* articles arguing for application = 55 *(85%)*	*N* citing neuroimaging = 52 *(80%)*
Anderson and Kiehl (2013)	CR	Y
Anderson and Kiehl (2014)	TR	Y
Baccarini and Malatesti (2017)	BIO	Y
Baskin-Sommers and Newman (2012)	TR	Y
Berryessa (2016)	CR	Y
Blair (2008)	CR	Y
Brink and Nelkin (2013)	CR	Y
Canavero (2014)	BIO	Y
Chialant et al. (2016)	TR	Y
Ciocchetti (2003)	CR	N
Corrado (2013)	PUN	N
DeLisi et al. (2009)	TR	Y
De Ridder et al. (2009)	BIO	Y
Dressing et al. (2008)	PUN	Y
Duff (2010)	CR	N
Fabian (2010)	CR	Y
Fine and Kennett (2004)	CR	Y[b]
Finlay (2011)	CR	N
Fischette (2004)	CR	N
Focquaert et al. (2013)	CR	Y
Focquaert et al. (2015)	CR	Y
Fox et al. (2013)	CR	Y
Freedman and Verdun-Jones (2010)	CR	Y
Fumagalli and Priori (2012)	CR	Y
Glannon (2008)	CR	Y
Glenn and Raine (2009a)	PUN	Y
Glenn and Raine (2009b)	TR, PREV	Y
Glenn and Raine (2014)	CR	Y
Glenn et al. (2011)	CR	Y
Glenn et al. (2013)	CR7	Y
Haji (2003)	PUN	N
Haji (2017)	CR	N
Hauser (2016)	CR	Y
Jurjako and Malatesti (2018a)	NLA	Y
Jurjako and Malatesti (2018b)	NLA	Y
Knabb et al. (2009)	CR	Y

(continued)

Table 6.1 (continued)

Authors	Primary conclusion[a]	Neuroimaging Y/N
Koenigs and Newman (2013)	CR	Y
Levy (2007)	CR	Y
Levy (2011)	PUN, PREV	Y
Levy (2014)	CR	N
Ling and Raine (2018)	CR	Y
Litton (2008)	UN	N
Litton (2013)	UN	N
Maibom (2008)	NLA	N
Marazziti et al. (2013)	CR	Y
Mei-Tal (2002)	CR	Y
Mendez (2009)	CR	Y
Morse (2008)	CR	N
Nadelhoffer (2015)	CR	Y
Nadelhoffer et al. (2012)	PRED	Y
Nair and Weinstock (2008)	CR	Y
O'Neill (2004)	PUN	Y
Ortega-Escobar et al. (2017)	CR	Y
Patrick et al. (2012)	NLA	Y
Poldrack et al. (2018)	PRED, PUN, etc.	Y
Reidy et al. (2015)	PREV	Y
Sartori et al. (2011)	TR	Y
Schopp and Slain (2000)	NLA	N
Shaw (2016)	CR	Y
Sifferd and Hirstein (2013)	CR	Y
Skeem et al. (2011)	NLA	Y
Umbach et al. (2015)	CR	Y
Vierra (2016)	CR	Y
Vitacco et al. (2013)	NLA	Y
Weber et al. (2008)	CR	Y

[a]CR = at least some bearing on criminal responsibility (no criminal responsibility, reduced responsibility, etc.); *BIO* neurobiological interventions, *TR* treatment/assessment, *PUN* punishment and/or sentencing, *PREV* prevention (including preventive detention), *PRED* prediction/neuroprediction, *NLA* data do not support legal application, *UN* undecided
[b]Cited data from individuals with prefrontal injuries, but not psychopaths

literature identified two 'true' nulls (and of these, one was only to distinguish successful from non-successful psychopaths). The applied literature mostly cited findings of one type: reduced structure or function (88% and 86% of all neuroimaging effects described, respectively). In other words, the applied studies' summaries of the neuroimaging data were consistently inaccurate (see Table 6.3).[2] The applied

---

[2] Null findings have been present since 2001. Their absence from the applied literature, therefore, is not a function of their unavailability for early publications.

Table 6.2 Neuroimaging effects cited in applied literature: Amygdala

Authors	Studies cited sMRI fMRI – Italicized	Effect	Suggested application
Anderson and Kiehl (2013)	Kiehl et al. (2001) Birbaumer et al. (2005) Veit et al. (2002) Harenski et al. (2010) Dolan and Fullam (2009) Müller et al. (2003)	– – – – –, + +	Legal responsibility
Anderson and Kiehl (2014)	Ermer et al. (2012) Yang et al. (2010) Kiehl et al. (2001) Birbaumer et al. (2005) Veit et al. (2002) Harenski et al. (2010) Dolan and Fullam (2009)	– – – – – – –	Treatment
Baskin-Sommers and Newman (2012)	Birbaumer et al. (2005) Glenn et al. (2009) Kiehl et al. (2001) Müller et al. (2003)	– – – +	Treatment
Blair (2008)	Kiehl et al. (2001) Birbaumer et al. (2005)	– –	Reduced responsibility
Chialant et al. (2016)	Kiehl et al. (2001)	–	Punishment, prevention, treatment
De Ridder et al., (2009)	Kiehl et al. (2001)	–	Neurosurgery
Fabian (2010)	Kiehl et al. (2001)	–	Mitigation, moral culpability, punishment
Focquaert et al. (2013)	Yang et al. (2009) Kiehl et al. (2001) Birbaumer et al. (2005)	– – –	Punishment
Focquaert et al. (2015)	Yang et al. (2009) Kiehl et al. (2001) Glenn et al. (2009)	– – –	Criminal responsibility, treatment
Fumagalli and Priori (2012)	Yang et al. (2006): Conference abstract Müller et al. (2003) Schneider et al. (2000) Kiehl et al. (2001) Veit et al. (2002) Birbaumer et al. (2005) Glenn et al. (2009) Harenski et al. (2010)	– + + – – – – –	Not specified: "Legal and clinical implications"
Glenn and Raine (2009)	Yang et al. (2006) Conference abstract Kiehl et al. (2001) Birbaumer et al. (2005) Veit et al. (2002) Glenn et al. (2009)	– – – – –	Pharmacological treatment, transcranial magnetic stimulation, crime prediction, punishment

(continued)

Table 6.2 (continued)

Authors	Studies cited sMRI fMRI – Italicized	Effect	Suggested application
Glenn and Raine (2014)	Yang et al. (2009) Kiehl et al. (2001) Birbaumer et al. (2005) Glenn et al. (2009) Müller et al. (2003)	− − − − +	Treatment, punishment, responsibility
Glenn et al. (2011)	Yang et al. (2009) Birbaumer et al. (2005) Kiehl et al. (2001) Glenn et al. (2009)	− − − −	No criminal responsibility
Hauser (2016)	Yang et al. (2009) Yang et al. (2010) Kiehl et al. (2001)	− − −	Not specified: Responsibility, punishment etc.
Knabb et al. (2009)	Kiehl et al. (2001)	−	Criminal responsibility, crime prediction
Koenigs and Newman (2013)	Yang et al. (2009) Kiehl et al. (2001) Birbaumer et al. (2005) Müller et al. (2003) Glenn et al. (2009) Deeley et al. (2006)	− − − + − 0	Not specified: Culpability, crime prediction, treatment
Ling and Raine (2018)	Yang et al. (2009) Ermer et al. (2012) Glenn et al. (2009) Birbaumer et al. (2005) Kiehl et al. (2001) Schultz et al. (2016)	− − − − − +	Crime prediction, treatment, punishment
Marazziti et al. (2013)	Veit et al. (2002)	−	Not specified: possibly relevant to responsibility
Mendez (2009)	Veit et al. (2002)	−	Legal culpability, treatment
Nadelhoffer et al. (2012)	Veit et al. (2002) Kiehl et al. (2001) Kiehl et al. (2004)	− − −	Violence prediction
Nadelhoffer (2015)	Veit et al. (2002) Kiehl et al. (2001) Kiehl et al. (2004)	− − −	Punishment
Ortega-Escobar et al. (2017)	Boccardi et al. (2011) Yang et al. (2009) Kiehl et al. (2001) Glenn et al. (2009) Birbaumer et al. (2005) Decety et al. (2014)	+ − − − − 0[a]	Not specified: criminal responsibility

(continued)

Table 6.2 (continued)

Authors	Studies cited sMRI fMRI – Italicized	Effect	Suggested application
Poldrack et al. (2018)	Ermer et al. (2012)	−	Crime prediction, sentencing, treatment, prevention
	Yang et al. (2010)	−	
	Veit et al. (2002)	−	
	Birbaumer et al. (2005)	−	
	Harenski et al. (2010)	−	
	Glenn et al. (2009)	−	
	Decety et al. (2013)	−	
	Motzkin et al. (2011)	−	
Reidy et al. (2015)	Yang et al. (2009)	−	Prevention, intervention
	Birbaumer et al. (2005)	−	
Sartori et al. (2011)	Birbaumer et al. (2005)	−	Mens Rea
	Glenn et al. (2009)	−	
	Müller et al. (2003)	+	
Sifferd and Hirstein (2013)	Birbaumer et al. (2005)	−	Criminal responsibility, treatment
Umbach et al. (2015)	Yang et al. (2009)	−	Punishment, crime prediction, intervention
	Boccardi et al. (2011)	+	
	Ermer et al. (2012)	−	
	Yang et al. (2010):	0	
	Successful psychopaths	−	
	Unsuccessful psychopaths	−	
	Birbaumer et al. (2005)	−	
	Glenn et al. (2009)	+	
	Kiehl et al. (2001)		
	Müller et al. (2003)		
Weber et al. (2008)	Raine et al. (2000)	−	Prevention, treatment, responsibility
	Tiihonen et al. (2000)	−	
	Birbaumer et al. (2005)	−	

[a]Incorrectly reported

Table 6.3 Distribution of effects in neuroimaging studies vs applied literature

Type of effects				
sMRI	Original studies	*(% of total)*	Applied literature	*(% of total)*
Null (none)	13	*46%*	1	*4%*
Smaller size	9	*32%*	22	*88%*
Larger size	6	*21%*	2	*8%*
Total	28	*100%*	25	*100%*
fMRI				
Null (none)	5	*46%*	2	*2%*
Hypoactive	4	*36%*	71	*86%*
Hyperactive	2	*18%*	10	*12%*
Total	11	*100%*	83	*100%*

Note. N findings in applied literature exceed original studies N because applied studies cite original studies more than once and empirical literatures, at least with respect to the amygdalae and the PCL-R, describe different universes of data: one with a clear neurobiological signal that amounts to a good case for action, the other constituting mostly noise.

We found only two publications that unequivocally argued against legal application of the neurobiological evidence (Patrick et al., 2012; Skeem et al., 2011). These publications also gave the most accurate review of the data. It is also important to note that, while they did not cite amygdala data specifically, two of the most recent applied papers (Haji, 2017; Jurjako & Malatesti, 2018a, b) were generally more careful about empirical data than their predecessors. If this holds, it may signal a shift toward improved accuracy in data reporting.

6.2   Spin

What might account for the citing imbalance in the applied literature? One possibility is that null findings in particular are difficult to find in the original research, as they are often de-emphasized in order to increase the likelihood of publication. This problem, along with other problematic reporting practices, is often termed *spin* and is widely discussed in the medical literature (e.g., Dwan et al., 2013; Franco et al., 2014; Lazarus et al., 2015). Spin may have played a role here as well, as null findings are indeed difficult to find in original studies on psychopaths. Consider for example a study by de Oliveira-Souza et al. (2008) that found reduced grey matter volume in several areas of psychopaths' brains, but not in the amygdalae. The only way to locate the null finding was to compare the study's a priori regions of interest, which included the amygdalae, to tables showing areas with statistically significant differences, which did not include the amygdalae. Unsurprisingly, only the positive findings from that study appear in the applied literature.

Spin in the empirical research may not fully explain problems in applied research, however. Some applied writers are also experts in the research field and should be able to detect even poorly reported null findings. Furthermore, some are demonstrably aware of studies with null findings (i.e. they cite them in other publications) but fail to cite them in the applied literature.[3] Findings may even reverse direction depending on the type of publication. Two researchers ran a meta-analysis and found that while antisocial individuals showed abnormal pre-frontal cortex function and structure, psychopathy scores did not moderate this effect (Yang & Raine, 2009). One of the authors subsequently argued that psychopaths should not be held criminally responsible in part because the meta-analysis had found "significant reductions in both structure and function in emotion-related areas [prefrontal cortex] in psychopathic individuals" (Glenn et al., 2011, p. 303).

---

[3] For example, Raine does not mention a null finding in de Oliveira-Souza et al. (2008) in several applied publication, but notes it in an empirical publication (Pardini et al., 2014).

Cases like these suggest that the applied literature on psychopathy may have an additional bias. As a rough test of this proposition, we examined citations of studies in which the direction of findings is clearly stated. We examined two studies published almost simultaneously by two research groups with overlapping personnel and samples, and the same imaging method (structural MRI), but opposite findings. One study was titled *Increased Volume of the Striatum in Psychopathic Individuals* (Glenn et al., 2010a, b), and the other *No Volumetric Differences in the Anterior Cingulate of Psychopathic Individuals* (Glenn et al., 2010a, b). The latter was the only null finding in the field so clearly stated. The positive finding was cited 55 times on PsycINFO, 131 times in Google Scholar, and 40 times in PubMed. In the applied literature it was cited five times (Anderson & Kiehl, 2012; Glenn & Raine, 2014; Ling & Raine, 2018; Poldrack et al., 2018; Reidy et al., 2015). Overall, the citations for the null finding numbered 5 in PsychINFO, 34 in Google Scholar, and 4 in PubMed. Applied literature did not cite it at all. To test whether this was because reviewers were more interested in the striatum than the cingulate cortex, we also examined a study that found reduced cingulate cortex volume in psychopaths (Boccardi et al., 2011). That study was cited 15 times in PsycINFO and 122 times in Google Scholar (PubMed did not list it). Applied literature cited it 4 times (Anderson & Kiehl, 2012, 2014; Hauser, 2016; Ortega-Escobar et al., 2017). In other words, citing appears to be more a function of positive findings than their neuroanatomical location.

6.3 Neurohype

In short, it appears that the applied literature is subject to two types of spin, at least when it comes to the amygdala: general reporting bias toward positive findings in the empirical studies, plus spin unique to applied publications. Data exaggeration of this type does not seem to be limited to neurobiology, however. For example, a study by Blair (1995) found differences in 10 psychopaths' and 10 non-psychopaths' performance in a test of moral reasoning. Blair took the data to mean that the psychopaths were incapable of this type of reasoning, a conclusion several philosophers used as supporting evidence for why psychopaths should enjoy reduced moral responsibility. Noting a critical methodological flaw in the study, a research group set out to replicate the findings, twice (minus the flaw and with a much larger sample sizes) and found no effect (Aharoni et al., 2012, 2014). However, philosophers mostly continued to cite the original study and maintained their original conclusion (see Jalava & Griffiths, 2017a, b).

But if all of the above is correct, the question is why the hype, and why in the direction of an apparent dysfunction? One possibility is that the philosophers and legal theorists rely on the scientific literature in good faith, and so miss the nulls along with almost everyone else. This, however, does not explain why some of these writers go to great and apparently independent lengths to describe the empirical studies that support their arguments, but miss others that do not. Nor does it explain why social scientists – some of whom are leading psychopathy researchers – miss them as well.

Another possibility for the spin is systemic: In the applied field, just like in the empirical field, strong data and strong arguments are more likely to find publication. Data, when strong, are rich source material for theories and recommendations, while there are only so many ways of saying the data are weak and not ready for application. Examining the data critically, in other words, comes with an opportunity cost.

Finally, the neurobiology of psychopathy is an appealing discourse because it appears to offer a straightforward explanation for a set of questions that are in reality very difficult to answer. The ultimate questions such as what causes crime, how personality is related to crime, and how neurobiology is related to personality, are complex and, most likely, require expertise in multiple scientific fields to answer. Most people would demand additional information before answering a question like: "Bob, a psychopath, shot Bill. What should we do with Bob?" However, by shifting the discourse from the specific (Bob) to the general (psychopaths) and adding a premise (psychopaths have abnormal brains) applied writers can propose a straightforward answer without having to tackle the hard questions. Kahneman termed this "substitution bias" (Kahneman, 2011). One way of dealing with complex questions, Kahneman (2011) argued, is to substitute easier ones for them. An easier question in this context is what neurobiological events correlate with psychopathy? This is often followed by another, linguistic, substitution: the term "correlates" is replaced with terms like "underlies", "is implicated in" or "plays a role in", to imply causal-like knowledge without the burden of proof that comes with an actual causal claim. Claims so construed pass as statements that something – in this case the behavior of psychopaths – is in some unspecified way *understood* (for a general treatment of this issue in the neurosciences, see Krakauer et al., 2017). The result often sounds something like this:

> Elucidation of the neural correlates of psychopathy may lead to improved management and treatment of the condition. Although some methodological issues remain, the neuroimaging literature is generally converging on a set of brain regions and circuits that are consistently implicated in the condition: the orbitofrontal cortex, amygdala, and the anterior and posterior cingulate and adjacent (para)limbic structures. We discuss these findings in the context of extant theories of psychopathy and highlight the potential legal and policy implications of this body of work (Anderson & Kiehl, 2012, p. 52).

Note the transition from "correlates" to brain regions "implicated" to an understanding of psychopathy sufficient for putting it to "legal and policy" use.

Substitutions like these change the rules not only for how we speak about psychopaths but also for who can talk about them. A philosopher, say, can find out what correlates with psychopathy and substitute that correlation for an understanding of why any given psychopath did what he or she did. The philosopher can then concentrate on the more manageable question of "implicated" brain regions and their relevance to such things as criminal responsibility. The resulting just-so account creates a concrete and intuitive dysfunction around which applied arguments can be built, a process similar using the game of Monopoly to explain why wealth concentrates in the real world.

6.4 Consequences of Neurohype

Neurohype has several potential consequences. One is reputational. Imagine the amygdala as a story and the applied writers as a news organization. In this case, the organization regularly gets the story wrong. No possible reason for this – bias, desire for publishable stories, uncritical trust in their sources – looks good for the organization, and it is reasonable to ask whether it can be trusted to get other stories right.

The applied field – though with exceptions – also reinforces a backward logic to understanding the relationship between brain and behavior. The important legal and clinical questions concern dysfunctions in psychopaths' cognition, emotion, and/or behavior. These dysfunctions, should they exist, may have neurobiological causes, but dysfunctions are not the same thing as their causes. By drawing attention to neurobiological abnormalities in the absence of other abnormalities, neurohype obscures the reason why psychopathy could be relevant to responsibility in the first place. Since psychopaths' cognition, emotion, and behavior, as discussed above, does not appear consistently different from non-psychopaths, applied writers are left to arrange the data in creative ways. For instance, one study found reduced amygdala activity in psychopaths while processing moral dilemmas (Glenn et al., 2009). Several applied writers cite the finding as evidence that psychopaths have problems in moral processing, and discuss other studies that show a link between the amygdala and moral emotions. However, the writers neglect to mention that psychopaths in the original study were indistinguishable from non-psychopaths in their actual responses to the moral dilemmas (the majority of fMRI studies in fact show no behavioral differences; Griffiths & Jalava, 2017).

This is also relevant to arguments about treatment, especially neurosurgery. Some writers suggest that treatment should be informed by psychopaths' amygdala deficits. This raises the question of what a treatment for a condition with no consistent and significant cognitive, emotional or behavioral dysfunction – unless one counts psychopathy itself as a dysfunction – and no known cause would hope to accomplish: a person with increased or reduced amygdala activity or someone with activity identical to non-psychopaths? Since the empirical literature shows psychopaths to fit all three profiles, neurobiologically-informed treatment should approach randomness.

Most importantly, however, the point of applied literature is to change policy and practice. If these are determined not by how psychopaths' brains actually are but how writers choose and report data about them, law and policy would arguably be more biased with neuroscience than without it.

6.5 Conclusion and Recommendations

Philosophers, social scientists, and lawyers have long argued for significant change in how we deal with psychopaths. This change, they argue, is justified on empirical – typically neurobiological – grounds. The arguments are on the whole subtle and substantive, but the data they rely on, at least with regards to the amygdala, are much weaker than the writers propose. The literature, in other words, shows neuro-hype. Applied arguments require a higher burden of proof than arguments in basic science. Therefore, if our findings generalize to brain structures beyond the amygdala, there is little reason to consider neurobiology in determining how we deal with psychopaths in the criminal justice system. This argument, though, comes with a caveat: If the data change, this conclusion should change as well.

There are a few simple ways to counteract neurohype. First, authors proposing application could frame their arguments as conditional on data, in the form of, say, "if the data are replicable and specific, then…". This would pre-empt the possibility of an entire body of work becoming obsolete if the data do not turn out as expected. Second, authors and editors might consider their arguments not as academic exercises but as something analogous to a physician prescribing medication. Here, the patients' health functions as checks and balances against carelessness or overstatement. Raising the stakes for the applied literature might likewise increase accuracy in data reporting.

More broadly, applied writers should pay closer attention not only to published data but also to the act of publishing that data. In the medical and social sciences it is becoming increasingly clear that published data are often skewed toward positive findings. Some researchers have begun to take steps to address this problem by, for example, estimating the number of unpublished null-findings needed to alter overall findings in meta-analyses. This sort of care – or even a basic recognition that spin and neurohype exist – is largely absent in applied psychopathy literature.

Finally, applied writers should pay closer attention to incentives. If a social scientist with a theory of their own reports little or no counter-evidence to it, it is likely that their evidence is too good to be true. Even if a neurobiological phenomenon did in reality correlate with a psychopathy diagnosis, null findings should still occur by chance. A field with no or very few null findings, therefore, looks, and is, unrealistic.

References

Aharoni, E., Sinnott-Armstrong, W., & Kiehl, K. A. (2012). Can psychopathic offenders discern moral wrongs? A new look at the moral/conventional distinction. *Journal of Abnormal Psychology, 121*(2), 484–497.

Aharoni, E., Sinnott-Armstrong, W., & Kiehl, K. A. (2014). What's wrong? Moral understanding in psychopathic offenders. *Journal of Research in Personality, 53*, 175–181.

Anderson, N. E., & Kiehl, K. A. (2012). The psychopath magnetized: Insights from brain imaging. *Trends in Cognitive Sciences, 16*(1), 52–60.

Anderson, N. E., & Kiehl, K. A. (2013). Functional neuroimaging and psychopathy. In K. A. Kiehl & W. P. Sinnott-Armstrong (Eds.), *Handbook on psychopathy and law* (pp. 131–149). Oxford University Press.

Anderson, N. E., & Kiehl, K. A. (2014). Psychopathy: Developmental perspectives and their implications for treatment. *Restorative Neurology and Neuroscience, 32*(1), 103–117.

Baccarini, E., & Malatesti, L. (2017). The moral bioenhancement of psychopaths. *Journal of Medical Ethics, 43*(10), 697–701.

Baskin-Sommers, A. R., & Newman, J. P. (2012). Cognition-emotion interactions is psychopathy: Implications for theory and practice. In H. H. Häkkänen-Nyholm & J. Nyholm (Eds.), *Psychopathy and law: A practitioner's guide* (pp. 79–97). Wiley.

Berryessa, C. M. (2016). Behavioral and neural impairments of frontotemporal dementia: Potential implications for criminal responsibility and sentencing. *International Journal of Law and Psychiatry, 46*, 1–6.

Bhattacharjee, Y. (2018, January). The science behind psychopaths and extreme altruists. *National Geographic Magazine.* Retrieved from https://www.nationalgeographic.com/magazine/2017/08/science-good-evil-charlottesville/

Birbaumer, N., Veit, R., Lotze, M., Erb, M., Hermann, C., Grodd, W., & Flor, H. (2005). Deficient fear conditioning in psychopathy: A functional magnetic resonance imaging study. *Archives of General Psychiatry, 62*(7), 799–805.

Blair, R. J. R. (1995). A cognitive developmental approach to morality: Investigating the psychopath. *Cognition, 57*, 1–29.

Blair, R. J. R. (2003). Neurobiological basis of psychopathy. *The British Journal of Psychiatry, 182*(1), 5–7.

Blair, R. J. R. (2008). The cognitive neuroscience of psychopathy and implications for judgments of responsibility. *Neuroethics, 1*(3), 149–157.

Blair, R. J. R. (2010). Neuroimaging of psychopathy and antisocial behavior: A targeted review. *Current Psychiatry Reports, 12*(1), 76–82.

Boccardi, M., Frisoni, G. B., Hare, R. D., Cavedo, E., Najt, P., Pievani, M., … Vaurio, O. (2011). Cortex and amygdala morphology in psychopathy. *Psychiatry Research: Neuroimaging, 193*(2), 85–92.

Brink, D. O., & Nelkin, D. (2013). Fairness and the architecture of responsibility. *Oxford Studies in Agency and Responsibility, 1*, 284–313.

Brook, M., Brieman, C. L., & Kosson, D. S. (2013). Emotion processing in psychopathy checklist—Assessed psychopathy: A review of the literature. *Clinical Psychology Review, 33*(8), 979–995.

Canavero, S. (2014). Criminal minds: Neuromodulation of the psychopathic brain. *Frontiers in Human Neuroscience, 8*, 1–3.

Chialant, D., Edersheim, J., & Price, B. H. (2016). The dialectic between empathy and violence: An opportunity for intervention? *The Journal of Neuropsychiatry and Clinical Neurosciences, 28*(4), 273–285.

Ciocchetti, C. (2003). The responsibility of the psychopathic offender. *Philosophy, Psychiatry, & Psychology, 10*(2), 175–183.

Corrado, M. L. (2013). Some notes on preventive detention and psychopathy. In K. Kiehl & W. P. Sinnott-Armstrong (Eds.), *Handbook on psychopathy and law* (pp. 346–357). Oxford University Press.

Cummings, M. A. (2015). The neurobiology of psychopathy: Recent developments and new directions in research and treatment. *CNS Spectrums, 20*(3), 200–206.

de Oliveira-Souza, R., Hare, R. D., Bramati, I. E., Garrido, G. J., Ignácio, F. A., Tovar-Moll, F., & Moll, J. (2008). Psychopathy as a disorder of the moral brain: Fronto-temporo-limbic grey matter reductions demonstrated by voxel-based morphometry. *NeuroImage, 40*(3), 1202–1213.

De Ridder, D., Langguth, B., Plazier, M., & Menovsky, T. (2009). Moral dysfunction: Theoretical model and potential neurosurgical treatments. In J. Verplaetse, J. Schrijver, S. Vanneste, & J. Braeckman (Eds.), *The moral brain* (pp. 155–183). Springer.

Decety, J., Chen, C., Harenski, C., & Kiehl, K. A. (2013). An fMRI study of affective perspective taking in individuals with psychopathy: Imagining another in pain does not evoke empathy. *Frontiers in Human Neuroscience, 7*, 489.

Decety, J., Skelly, L., Yoder, K. J., & Kiehl, K. A. (2014). Neural processing of dynamic emotional facial expressions in psychopaths. *Social Neuroscience, 9*(1), 36–49.

Del Casale, A., Kotzalidis, G. D., Rapinesi, C., Di Pietro, S., Alessi, M. C., Di Cesare, G., … Ferracuti, S. (2015). Functional neuroimaging in psychopathy. *Neuropsychobiology, 72*(2), 97–117.

DeLisi, M., Umphress, Z. R., & Vaughn, M. G. (2009). The criminology of the amygdala. *Criminal Justice and Behavior, 36*(11), 1241–1252.

Denno, D. W. (2015). The myth of the double-edged sword: An empirical study of neuroscience evidence in criminal cases. *Boston College Law Review, 56*, 493–551.

Deeley, Q., Daly, E., Surguladze, S., Tunstall, N., Mezey, G., Beer, D., & Clarke, A. (2006). Facial emotion processing in criminal psychopathy. *The British Journal of Psychiatry, 189*(6), 533–539.

Dolan, M. C., & Fullam, R. S. (2009). Psychopathy and functional magnetic resonance imaging blood oxygenation level-dependent responses to emotional faces in violent patients with schizophrenia. *Biological Psychiatry, 66*(6), 570–577.

Dressing, H., Sartorius, A., & Meyer-Lindenberg, A. (2008). Implications of fMRI and genetics for the law and the routine practice of forensic psychiatry. *Neurocase, 14*(1), 7–14.

Duff, A. (2010). Psychopathy and answerability. In L. Malatesti & J. McMillan (Eds.), *Responsibility and psychopathy: Interfacing law, psychiatry, and philosophy* (pp. 199–212). Oxford University Press.

Dwan, K., Gamble, C., Williamson, P. R., & Kirkham, J. J. (2013). Systematic review of the empirical evidence of study publication bias and outcome reporting bias—An updated review. *PLoS One, 8*(7), e66844.

Ermer, E., Cope, L. M., Nyalakanti, P. K., Calhoun, V. D., & Kiehl, K. A. (2012). Aberrant paralimbic gray matter in criminal psychopathy. *Journal of Abnormal Psychology, 121*(3), 649–658.

Fabian, J. M. (2010). Neuropsychological and neurological correlates in violent and homicidal offenders: A legal and neuroscience perspective. *Aggression and Violent Behavior, 15*(3), 209–223.

Fine, C., & Kennett, J. (2004). Mental impairment, moral understanding and criminal responsibility: Psychopathy and the purposes of punishment. *International Journal of Law and Psychiatry, 27*, 425–443.

Finlay, S. (2011). The selves and the shoemaker: Psychopaths, moral judgment, and responsibility. *The Southern Journal of Philosophy, 49*, 125–133.

Fischette, C. (2004). Psychopathy and responsibility. *Virginia Law Review, 90*, 1423–1485.

Focquaert, F., Glenn, A. L., & Raine, A. (2013). Free will, responsibility, and the punishment of criminals. In T. A. Nadelhoffer (Ed.), *The future of punishment* (pp. 247–274). Oxford University Press.

Focquaert, F., Glenn, A. L., & Raine, A. (2015). Psychopathy and free will from a philosophical and cognitive neuroscience perspective. In W. Glannon (Ed.), *Free will and the brain: Neuroscientific, philosophical, and legal perspectives* (pp. 103–124). Cambridge University Press.

Fox, A. R., Kvaran, T. H., & Fontaine, R. G. (2013). Psychopathy and culpability: How responsible is the psychopath for criminal wrongdoing? *Law & Social Inquiry, 38*(1), 1–26.

Franco, A., Malhotra, N., & Simonovits, G. (2014). Publication bias in the social sciences: Unlocking the file drawer. *Science, 345*(6203), 1502–1505.

Freedman, L. F., & Verdun-Jones, S. N. (2010). Blaming the parts instead of the person: Understanding and applying neurobiological factors associated with psychopathy. *Canadian Journal of Criminology and Criminal Justice, 52*(1), 29–53.

Fumagalli, M., & Priori, A. (2012). Functional and clinical neuroanatomy of morality. *Brain, 135*(7), 2006–2021.

Gao, Y., & Raine, A. (2010). Successful and unsuccessful psychopaths: A neurobiological model. *Behavioral Sciences & the Law, 28*(2), 194–210.

Gao, Y., Glenn, A. L., Schug, R. A., Yang, Y., & Raine, A. (2009). The neurobiology of psychopathy: A neurodevelopmental perspective. *The Canadian Journal of Psychiatry, 54*(12), 813–823.

Gaudet, L. M., Lushing, J. R., & Kiehl, K. A. (2014). Functional magnetic resonance imaging in court. *AJOB Neuroscience, 5*(2), 43–45.

Glannon, W. (2008). Moral responsibility and the psychopath. *Neuroethics, 1*, 158–166.

Glenn, A. L., & Raine, A. (2008). The neurobiology of psychopathy. *Psychiatric Clinics of North America, 31*(3), 463–475.

Glenn, A. L., & Raine, A. (2009a). The immoral brain. In J. Verplaetse, J. de Schrijver, S. Vanneste, & J. Braeckman (Eds.), *The moral brain: Essays on the evolutionary and neuroscientific aspects of morality* (pp. 45–67). Springer.

Glenn, A. L., & Raine, A. (2009b). Psychopathy and instrumental aggression: Evolutionary, neurobiological, and legal perspectives. *International Journal of Law and Psychiatry, 32*(4), 253–258.

Glenn, A. L., & Raine, A. (2014). *Psychopathy: An introduction to biological findings and their implications*. NYU Press.

Glenn, A. L., Raine, A., & Schug, R. A. (2009). The neural correlates of moral decision-making in psychopathy. *Molecular Psychiatry, 14*, 5–6.

Glenn, A. L., Raine, A., Yaralian, P. S., & Yang, Y. (2010a). Increased volume of the striatum in psychopathic individuals. *Biological Psychiatry, 67*(1), 52–58.

Glenn, A. L., Yang, Y., Raine, A., & Colletti, P. (2010b). No volumetric differences in the anterior cingulate of psychopathic individuals. *Psychiatry Research: Neuroimaging, 183*(2), 140–143.

Glenn, A. L., Raine, A., & Laufer, W. S. (2011). Is it wrong to criminalize and punish psychopaths? *Emotion Review, 3*(3), 302–304.

Glenn, A. L., Laufer, W. S., & Raine, A. (2013). Author reply: Vitacco, Erickson, and Lishner: Holding psychopaths morally and criminally culpable. *Emotion Review, 5*(4), 426–427.

Griffiths, S. Y., & Jalava, J. V. (2017). A comprehensive neuroimaging review of PCL-R defined psychopathy. *Aggression and Violent Behavior, 36*, 60–75.

Hagerty, B. B. (2017, June). When your child is a psychopath. *The Atlantic Monthly, 319*(5), 78.

Haji, I. (2003). The emotional depravity of psychopaths and culpability. *Legal Theory, 9*(1), 63–82.

Haji, I. (2017). Empathy and legal responsibility. In H. Maibom (Ed.), *The Routledge handbook of the philosophy of empathy* (pp. 253–263). Routledge.

Hare, R. D. (2003). *Hare psychopathy checklist: Revised (PCL-R)*. Multi-Health Systems, Inc.

Harenski, C. L., Harenski, K. A., Shane, M. S., & Kiehl, K. A. (2010). Aberrant neural processing of moral violations in criminal psychopaths. *Journal of Abnormal Psychology, 119*(4), 863–874.

Hart, S. D., Cox, D. N., & Hare, R. D. (1995). *Hare psychopathy checklist: Screening version (PCL: SV)*. Multi-Heath Systems.

Hauser, L. L. (2016). Forensic implications of neuroscientific advancements. *Journal of the American Academy of Psychiatry and the Law Online, 44*(2), 193–197.

Herba, C. M., Hodgins, S., Blackwood, N., Kumari, V., Naudts, K. H., & Phillips, M. (2007). The neurobiology of psychopathy: A focus on emotion processing. In H. Herve & J. Yuille (Eds.), *Psychopathy: Theory, research and social implications* (pp. 251–283). Lawrence Erlbaum.

Hoppenbrouwers, S. S., Bulten, B. H., & Brazil, I. A. (2016). Parsing fear: A reassessment of the evidence for fear deficits in psychopathy. *Psychological Bulletin, 142*(6), 1–28.

Jalava, J., & Griffiths, S. (2017a). Philosophers on psychopaths: A cautionary tale in interdisciplinarity. *Philosophy, Psychiatry, & Psychology, 24*(1), 1–12.

Jalava, J., & Griffiths, S. (2017b). Call me irresponsible: Is Psychopaths' responsibility a matter of (data) preference? *Philosophy, Psychiatry, & Psychology, 24*(1), 21–24.

Jurjako, M., & Malatesti, L. (2018a). Neuropsychology and the criminal responsibility of psychopaths: Reconsidering the evidence. *Erkenntnis, 83*(5), 1003–1025.

Jurjako, M., & Malatesti, L. (2018b). Psychopathy, executive functions, and neuropsychological data: A response to Sifferd and Hirstein. *Neuroethics, 11*(1), 55–65.

Kahneman, D. (2011). *Thinking, fast and slow*. Farrar, Straus & Giroux.

Kiehl, K. A. (2006). A cognitive neuroscience perspective on psychopathy: Evidence for paralimbic system dysfunction. *Psychiatry Research, 142*(2–3), 107–128.

Kiehl, K. A. (2014). *The psychopath whisperer: The science of those without conscience*. Crown.

Kiehl, K. A., Smith, A. M., Hare, R. D., Mendrek, A., Forster, B. B., Brink, J., & Liddle, P. F. (2001). Limbic abnormalities in affective processing by criminal psychopaths as revealed by functional magnetic resonance imaging. *Biological Psychiatry, 50*(9), 677–684.

Kiehl, K. A., Smith, A. M., Mendrek, A., Forster, B. B., Hare, R. D., & Liddle, P. F. (2004). Temporal lobe abnormalities in semantic processing by criminal psychopaths as revealed by functional magnetic resonance imaging. *Psychiatry Research: Neuroimaging, 130*(1), 27–42.

Knabb, J. J., Welsh, R. K., Ziebell, J. G., & Reimer, K. S. (2009). Neuroscience, moral reasoning, and the law. *Behavioral Sciences & the Law, 27*(2), 219–236.

Koenigs, M., & Newman, J. P. (2013). The decision-making impairment in psychopathy: Psychological and neurobiological mechanisms. In K. A. Kiehl & W. P. Sinnott-Armstrong (Eds.), *Handbook on psychopathy and law* (pp. 93–106). Oxford University Press.

Koenigs, M., Baskin-Sommers, A., Zeier, J., & Newman, J. P. (2011). Investigating the neural correlates of psychopathy: A critical review. *Molecular Psychiatry, 16*(8), 792–799.

Krakauer, J. W., Ghazanfar, A. A., Gomez-Marin, A., MacIver, M. A., & Poeppel, D. (2017). Neuroscience needs behavior: Correcting a reductionist bias. *Neuron, 93*(3), 480–490.

Lazarus, C., Haneef, R., Ravaud, P., & Boutron, I. (2015). Classification and prevalence of spin in abstracts of non-randomized studies evaluating an intervention. *BMC Medical Research Methodology, 15*(1), 1–8.

Levy, N. (2007). The responsibility of the psychopath revisited. *Philosophy, Psychiatry, & Psychology, 14*(2), 129–138.

Levy, K. (2011). Dangerous psychopaths: Criminally responsible but not morally responsible, subject to criminal punishment and to preventive detention. *San Diego Law Review, 48*, 1299–1395.

Levy, N. (2014). Psychopaths and blame: The argument from content. *Philosophical Psychology, 27*(3), 351–367.

Lilienfeld, S. O., Aslinger, E., Marshall, J., & Satel, S. (2018). Neurohype: A field guide to exaggerated brain-based claims. In L. S. M. Johnson & K. S. Rommelfanger (Eds.), *Routledge handbooks in applied ethics. The Routledge handbook of neuroethics* (pp. 241–261). Routledge/Taylor & Francis Group.

Ling, S., & Raine, A. (2018). The neuroscience of psychopathy and forensic implications. *Psychology, Crime & Law, 24*(3), 296–312.

Litton, P. J. (2008). Responsibility status of the psychopath: On moral reasoning and rational self-governance. *Rutgers Law Journal, 39*, 349–392.

Litton, P. J. (2013). Criminal responsibility and psychopathy. Do psychopaths have a right to excuse? In K. Kiehl & W. Sinnott-Armstrong (Eds.), *Handbook on psychopathy and law* (pp. 275–296). Oxford University Press.

Maibom, H. L. (2008). The mad, the bad, and the psychopath. *Neuroethics, 1*(3), 167–184.

Marazziti, D., Baroni, S., Landi, P., Ceresoli, D., & Dell'Osso, L. (2013). The neurobiology of moral sense: Facts or hypotheses? *Annals of General Psychiatry, 12*(1), 1–12.

Marshall, J., Watts, A. L., & Lilienfeld, S. O. (2016). Do psychopathic individuals possess a misaligned moral compass? A meta-analytic examination of psychopathy's relations with moral judgment. *Personality Disorders: Theory, Research, and Treatment, 9*(1), 1–11.

Mei-Tal, M. (2002). The criminal responsibility of psychopathic offenders. *Israel Law Review, 36*(02), 103–121.

Mendez, M. F. (2009). The neurobiology of moral behavior: Review and neuropsychiatric implications. *CNS Spectrums, 14*(11), 608–620.

Morse, S. J. (2008). Psychopathy and criminal responsibility. *Neuroethics, 1*(3), 205–212.

Motzkin, J. C., Newman, J. P., Kiehl, K. A., & Koenigs, M. (2011). Reduced prefrontal connectivity in psychopathy. *The Journal of Neuroscience, 31*(48), 17348–17357.

Müller, J. L., Sommer, M., Wagner, V., Lange, K., Taschler, H., Röder, C. H., ... & Hajak, G. (2003). Abnormalities in emotion processing within cortical and subcortical regions in criminal psychopaths: Evidence from a functional magnetic resonance imaging study using pictures with emotional content. *Biological Psychiatry, 54*(2), 152–162.

Nadelhoffer, T. (2015). Pure retributivism and the problem of psychopathy: A preliminary investigation/Retributivismo puro eo Problema da Psicopatia: Uma Investigação Preliminar. *Revista Direito, Estado e Sociedade, 47,* 156–204.

Nadelhoffer, T., Bibas, S., Grafton, S., Kiehl, K. A., Mansfield, A., Sinnott-Armstrong, W., & Gazzaniga, M. (2012). Neuroprediction, violence, and the law: Setting the stage. *Neuroethics, 5*(1), 67–99.

Nair, M. S., & Weinstock, R. (2008). Psychopathy, diminished capacity and responsibility. In A. R. Felthous & H. Sass (Eds.), *The international handbook of psychopathic disorders and the law: Laws and policies, volume II* (pp. 275–301). John Wiley & Sons Ltd..

O'Neill, M. E. (2004). Irrationality and the criminal sanction. *Supreme Court Economic Review, 12,* 139–180.

Ortega-Escobar, J., Alcázar-Córcoles, M. Á., Puente-Rodríguez, L., & Penaranda-Ramos, E. (2017). Psychopathy: Legal and neuroscientific aspects. *Anuario de Psicología Jurídica, 27*(1), 57–66.

Pardini, D. A., Raine, A., Erickson, K., & Loeber, R. (2014). Lower amygdala volume in men is associated with childhood aggression, early psychopathic traits, and future violence. *Biological Psychiatry, 75*(1), 73–80.

Patrick, C. J., Venables, N. C., & Skeem, J. (2012). Psychopathy and brain function: Empirical findings and legal implications. In H. H. Häkkänen-Nyholm & J. Nyholm (Eds.), *Psychopathy and law: A practitioner's guide* (pp. 39–77). John Wiley & Sons, Ltd..

Poeppl, T. B., Donges, M. R., Mokros, A., Rupprecht, R., Fox, P. T., Laird, A. R., … Eickhoff, S. B. (2018). A view behind the mask of sanity: Meta-analysis of aberrant brain activity in psychopaths. *Molecular Psychiatry, 24*(3), 463–470.

Poldrack, R. A., Monahan, J., Imrey, P. B., Reyna, V., Raichle, M. E., Faigman, D., & Buckholtz, J. W. (2018). Predicting violent behavior: What can neuroscience add? *Trends in Cognitive Sciences, 22*(2), 111–123.

Pujara, M., & Koenigs, M. (2014). Neuroimaging studies of psychopathy. In R. A. J. O. Dierckz, A. Otte, E. F. J. de Vries, A. van Waarde, & J. A. den Boer (Eds.), *PET and SPECT in psychiatry* (pp. 657–674). Berlin.

Purdie, J. (Writer/Director). (2002). *Psychopath* [video file]. Retrieved from http://www.youtube.com/watch?v=qm1S_n0V5kk&feature=related

Raine, A., Lencz, T., Bihrle, S., LaCasse, L., & Colletti, P. (2000). Reduced prefrontal gray matter volume and reduced autonomic activity in antisocial personality disorder. *Archives of General Psychiatry, 57*(2), 119–127.

Reidy, D. E., Kearns, M. C., DeGue, S., Lilienfeld, S. O., Massetti, G., & Kiehl, K. A. (2015). Why psychopathy matters: Implications for public health and violence prevention. *Aggression and Violent Behavior, 24,* 214–225.

Ridley, J. (2018, March 7). How to tell if your child is a future psychopath. *New York Post.* Retrieved from https://nypost.com/2018/03/07/how-to-tell-if-your-child-is-a-future-psychopath/

Rogers, R. (1995). *Diagnostic and structured interviewing.* Psychological Assessment Resources.

Santana, E. J. (2016). The brain of the psychopath: A systematic review of structural neuroimaging studies. *Psychology & Neuroscience, 9*(4), 420–443.

Sartori, G., Pellegrini, S., & Mechelli, A. (2011). Forensic neurosciences: From basic research to applications and pitfalls. *Current Opinion in Neurology, 24*(4), 371–377.

Schneider, F., Habel, U., Kessler, C., Posse, S., Grodd, W., & Müller-Gärtner, H. W. (2000). Functional imaging of conditioned aversive emotional responses in antisocial personality disorder. *Neuropsychobiology, 42*(4), 192–201.

Schopp, R. F., & Slain, A. J. (2000). Psychopathy, criminal responsibility, and civil commitment as a sexual predator. *Behavioral Sciences & the Law, 18*(2–3), 247–274.

Schultz, D. H., Balderston, N. L., Baskin-Sommers, A. R., Larson, C. L., & Helmstetter, F. J. (2016). *Psychopaths show enhanced amygdala activation during fear conditioning. Frontiers in Psychology, 7.*

Sethi, A., Sarkar, S., Dell'Acqua, F., Viding, E., Catani, M., Murphy, D. G., & Craig, M. C. (2018). Anatomy of the dorsal default-mode network in conduct disorder: Association with callous-unemotional traits. *Developmental Cognitive Neuroscience, 30*, 87–92.

Shaw, E. (2016). Psychopathy, moral understanding and criminal responsibility. *European Journal of Current Legal Issues, 22*(2), 1–25.

Sifferd, K. L., & Hirstein, W. (2013). On the criminal culpability of successful and unsuccessful psychopaths. *Neuroethics, 6*(1), 129–140.

Skeem, J. L., Polaschek, D. L., Patrick, C. J., & Lilienfeld, S. O. (2011). Psychopathic personality: Bridging the gap between scientific evidence and public policy. *Psychological Science in the Public Interest, 12*(3), 95–162.

Smith, S. F., & Lilienfeld, S. O. (2015). The response modulation hypothesis of psychopathy: A meta-analytic and narrative analysis. *Psychological Bulletin, 141*(6), 1145–1177.

Stratton, J., Kiehl, K. A., & Hanlon, R. E. (2015). The neurobiology of psychopathy. *Psychiatric Annals, 45*(4), 186–194.

The People of the State of Illinois v. Brian J. Dugan. 2009. No. #05 CF 3491. In the Circuit Court of Du Page County for the Eighteenth Judicial Circuit of Illinois.

Tiihonen, J., Hodgins, S., Vaurio, O., Laakso, M., Repo, E., Soininen, H., Aronen, H. J., Nieminen, P., & Savolainen, L. (2000). Amygdaloid volume loss in psychopathy. Society for Neuroscience abstracts, 2017.

Umbach, R., Berryessa, C. M., & Raine, A. (2015). Brain imaging research on psychopathy: Implications for punishment, prediction, and treatment in youth and adults. *Journal of Criminal Justice, 43*(4), 295–306.

University Of Southern California. (2004, March 11). USC study finds faulty wiring in psychopaths. *ScienceDaily*. Retrieved September 29, 2018 from www.sciencedaily.com/releases/2004/03/040311072248.htm

Veit, R., Flor, H., Erb, M., Hermann, C., Lotze, M., Grodd, W., & Birbaumer, N. (2002). Brain circuits involved in emotional learning in antisocial behavior and social phobia in humans. *Neuroscience Letters, 328*(3), 233–236.

Vierra, A. (2016). Psychopathy, mental time travel, and legal responsibility. *Neuroethics, 9*(2), 129–136.

Vitacco, M. J., Erickson, S. K., & Lishner, D. A. (2013). Comment: Holding psychopaths morally and criminally culpable. *Emotion Review, 5*(4), 423–425.

Vul, E., Harris, C., Winkielman, P., & Pashler, H. (2009). Puzzlingly high correlations in fMRI studies of emotion, personality, and social cognition. *Perspectives on Psychological Science, 4*(3), 274–290.

Weber, S., Habel, U., Amunts, K., & Schneider, F. (2008). Structural brain abnormalities in psychopaths—A review. *Behavioral Sciences & the Law, 26*(1), 7–28.

Yang, Y., & Raine, A. (2009). Prefrontal structural and functional brain imaging findings in antisocial, violent, and psychopathic individuals: A meta-analysis. *Psychiatry Research: Neuroimaging, 174*(2), 81–88.

Yang, Y., Raine, A., Narr, K., Lencz, T., & Toga, A. (2006). Amygdala volume reduction in psychopaths. *Society for Research in psychopathology annual meeting*. Society for Research in Psychopathology.

Yang, Y., Raine, A., Narr, K. L., Colletti, P., & Toga, A. W. (2009). Localization of deformations within the amygdala in individuals with psychopathy. *Archives of General Psychiatry, 66*(9), 986–994.

Yang, Y., Raine, A., Colletti, P., Toga, A. W., & Narr, K. L. (2010). Morphological alterations in the prefrontal cortex and the amygdala in unsuccessful psychopaths. *Journal of Abnormal Psychology, 119*(3), 546–554.

Part II
The Plausibility and Validity of Psychopathy

Chapter 7
In Fieri Kinds: The Case of Psychopathy

Zdenka Brzović and Predrag Šustar

Abstract We examine the philosophical and empirical issues related to the question whether psychopathy can be considered a psychiatric natural kind. Natural kinds refer to categories that are privileged because they the capture certain real divisions in nature. Generally, in philosophical debates regarding psychiatry, there is much scepticism about the possibility that psychiatric categories track natural kinds. We outline the main positions in the debate about natural kinds in psychiatry and examine whether psychopathy can be considered as a natural kind on any of the proposed accounts. By examining the scientific literature on psychopathy, we draw two main conclusions: (1) the empirical data currently do not support the view that this condition is unified enough to be considered a natural kind; (2) the construct of psychopathy plays useful roles both in the context of scientific research and in forensic and clinical settings. These considerations bring us to our tentative conclusion about psychopathy as a kind in the making, a sort of "under construction" category with potential for improvement and refinement with further research.

Keywords Psychiatric kinds · Natural kinds · Theoretical constructs · Epistemic values · Cluster kinds · Psychopathy

7.1 Introduction

This introductory chapter and the other chapters in Part II of the book examine the scientific status of psychopathy in the sense of a psychiatric classification, both from scientific and philosophical viewpoints. A frequently shared assumption by scientists, philosophers and the general public is that scientific classifications are to be considered as privileged ones in comparison to many other groupings that we

Z. Brzović (✉) · P. Šustar
Department of Philosophy, Faculty of Humanities and Social Sciences, University of Rijeka, Rijeka, Croatia
e-mail: zdenka@uniri.hr; psustar@ffri.uniri.hr

L. Malatesti et al. (eds.), *Psychopathy*, History, Philosophy and Theory of the Life Sciences 27, https://doi.org/10.1007/978-3-030-82454-9_7

might come up with. This, in turn, means that we need to provide certain criteria for what counts as a scientific classification and what makes it privileged.

Philosophical debates regarding these issues can be summarized through the question about whether certain scientific categories correspond to *natural kinds*. Ideally, philosophers aim at setting up a list of requirements or constraints about what establishes a determined grouping as a natural one. The idea is that scientific classifications as the paradigmatic instances of natural kinds should fulfil those requirements, while other, non-scientific classifications or even pseudoscientific ones will not make the cut. As we will see in the next section, philosophers typically start out with a very strict account of natural kinds and, then, loosen up the criteria when it becomes obvious that hardly any scientific classification (if any at all) would qualify for the natural kinds status.

The tendency of relaxing the criteria for what constitutes natural kinds is well illustrated through the debate on psychiatric kinds. In psychiatry, there are numerous issues related to quick and easy criteria for what constitutes a relevant kind. Also, there is pressure to have at least minimal threshold for which kinds are to be considered as relevant both in the research-oriented projects and in more practical contexts. As is probably clear to anyone coming across the book, to be classified or diagnosed as a psychopath is no laughing matter. The least we can ask in that regard is that the criteria for such a classification be carefully established, if possible.

In what follows, we present an outline of the philosophical debate on natural kinds in psychiatry. Then, we examine how the accounts in question can be applied to the particular case of psychopathy. Moreover, we try to position the contributions to this part of the book into the on-going wider debate on the status of psychopathy as a natural kind. At the end, we summarize the core claims of the papers grouped in this part, thereby primarily referring to the kinds and scientific classifications debate in the current philosophy of the life sciences.

7.2 Natural Kinds in Psychiatry

Debates about natural kinds in psychiatry are often conducted in terms of the question whether psychiatric conditions are *real* (see, for instance, Beebee & Sabbarton-Leary, 2010; Kendler, 2015, 2016; Kohne, 2015). Essentialism about natural kinds is often taken as the perfect candidate account for a realist take on natural kinds, because it provides clear criteria for discovering clear-cut boundaries between kinds, signalling that we are indeed carving up the purported nature's joints. Namely, the possession of essence, an important intrinsic, deep-level property, shared by all and only kind members responsible for all of their other common properties, allows us to establish clear-cut boundaries between members and non-members of a kind.

The problem, however, is that in psychiatry, we do not encounter essentialist classifications. Not only, we do not encounter discrete, essentialist categories, which seems to be the case in other special sciences as well,[1] but psychiatric classifications also show some additional interesting problems of their own. Namely, the unsuitability of essentialism for capturing classifications in the mature sciences usually leads to rejection of essentialism and the view that a different, less strict, account of natural kinds ought to be endorsed. The underlying idea here is that in the mature sciences, we have well-established theories with cleared up concepts and categories that have withheld serious empirical scrutiny and, as such, they represent the best available candidates for natural kinds. However, the maturity of psychiatry in that regard has been extensively questioned. For instance, Thomas Insel, former director of the National Institute of Mental Health (NIMH), in an NIMH 2013 blog post, reports that psychiatric categories, he refers primarily to the DSM ones, are not based on any objective laboratory measure, but rather on a consensus about clusters of clinical symptoms. He states that "[I]n the rest of medicine, this would be equivalent to creating diagnostic systems based on the nature of chest pain or the quality of fever." (Insel, 2013).

The authors of DSM-5 acknowledge that current diagnostic criteria for disorders do not necessarily identify a homogenous group of patients that can reliably be characterized with all the various types of validators. They state, however, that „[U]ntil incontrovertible etiological or pathophysiological mechanisms are identified to fully validate specific disorders or disorder spectra, the most important standard for the DSM-5 disorder criteria will be their clinical utility for the assessment of clinical course and treatment response of individuals grouped by a given set of diagnostic criteria." (American Psychiatric Association, 2013, p. 20).

Another, more general issue that might be raised is that in psychiatry, but also in psychological sciences more generally, we are dealing with constructs which are theoretical abstractions derived from theory, as is nicely elaborated by Sellbom et al. (this volume). As such, they are not observable, but need to be inferred through operationalization. One might argue that this is problematic since we lack direct access to prospective psychological kinds. Moreover, psychologists, when assessing construct validity, assess how well the test measures that which it claims to measure and do not enter into further discussions whether there is an underlying reality to their classifications, i.e. whether the proposed constructs track some real divisions in nature.[2] Even though psychological constructs do at first sight seem more problematic than entities referred to in scientific theories in other disciplines, it is hard to come up with a principled reason why this should be the case. Namely, the observability of theoretical constructs is an issue for other sciences as well, and should not automatically be considered as ground for scepticism about psychiatric kinds.

---

[1] Chemistry is often invoked as a domain of paradigmatic essentialist kinds, but even this has recently been under scrutiny.

[2] However, in Sect. 7.3 we will see that constructs that exhibit a high co-instantiation of clustering traits which can be construed as discontinuous from other clusters are characterized by behavioural scientists as taxons, something which can be taken as corresponding to a natural kinds, at least on some interpretations.

We outline the main philosophical reactions to these reported problems concerning psychiatric classifications. We distinguish three main reactions to such difficulties: (1) there are no natural kinds in psychiatry; (2) there are natural kinds in psychiatry and psychiatric conditions (or many of them) correspond to natural kinds when we adopt a less strict account of natural kinds; (3) the strategy of examining current scientific psychiatric practice in search for natural kinds is mistaken, because the practice itself is problematic and needs to be thoroughly reassessed. In the next section, we address (1) and (2) in more detail, since they represent the prevailing reactions in the corresponding discussions. Reaction (3), represented here by Maraun (this volume), is, in his own words, an "iconoclastic" one. We now briefly mention Maraun's objections to the current psychiatric classificatory practice and point to the reader the corresponding text.

Maraun criticizes the strategy used by the psychiatric practice termed "General Construct Validity" (GCV). GCV distinguishes between (1) theoretical constructs as abstract hypothetical devices derived from a theory, and (2) their measurement that, at best, amounts to an empirical approximation to the construct inferred through observable quantitative indicators (for an informative summary with regard to this dichotomy, see Sellbom et al., this volume). Maraun objects that the introduction of the notion of *construct* is problematic because it distorts the relation between a *term* and a *class of referents* that should be easily fixed by a clear and unambiguous rule for the term use. The faulty presuppositions here are, in his view, that the meaning of the term can be brought into focus through empirical research, and that conceptual confusion is a commonplace in science. In Maraun's view, clarity is a hallmark of fruitful scientific enterprise, and the rules of ascription for technical scientific terms should be given according to the necessary and sufficient conditions approach.

We have argued elsewhere that many scientific categories are introduced as vague notions serving a unifying role in the scientific research and have proposed a similar role for the category of psychopathy (see Brzović et al., 2017). However, we do not discuss these issues any further in this setting. The main point that we wish to emphasize here is the following one: there are two routes for assessing psychiatric categories. *First* one is by endorsing the current scientific standards and practices in psychiatry and examining whether a particular category, psychopathy in our case, fulfils certain criteria for establishing a good or relevant scientific kind. *Second* one, exemplified by Maraun's approach is an external criticism of the scientific practice. The latter strategy starts out with some prior assumptions regarding what constitutes a good scientific category, and then examines whether certain scientific practice corresponds to that model.

7.2.1   *There Are No Natural Kinds in Psychiatry*

A common strategy in arguing for the claim that, in principle, there are no natural kinds in psychiatry, invokes the fact that psychiatric kinds are not mind-independent. Three main reasons have been offered that preclude the mind-independence of psychiatric kinds: (1) in determining the boundaries of psychiatric conditions, our decisions play important roles; (2) what gets classified in psychiatry is necessarily mind-dependent. Finally, potentially the most devastating reason: (3) that the mind-dependence involved in psychiatric classification is of a problematic, non-epistemic type. That is, the decisions about the correct classificatory systems in psychiatry involve evaluative judgments heavily influenced by social norms. In what follows, we examine the above reasons in more detail.[3]

Let us start with (1). The basic claim here is that there are no natural kinds in psychiatry because psychiatric kinds are not *real*, but, rather, conventional. Nonetheless, since there are important and interesting classifications that psychiatrists refer to, there are proposals for distinguishing such (partly) conventional categories as relevant for the biomedical contexts. For instance, Zachar (2000, 2002, 2015) introduces the distinction between *natural* kinds that represent what exists independently from our classifications, and *practical* kinds, which represent phenomena as a result of various decisions that we make in the act of scientific classifying. He summarizes this by a catchy motto: 'natural kinds are made by the world and practical kinds are made by us'.

As for reason (2), some authors have claimed that human (and social) kinds are automatically disqualified as candidates for natural kinds, because they are inevitably mind-dependent (see, for instance, Held, 2017). However, a distinction should be made between the fact that the entities classified into kinds are mind-dependent, for example, things such as mental states, behaviours, and similar phenomena, and the fact that our classifications hold in virtue of mind-dependent facts. That is, it does not follow that classifications of mind-dependent entities are necessarily mind-dependent. One can argue that we can discover the correct way to classify certain mind-dependent entities, which holds independently of us, the classifiers. In that regard, Zachar rightly emphasizes the importance of our decisions for

---

[3] Another potential reason for why psychiatric categories cannot be considered as natural kinds is put forward by Ian Hacking. He refers to the so-called "looping effect", associated with human kinds (see Hacking, 1996). In his view, human kinds are characterized by the looping effects, i.e., the classifications of humans into kinds can change what gets classified (either because we become aware of the classification in question or because we generally respond to being treated differently, even without knowing about the classification). This, in turn, can loop back to change the classification in question. Hacking does not specify what condition traditionally associated with natural kinds is incompatible with the looping of human kinds. Rather, he just claims that natural kinds are supposed to be "indifferent" (Hacking, 1999), i.e., they do not change in response to our classificatory efforts. We will not discuss here whether this objection to the natural kinds status can be subsumed under the mind-dependence reason against psychiatric natural kinds, but does appear close to reason (2).

circumscribing psychiatric kinds. In other words, if what constitutes a psychiatric kind is partly a result of our decision, then they are indeed mind-dependent.

That said, many authors do not take mind-dependence to be a problem for realism about natural kinds. For instance, Ereshefsky and Reydon argue that natural kinds need not be mind-dependent (see Ereshefsky, 2018; Ereshefsky & Reydon, 2021). Magnus and Khalidi defend an even stronger claim that realism about natural kinds needs not be understood in terms of mind-independence criterion (Muhammad Ali Khalidi, 2016; see Muhammed Ali Khalidi, 2013; Magnus, 2012). These arguments, however, rest on a dichotomy between realist and conventionalist views, where conventionalism (or antirealism) comes down to the extreme position that nothing in the world constraints our categories, i.e., they are all equally arbitrary. Yet, this is not the standard understanding of the opposition between realism and conventionalism. The standard characterization of conventionalism about natural kinds can be summarized as the view according to which nature does not come in pre-packaged clear-cut categories that we just have to discover or recognize.[4] This is the case, because there are no clear divisions ("chasms" or "gaps") between natural phenomena, but a continuum characterized in terms of degrees of similarity and difference, or we are not able to grasp such divisions. In any case it follows that we are supposed to decide where to draw the line between them (see, for instance, Koslicki, 2008).

We will not dwell further on the issues regarding realism and mind-independence in the debate under consideration. The interesting claim that we can draw from this is that psychiatric categories can be considered as natural kinds, even though they partly depend on our decisions about where to draw the line on a continuum of symptoms, behaviours, brain states, responses to medications, or whatever else is relevant for psychiatric classifications. In what follows, we will show that all the accounts of natural kinds that can be applied to the case of psychiatric conditions accept the influence of classifiers in determining what constitutes a relevant natural kind.[5] Once we accept this as a non-problematic fact, another pair of issues crops up: (i) to what degree our interests and aims can determine whether we will consider a psychiatric condition to correspond to a natural kind; (ii) whether only epistemic interests are relevant, or practical ones may also enter into consideration.

These issues, especially the latter one, bring us to the third reason that has been put forward in favour of the view that there are no natural kinds in psychiatry, namely, that the mind-dependence involved in psychiatric classification is of a problematic, non-epistemic type. Critics of psychiatric classification have claimed that it is *conventionalist* in a stronger sense then discussed above, i.e., based on

---

[4] Essentialism is a typical view of this sort.

[5] Even if there are psychiatric conditions where we can circumscribe their natural boundaries, i.e., where no human decisions are required as to where to draw the line between different symptoms, behaviors constituting the condition, this does not undermine the fact that most psychiatric conditions are not of this type. Hence, our decisions where to draw the line become relevant. Accordingly, any account that aims at capturing a relevant amount of psychiatric categories will allow for our influence within the classifying activity.

non-epistemic values, resulting in problematic categories demarcated based on deviance from social norms. The underlying assumption here is that classifications are based on non-epistemic values when they do not fulfil the epistemic ones, but we still entertain them for practical purposes. For instance, deviance from social norms is problematic for the society, so this might serve as an incentive to declare such behaviour a type of mental illness. Such strongly conventionalist categories are deeply problematic because they are not grounded in characteristics of the subjects classified but in someone's decisions about not condoning certain types of behaviour.

Mayes and Horwitz (2005) describe the introduction of DSM-III, the first of the symptom-based, non-causal classificatory standards, as the result of "a vehement political struggle for professional status and direction" (2005, p. 266). Furthermore, many psychiatric categories have been appropriated from folk terms and classifications, potentially carrying along lay beliefs, and even misconceptions about the condition. For example, Sellbom et al. (this volume) interestingly point out how myths, misconceptions and fallacies are not specific only to lay people's understanding of psychopathy, but also entrenched among many mental health professionals.

Now, thus far, we have entertained the view that influence of our interests and values does not preclude psychiatric categories to correspond to natural kinds. However, the standard presupposition is that the interests and values in question are *epistemic* ones.[6] However, the question arises whether influence of *non-epistemic* interests and aims can also be compatible with the natural kind status of psychiatric categories.

Certain amount influence of non-epistemic or contextual values, such as the political struggles involved in the creation of DSM-III, does not necessarily preclude the objectivity and scientific relevance of those categories. We might still think of those categories as objective even if the reason for introducing a certain type of classificatory system is partly ideologically motivated. Take a hypothetical case where a new classificatory system is proposed because the old results in great expenses. Let us say that the old system was inefficient in treating individuals that it classifies as having a certain condition. Here, even if motivation for introducing the new one is a cost-benefit analysis and not concern with public health, it could be endorsed if the new system is more efficient. Efficiency is also a practical, non-epistemic aim, but we suppose that it is a product of the fact that a classification satisfies epistemic aims as well.

Perhaps we can even allow some degree of non-epistemic values to influence the decision about where exactly to draw the line between certain conditions as long as the conditions themselves are based on some objective scientific considerations and the ambiguity consists in where their boundaries are. However, the important

---

[6] Epistemic (or cognitive values) such as predictive accuracy, scope, unification, explanatory power, simplicity, coherence with other accepted theories, etc. Non-epistemic (non-cognitive or contextual) values are moral, personal, social, political and cultural ones (see Reiss, & Sprenger, 2014).

question is how much of non-epistemic values is *too much*? It is clear that the categories under consideration cannot be entirely based on non-epistemic values. For instance, ideologically motivated. In that case, classifications would be social constructions without anything independent of us to enable grounding. In such cases, it is clear that psychiatric categories certainly do not qualify as natural kinds. Showing that they are completely the product of our decisions to group certain people in a certain way without having any basis in the causal structure of the world, as was the case with certain historical classifications such as drapetomania or hysteria, would be a strong case against psychiatric classifications and the scientific status of psychiatry.

It appears, however, that in most cases, psychiatric classifications are not entirely socially constructed. Rather, there are often determined grounds at display for psychiatric classifications, but the interesting question is how causally grounded they need to be in order to consider them as natural kinds. Namely, we can come up with many classifications that are, to a certain extent, grounded in nature, but we would not straightforwardly consider them as natural kinds. To the issue concerned with the right amount of grounding for psychiatric kinds as candidates for natural kinds, we turn in the next section. We examine positions that hold that psychiatric categories are natural kinds, if we adopt a less strict account of natural kinds than essentialism.

7.2.2 Natural Kinds in Psychiatry: Cluster Kinds, Promiscuous Kinds or Grounded Functionality of Kinds?

This part of the debate on psychiatric kinds closely resembles the analogous debate in the philosophy of biology and other special sciences: we look at scientific classifications in the prospective field, notice that they cannot be captured by an essentialist account and, then, look for alternative accounts that would be suitable for this purpose. The first candidate is the cluster account of natural kinds, introduced by Richard Boyd (1991) under the term "Homeostatic Properties Cluster" (HPC). The main idea is that there are clusters of co-occurring properties shared by members of a kind, unified by a common cause or a mechanism. Cluster accounts are often taken as the most serious candidates for capturing kinds in the special sciences, psychiatry included (see, for instance, Kendler et al., 2011; Tsou, 2016).

There are two potential problems in applying the HPC account on the psychiatric kinds here at issue. A *first* problem relates to the notion of underlying mechanisms responsible for clustering of properties. We have seen earlier that psychiatric categories often do not refer to causes of various constellations of symptoms making up a certain condition. Some authors have argued that clusters of stable properties are enough for a grouping to be considered as a natural kind, without worrying about the underlying causes of stability (see Slater, 2015). Thus, we might want to argue that some cluster accounts (such as Slater's "Stable Property Clusters" or SPC) are applicable to psychiatric kinds. The main issue, however, concerns the question whether psychiatric conditions are homogenous enough to be captured by the cluster accounts, as we will see in the case of psychopathy.

Such concerns have led to formulations of more encompassing, or "relaxed" (Varga, 2018)[7] accounts circumscribing conditions that cannot be characterized in terms of clusters of common properties, i.e., cases where no substantial correlation between properties (or traits) characterizing a certain condition can be found. The most prominent of such relaxed accounts is "promiscuous realism" (Dupré, 1981, 1993). Its central claim is that the world is immensely complex and that we can categorize it in countless cross-cutting ways. It follows that an individual is objectively a member of many distinct kinds and there is no *the* natural kind to which it belongs. We can see here that kinds are not perceived as special, privileged categories, but, rather, groupings that can be unified by a minimal grounding – such as sharing a common property or properties relevant for our classificatory interests. If we apply this account to the case of psychiatric kinds, the fact that psychiatric conditions are not unified enough does not necessarily represent a relevant obstacle, as long as they serve the goals that motivate our inquiry. Rachel Cooper (2005) applies promiscuous realism to the psychiatric categories, and argues that natural kinds pick out entities with similar determining properties. The grades of similarity required are relative to the aims of the scientific inquiry. Also, Cooper talks about *genuine* properties as the relevant ones, because they "endow entities with particular causal powers, and ground objective similarities." (p. 52). This constraint distinguishes promiscuous realism from even more relaxed accounts such as the one defended by one of the contributors to this section Thomas Reydon, the *Grounded Functionality Account* to which we turn to next.

The account of natural kinds he puts forward together with Marc Ereshefsky (in press) introduces two constraints on what can be considered as natural classifications: (1) they should serve either the epistemic or non-epistemic functions they are posited for, and (2) they should serve them because they are grounded in the world. We will go back to requirement (1) in our brief survey of psychopathy as a potential scientific kind in the next section, but let us shortly focus on (2). What is interesting here is that the grounding in question allows that scientific classifications need not be causal kinds, i.e., members of natural kinds need not share a common property or underlying mechanism. Moreover, it appears that it is possible that they do not possess any common set of causal properties. A candidate for natural kinds, described in an earlier paper by Ereshefsky and Reydon (2015), relates to the case of functional kinds, where entities classified into a kind can have various structures and properties.[8] This makes their account even more

---

[7] It needs to be noted, however, that Varga calls "relaxed" all the non-essentialist accounts of kinds, but here we take as relaxed only those that are less demanding than the cluster accounts. The reasoning behind this is that in psychiatry many disorders do not fit the cluster demand. Thus, the "relaxing" condition should go beyond the cluster accounts.

[8] The received view is that functional kinds are not candidates for natural kinds (see, for, instance Fodor, 1974), but as we can see in this strand of the debate, there are exceptions to such a view.

encompassing than promiscuous realism. Ereshefsky's and Reydon's grounding condition states that the kinds in question ought to reflect *an aspect* of the world, rather than *merely* human interests and practices. Thus, minimally, the grouping is natural if it is not entirely arbitrary. Before addressing the issue of how psychopathy's status as a potential natural kind fares against these proposed accounts of natural kinds, let us briefly get back to the question of realism of psychiatric kinds.

All the proposed accounts that aim to loosen up the strict essentialist criteria in order to capture relevant scientific classifications share the presupposition stating that kinds will be considered as natural partly depends on our epistemic aims, decisions, and scientific practices.[9] The Grounded Functionality Account, however, even allows for non-epistemic interests and values to play an important role in determining which kinds are rightly considered as natural. The interesting issue here is concerned with an apparent reciprocity between the degree of "relaxedness" of the proposed account and the burden it puts on the contribution of our interests and values in determining which kinds are natural. Thus, proposals with minimal grounding requirements, if they do not wish to end up as a mere science reporting,[10] need to carefully filter out what it means that certain scientific classifications fulfil our interests, what interests exactly, and to what degree, in order to count as natural kinds. In other words, loosening up the "metaphysical" burden has to be accompanied by an elaborated view of what it means that a determined classification is natural, without thereby collapsing into a downright social construct.

In what follows, we examine whether there are sufficient grounds for considering psychopathy as a psychiatric natural kind. We briefly survey the current state of psychopathy research, in particular, (1) whether there is a consensus with regard to how we understand, conceptualize, and measure this condition (for more on this, see Sellbom et al., this volume), and (2) whether the classification is unified enough to be considered as a natural kind. Thus, in what follows, we examine how the scientific debates regarding the scientific legitimacy of the construct of psychopathy can be linked to the three non-essentialist accounts of natural kinds mentioned above. That is, we assess whether one or more of these accounts can be applied to the case of psychopathy and, if so, what this means for the natural kind status of psychopathy.

---

[9] At least as they are standardly presented in the literature.

[10] As we will see in the next section, and also in the chapters in this part, as far as the conditions as complex as psychopathy are concerned, even science reporting becomes a challenging task, since there is no clear scientific consensus concerned with the status of psychopathy as a scientific category.

7.3   Psychopathy as a Psychiatric Kind

At the most general level, we approach the question whether psychopathy can be considered a natural kind in psychiatry by examining how unified of a phenomenon it is, and if it allows us to formulate sound inductive generalizations about the people classified as psychopaths. More specifically, in the literature four options have been examined whether psychopathy fulfils the conditions for a natural kind: (1) an HPC cluster kind; (2) an SPC cluster kind; (3) a promiscuous realist kind and (4) grounded functionality kind. Option (1) is the most demanding but it can be taken as too strict. If, as we have stated earlier, most DSM categories do not fulfil the condition of tracking common underlying causes or mechanisms responsible for the symptoms characterizing them, then it might seem too demanding that psychopathy should satisfy them nonetheless.[11]

The first two options are most relevant when it comes to examining empirical data and assessing whether they support the view that psychopathy is a unified category. This is because they provide relatively strict grounding conditions for what constitutes a kind. With options (3) and (4) as we will see, the grounding condition is relatively minimal and the burden is put on specifying which of our interests, values and aims are relevant for establishing psychopathy as a natural kind. Let us start first with option (2) and examine whether psychopathy can be considered as a condition characterized by a cluster of common symptoms.

To get a more precise formulation of the clustering requirement, we rely on Slater's (2015) elaboration of it within his SPC account. For a cluster to be considered stable it needs to allow reliable inferences from observations of certain features to instantiations of other features. Arguably the most reliable inferences are based on categories that are deemed to be discrete. For instance, a high correlation between traits is expected to increase the probability that they will be co-instantiated. When this co-instantiation or clustering of traits can be justifiably construed as discontinuous from other clusters, then behavioural scientists talk about the existence of a "discrete category", "taxon", a "type", a "species", a "disease entity", a "nonarbitrary category" or a "*natural kind*" (see Meehl, 1992). Now, this is exactly what interests us here: are traits connected to psychopathy correlated in such a way to form a taxon or a discrete category.

It seems that psychopathy should not be construed as a taxon. Although early studies suggested that psychopathy might be thought of as a taxon (Harris et al., 1994, 2007), they have not been corroborated with further research. In fact, more recent studies indicate that psychopathic traits should be construed as dimensionally distributed in the general population, thus suggesting that there is no clear cut off line that would justify thinking about them as forming a discrete category (Edens et al., 2006, 2011; Murrie et al., 2007; Walters et al., 2011).

---

[11] One might argue, however, that the severity of social consequences involved in labeling someone a psychopath requires stronger grounding then for most psychiatric categories.

Nonetheless, even if psychopathy is not a taxon, this does not necessarily show that different measures of psychopathy do not delineate clusters of traits that might underlie reliable inferences. For instance, studies show that Robert Hare's Psychopathy Checklist-Revised (Hare, 2003), a widely used measure for testing psychopathy in forensic settings, captures several clusters of antisocial personality subtypes which differentially correlate with external criteria, such as anxiety, emotional expression, and criminal recidivism (Driessen et al., 2018). However, despite the influence of PCL-R in psychopathy research and practical application, recently it has been forcefully criticised on empirical grounds for its limitations in delineating a category of personality traits associated with impaired moral psychology that in turn predicts treatment outcomes and criminal behaviour (DeMatteo et al., 2020; Larsen et al., 2020; Olver et al., 2020). In other words, the complaint is that the reliability of PCL-R-based inferences in theoretical and practical contexts might be more limited than it would be expected if it delineated a natural kind.

However, even if we set aside these pending empirical issues that are internal to the application of PCL-R, we should keep in mind, as nicely elaborated by Sellbom et al.'s (this volume), that there are different measures of the construct of psychopathy and that we should not equate psychological constructs with their measurements (see also Cooke, 2018). This is important, because, no matter the usefulness of PCL-R in different contexts, it is still *a* measure of psychopathy whose main purpose is to delineate individuals with specific maladaptive tendencies within the forensic populations. This raises the question whether similar types of stable clusters of personality and behavioural traits are likely be found in other populations. In fact, there is substantial disagreement on this point between psychopathy researchers, because when we turn to other measures of psychopathy, we do not encounter substantial evidence for the stable cluster of properties view. The most telling in this respect is that the psychometric studies of the Psychopathic Personality-Inventory (PPI), another widely used psychopathy measure designed for subclinical (general) populations, provide evidence for thinking that psychopathy is composed of a manifold of personality and behavioural traits that do not enable reliable predictions because they are only loosely correlated (Lilienfeld, 2013).

So far, we have examined top-down approaches to determining whether psychopathy is a good scientific kind. We start from psychological and behavioural traits and then try to determine whether they capture reliable cluster kinds. Given the problems this approach has encountered, another option would be to adopt a more bottom-up approach that starts with neurobiological mechanisms associated with antisocial behaviour and sees whether in this way we might unravel some interesting results or common causes that might explain the behavioural and psychological clustering of psychopathic traits (Brazil et al., 2018; Jurjako et al., 2020). This approach is compatible with the HPC account of natural kinds (our option 1), and it might even be suggested by it. For instance, there are two main research frameworks that conceptualize psychopathy as a unitary construct and they purport to unravel the main common cause of psychopathy (however, see Groat & Shane, 2020 for third competitor to these research frameworks). Affect-based accounts associate psychopathy with deficits in experiencing, recognizing, and learning from affective stimuli (see, e.g., Blair et al., 2005).

Alternatively, the cognitive or attention-based accounts, construe psychopathic maladaptive behaviours as stemming from suboptimal allocating of attentional resources to cues that are relevant for successful performance as measured by experimental tasks (see, e.g. Koenigs & Newman, 2013). What is common to these two research frameworks is that both purport to explain the characteristic personality affective and behavioural features exhibited by psychopathic individuals. For an informative and exhaustive review of various theories belonging to these two camps, and empirical work done on maladaptive behaviour in psychopathy in the context of reinforcement learning and risky decision-making we refer you Glimmerveen et al. (this volume).

However, we might wonder about the feasibility of such a bottom-up approach to determining the causes of psychopathic traits, understood as a constellation of cognitive, affective and behavioural traits, if there is no consensus on whether they capture a unitary construct or with regards to their optimal measure. This brings us back to the methodological approach, criticized by Maraun (this volume), that the meaning of technical scientific terms can be brought into focus through empirical research. But it seems that our current lack of insight into the intricacies surrounding the psychopathy condition (like many other psychiatric conditions as well) justifies such an approach, even if it still has, in Kuhnian terms a preparadigmatic flavour to it (Kuhn, 1994).

Nonetheless, the more feasible idea of a thoroughly bottom-up approach would be to concentrate on one or several compatible measures of psychopathy and then search for common specific aetiology or underlying neurobiological causes responsible for the constellation of traits as identified with those measures. In this process, we could use the empirical data about the underlying aetiology to refine or even substitute currently used measures with improved ones (Brazil et al., 2018). In addition, in this process might help us to refine our conceptualizations of the construct of psychopathy and not just its measures that would enable us to successfully deal with the negative effects that drive the psychopathy research in the first place; and that is to find effective treatments for reducing maladaptive and antisocial behaviour as typically exhibited by psychopathic individuals.

Let us now turn to a different strategy, exemplified by the options (3) and (4) that starts not with the question about the properties that characterize kind members, but with the question about what purpose certain classification is supposed to fulfil and how well it serves this purpose. Reydon (this volume) offers a nice elaboration of such a strategy. He acknowledges that psychopathy would not count as a natural kind on most accounts of natural kinds. It does not figure as the basis of inferential statements or scientific explanations, meaning that it does fulfil the epistemic functions commonly associated with natural kinds. Reydon argues, however, that there are many other functions, both epistemic and non-epistemic ones that kinds perform in scientific practice, clinical practice and elsewhere. In his view, psychopathy is a good scientific kind because saying that someone is a psychopath locates that person somewhere in the space of distribution of behavioural traits, thereby serving important functions in the scientific research, clinical purposes and clinical intervention.

This is a legitimate strategy and we agree with Reydon that the notion of psychopathy serves useful purposes. We have argued elsewhere (Brzović et al., 2017) that there are several important purposes that psychopathy serves: it reliably predicts recidivism and violent behaviour (when characterized by the PCL-R), it narrows the scientific focus to groups of people within the broader category of ASPD and it has an important communicative role in forensic and clinical research due to the standardisation of assessment procedures. We associate these important functions with fulfilling the epistemic functions or aims, with practical functions being only derivative. Moreover, Reydon' strategy seems to be along the lines of proponents of the DSM-5 when they argue that until we are able to identify etiological or pathophysiological mechanisms of various disorders, the most important standard for judging psychiatric categories is their clinical utility (clinical course and treatment response of individuals grouped).

However, psychopathy is an extremely serious diagnosis and the standards for its assessment should be equally serious. This is especially salient given the severe consequences of such a diagnosis and the minimal treatment options. The strategy of keeping the useful (to a certain extent) scientific categories despite the lack of their grounding is perhaps less problematic in fields with less influence of lay characterizations, stereotypings and misconceptions. Given the potential of psychopathy research to be influenced by such problematic misconceptions as emphasized by Sellbom et al. (this volume), we need to be especially careful in examining whether the perceived usefulness is more than just confirmation of some previously acquired biases. While the massive amount of empirical research is underway in preventing such scenarios, our tentative conclusion would be that it is still premature to consider psychopathy a natural kind. It does, however, deserve a status of a category in the making, something that serves to unify scientific research, but still without sufficient homogeneity to deserve a status of a full-blown natural kind.

In the next section, we summarize the core claims of the papers grouped in this part.

7.4   Chapter Summaries

Michael D. Maraun, in the chapter entitled "Psychopathy and the Issue of Existence", singles out the following main questions: does psychopathy exist? ("Question I"); if it exists, what are its causes? ("Question II"). The questions of *existence* and *causality*, as targeted by scientific practice, the author argues, however, remain unresolvable in the context of psychopathy research. Hence, one of the most valuable outcomes of this analysis points to the fact that experimental investigation can thrive, recent genome biology is especially illustrative further example in this regard. Nevertheless, that kind of success within scientific practices seems to be "irrespective of the presence of *conceptual* confusion" (emphasis added) therein.

The chapter "Assessment of Psychopathy: Addressing Myths, Misconceptions, and Fallacies" by Martin Sellbom, Scott O. Lilienfeld, Robert D. Latzman, and Dustin B. Wygant conducts a meta-analysis of the broader psychopathology literature by targeting *general issues* in psychopathy assessment and, subsequently, a set of *additional issues*. As to the general issues, the authors confront some common conceptual "myths, misconceptions, and fallacies", which have been transmitted by the literature. The "myths" are concerned with the existence of so-called 'gold standards', primarily, PCL-R as *the* assessment tool or, even, as amounting to the psychopathy construct itself; the "misconceptions" relate to a number of theoretical issues affecting operationalization; and "fallacies" concerned with interpreting psychopathy scores. Apart from these general issues, the chapter also addresses some additional ones through the following questions: (i) are self-report measures inherently problematic for psychopathy assessment? (ii) should informant reports be ignored in adult psychopathy assessment? Namely, the authors here point out that exist a number of good reasons to take into account informant-report-based methodologies in the assessment of psychopathy, including the possibility that informant data may help to circumvent "blind spots" in psychopathic individuals' self-reporting; (iii) are brief measures appropriate for the assessment of psychopathy?

Now, as to *the natural kinds debate* more strictly, the chapter interestingly emphasizes the fact that "there is surprisingly little empirical foundation for the prevalent practice of using cut-off scores on psychopathy measures, let alone for the specific cut-offs recommended in psychopathy manuals". Thus, the corresponding analyses have clearly shown that psychopathy is not a "taxon, that is, a class (category) in nature". Accordingly, the authors conclude that the structure of psychopathy should be instead conceived on a continuum, not as an all-or-nothing affair. Or, as they characterize the conceptual shift in question, its nature appears to be "*dimensional*, meaning that the differences between psychopathic and non-psychopathic individuals are of degree, not of kind".

In the chapter "Psychopathy as a Scientific Kind: On Usefulness and Underpinnings", Thomas A.C. Reydon argues that whether psychopathy is a natural kind represents a non-starter for the corresponding debate, in particular when the question is addressed by the mainstream philosophical accounts of natural kinds. On the contrary, according to Reydon's and associates' *Grounded Functionality Account* (GFA), we should ask instead, in a reversed order with respect to the mainstream accounts of natural kinds, the following questions: *firstly*, what the grouping that the term denotes is good for, and, *secondly*, whether its role is grounded in the world. That reversal in approaching the debate enables to get around some considerable impasses, such as whether psychopathy represents a grouping in the world as it is *independently* of our decisions and other interventions. The conclusion of the GFA line of reasoning is that psychopathy constitutes a good scientific or, interchangeably, according to the author, natural kind.

However, the mainstream strand of the natural kinds debate rebuttal on that conclusion could be easily envisioned along the following lines: (i) psychopathy as a scientific or natural kind is not sufficiently successful in generating inferential statements, that is, empirical generalizations about the traits that all or a relevant majority of psychopaths exhibit or the supposed natural kind fails to act as an appropriate basis for scientific explanations. As such, it is not performing the standard epistemic functions that the mainstream accounts in this philosophical area assign to kind terms in the sciences. However, according to the author, there are other epistemic and, especially, non-epistemic functions that kind terms may perform in psychiatry and similar scientific practices. Thus, by overly insisting "on just one epistemic role entails missing much of how kind terms function in actual scientific practice, clinical practice, and elsewhere". The upshot of such a GFA application to the psychopathy construct comes not to the view that psychopathy does not relate to a kind of people or a kind of behaviors exhibited by some specific group of individual organisms, but of psychopathy as a "region on a *multidimensional space* of behaviors" (emphasis added). In that, in our view, there is a close similarly between the GFA outcomes and the dimensionality claim advocated in Sellbom's et al. chapter "Assessment of Psychopathy: Addressing Myths, Misconceptions, and Fallacies". Essentially, both new accounts join to the natural kinds debate by saying that a certain human individual organism is a psychopath, on these views, amounts to locating that individual at some point within the corresponding dimensional region. However, that conclusion brings us back to the difficulties that we have emphasized before.

References

American Psychiatric Association. (2013). *Diagnostic and statistical manual of mental disorders: DSM-5* (5th ed.). American Psychiatric Association.

Beebee, H., & Sabbarton-Leary, N. (2010). Are psychiatric kinds real? *European Journal of Analytic Philosophy, 6*(1), 11–27.

Blair, R. J. R., Mitchell, D., & Blair, K. (2005). *The psychopath: Emotion and the brain.* Blackwell.

Boyd, R. (1991). Realism, anti-foundationalism and the enthusiasm for natural kinds. *Philosophical Studies, 61*(1–2), 127–148.

Brazil, I. A., van Dongen, J. D. M., Maes, J. H. R., Mars, R. B., & Baskin-Sommers, A. R. (2018). Classification and treatment of antisocial individuals: From behavior to bio-cognition. *Neuroscience & Biobehavioral Reviews, 91,* 259–277. https://doi.org/10.1016/j.neubiorev.2016.10.010

Brzović, Z., Jurjako, M., & Šustar, P. (2017). The kindness of psychopaths. *International Studies in the Philosophy of Science, 31*(2), 189–211. https://doi.org/10.1080/02698595.2018.1424761

Cooke, D. J. (2018). Psychopathic personality disorder: Capturing an elusive concept. *European Journal of Analytic Philosophy, 14*(1), 15–32. https://doi.org/10.31820/ejap.14.1.1

Cooper, R. V. (2005). *Classifying madness: A philosophical examination of the diagnostic and statistical manual of mental disorders.* Springer.

DeMatteo, D., Hart, S. D., Heilbrun, K., Boccaccini, M. T., Cunningham, M. D., Douglas, K. S., Dvoskin, J. A., Edens, J. F., Guy, L. S., Murrie, D. C., Otto, R. K., Packer, I. K., & Reidy, T. J. (2020). Statement of concerned experts on the use of the Hare psychopathy checklist—

Revised in capital sentencing to assess risk for institutional violence. *Psychology, Public Policy, and Law, 26*(2), 133–144. https://doi.org/10.1037/law0000223

Driessen, J. M. A., Fanti, K. A., Glennon, J. C., Neumann, C. S., Baskin-Sommers, A. R., & Brazil, I. A. (2018). A comparison of latent profiles in antisocial male offenders. *Journal of Criminal Justice, 57*, 47–55. https://doi.org/10.1016/j.jcrimjus.2018.04.001

Dupré, J. (1981). Natural kinds and biological taxa. *The Philosophical Review, 90*(1), 66–90. https://doi.org/10.2307/2184373

Dupré, J. (1993). *The disorder of things: Metaphysical foundations of the disunity of science.* Harvard University Press.

Edens, J. F., Marcus, D. K., Lilienfeld, S. O., & Poythress, N. G. (2006). Psychopathic, not psychopath: Taxometric evidence for the dimensional structure of psychopathy. *Journal of Abnormal Psychology, 115*(1), 131–144. https://doi.org/10.1037/0021-843X.115.1.131

Edens, J. F., Marcus, D. K., & Vaughn, M. G. (2011). Exploring the Taxometric status of psychopathy among youthful offenders: Is there a juvenile psychopath taxon? *Law and Human Behavior, 35*(1), 13–24. https://doi.org/10.1007/s10979-010-9230-8

Ereshefsky, M. (2018). Natural kinds, mind Independence, and defeasibility. *Philosophy of Science, 85*(5), 845–856. https://doi.org/10.1086/699676

Ereshefsky, M., & Reydon, T. A. C. (2015). Scientific kinds. *Philosophical Studies, 172*(4), 969–986.

Ereshefsky, M., & Reydon, T. A. C. (2021). The grounded functionality account of natural kinds. In W. Bausman (Ed.), *From biological practice to scientific metaphysics.* University of Minnesota Press.

Fodor, J. A. (1974). Special sciences (or: The disunity of science as a working hypothesis). *Synthese, 28*(2), 97–115. https://doi.org/10.1007/BF00485230

Glimmerveen, J. C., Maes, J. H. R., & Brazil, I. A. (this volume). Psychopathy, maladaptive learning and risk-taking. In L. Malatesti, J. McMillan, & P. Šustar (Eds.), *Psychopathy. Its uses, validity, and status.* Springer.

Groat, L. L., & Shane, M. S. (2020). A motivational framework for psychopathy: Toward a reconceptualization of the disorder. *European Psychologist, 25*(2), 92–103. https://doi.org/10.1027/1016-9040/a000394

Hacking, I. (1996). The looping effects of human kinds. In D. Sperber, D. Premack, & A. J. Premack (Eds.), *Causal cognition* (pp. 351–383). Oxford University Press. https://doi.org/10.1093/acprof:oso/9780198524021.003.0012

Hacking, I. (1999). *The social construction of what?* Harvard University Press.

Hare, R. D. (2003). *The Hare psychopathy checklist revised* (2nd ed.). Multi-Health Systems.

Harris, G. T., Rice, M. E., & Quinsey, V. L. (1994). Psychopathy as a taxon: Evidence that psychopaths are a discrete class. *Journal of Consulting and Clinical Psychology, 62*(2), 387–397.

Harris, G. T., Rice, M. E., Hilton, N. Z., Lalumiére, M. L., & Quinsey, V. L. (2007). Coercive and precocious sexuality as a fundamental aspect of psychopathy. *Journal of Personality Disorders, 21*(1), 1–27. https://doi.org/10.1521/pedi.2007.21.1.1

Held, B. S. (2017). The distinction between psychological kinds and natural kinds revisited: Can updated natural-kind theory help clinical psychological science and beyond meet psychology's philosophical challenges? *Review of General Psychology, 21*(1), 82–94. https://doi.org/10.1037/gpr0000100

Insel, T. R. (2013). *Transforming diagnosis.* https://www.Nimh.Nih.Gov/; https://www.nimh.nih.gov/about/directors/thomas-insel/blog/2013/transforming-diagnosis.shtml

Jurjako, M., Malatesti, L., & Brazil, I. A. (2020). Biocognitive classification of antisocial individuals without explanatory reductionism. *Perspectives on Psychological Science, 15*(4), 957–972. https://doi.org/10.1177/1745691620904160

Kendler, K. S. (2015). Toward a limited realism for psychiatric nosology based on the coherence theory of truth. *Psychological Medicine, 45*(6), 1115–1118. https://doi.org/10.1017/S0033291714002177

Kendler, K. S. (2016). The nature of psychiatric disorders. *World Psychiatry, 15*(1), 5–12. https://doi.org/10.1002/wps.20292

Kendler, K. S., Zachar, P., & Craver, C. (2011). What kinds of things are psychiatric disorders? *Psychological Medicine, 41*(6), 1143–1150. https://doi.org/10.1017/S0033291710001844

Khalidi, M. A. (2013). *Natural categories and human kinds: Classification in the natural and social sciences*. Cambridge University Press.

Khalidi, M. A. (2016). Mind-Dependent Kinds. *Journal of Social Ontology, 2*(2), 223–246. https://doi.org/10.1515/jso-2015-0045

Koenigs, M., & Newman, J. P. (2013). The decision-making impairment in psychopathy. In K. A. Kiehl & W. P. Sinnott-Armstrong (Eds.), *Handbook on Psychopathy and Law* (pp. 93–106). Oxford University Press.

Kohne, A. C. J. (2015). The dubious "reality" of a psychiatric classification. *Tijdschrift voor Psychiatrie, 57*(6), 433–440.

Koslicki, K. (2008). Natural kinds and natural kind terms. *Philosophy Compass, 3*(4), 789–802. https://doi.org/10.1111/j.1747-9991.2008.00157.x

Kuhn, T. S. (1994). *The structure of scientific revolutions* (2nd ed., enlarged). Chicago University Press.

Larsen, R. R., Jalava, J., & Griffiths, S. (2020). Are Psychopathy Checklist (PCL) psychopaths dangerous, untreatable, and without conscience? A systematic review of the empirical evidence. *Psychology, Public Policy, and Law, 26*(3), 297–311. https://doi.org/10.1037/law0000239

Lilienfeld, S. O. (2013). Is psychopathy a syndrome? Commentary on Marcus, Fulton, and Edens. *Personality Disorders: Theory, Research, and Treatment, 4*(1), 85–86.

Magnus, P. D. (2012). *Scientific enquiry and natural kinds from planets to mallards*. Palgrave Macmillan.

Maraun, M. (this volume). Psychopathy and the issue of existence. In L. Malatesti, J. McMillan, & P. Šustar (Eds.), *Psychopathy. Its uses, validity, and status*. Springer.

Mayes, R., & Horwitz, A. V. (2005). DSM-III and the revolution in the classification of mental illness. *Journal of the History of the Behavioral Sciences, 41*(3), 249–267. https://doi.org/10.1002/jhbs.20103

Meehl, P. E. (1992). Factors and taxa, traits and types, differences of degree and differences in kind. *Journal of Personality, 60*(1), 117–174. https://doi.org/10.1111/j.1467-6494.1992.tb00269.x

Murrie, D. C., Marcus, D. K., Douglas, K. S., Lee, Z., Salekin, R. T., & Vincent, G. (2007). Youth with psychopathy features are not a discrete class: A Taxometric analysis. *Journal of Child Psychology and Psychiatry, 48*(7), 714–723.

Olver, M. E., Stockdale, K. C., Neumann, C. S., Hare, R. D., Mokros, A., Baskin-Sommers, A. R., Brand, E., Folino, J., Gacono, C., Gray, N. S., Kiehl, K. A., Knight, R. A., Leon-Mayer, E., Logan, M., Meloy, J. R., Roy, S., Salekin, R. T., Snowden, R., Thomson, N., … Yoon, D. (2020). Reliability and validity of the psychopathy checklist-revised in the assessment of risk for institutional violence: A cautionary note on DeMatteo et al. (2020). *Psychology, Public Policy, and Law, 26*(4), 490–510.

Reiss, J., & Sprenger, J. (2014). *Scientific objectivity (Stanford Encyclopedia of Philosophy)*. https://plato.stanford.edu/entries/scientific-objectivity/

Sellbom, M., Lilienfeld, S. O., Latzman, R. D., & Wygant, D. B. (this volume). Assessment of psychopathy: Addressing myths, misconceptions, and fallacies. In L. Malatesti, J. McMillan, & P. Šustar (Eds.), *Psychopathy. Its uses, validity, and status*. Springer.

Slater, M. H. (2015). Natural kindness. *The British Journal for the Philosophy of Science, 66*(2), 375–411. https://doi.org/10.1093/bjps/axt033

Tsou, J. Y. (2016). Natural kinds, psychiatric classification and the history of the DSM. *History of Psychiatry, 27*(4), 406–424.

Varga, S. (2018). "Relaxed" natural kinds and psychiatric classification. *Studies in History and Philosophy of Science Part C: Studies in History and Philosophy of Biological and Biomedical Sciences, 72*, 49–54. https://doi.org/10.1016/j.shpsc.2018.10.001

Walters, G. D., Marcus, D. K., Edens, J. F., Knight, R. A., & Sanford, G. M. (2011). In search of the psychopathic sexuality taxon: Indicator size does matter. *Behavioral Sciences & the Law, 29*(1), 23–39.

Zachar, P. (2000). Psychiatric disorders are not natural kinds. *Philosophy, Psychiatry, & Psychology, 7*(3), 167–182.

Zachar, P. (2002). The practical kinds model as a pragmatist theory of classification. *Philosophy, Psychiatry, & Psychology, 9*(3), 219–227.

Zachar, P. (2015). Psychiatric disorders: Natural kinds made by the world or practical kinds made by us? *World Psychiatry, 14*(3), 288–290. https://doi.org/10.1002/wps.20240

Chapter 8
Psychopathy and the Issue of Existence

Michael D. Maraun

Abstract It is elementary logic, that a precondition for a sentence to be a scientific proposition is that it have a truth value (the latter, potentially determinable with reference to empirical evidence); and a precondition that it have a truth value, is that it have a *sense*. It is argued, herein, that, in consequence of ambiguity attendant to the grounds of ascription of the focal term, *psychopathy* (and cognates), linguistic expressions relating to the issues of the existence- and causes- of psychopathy, are absent a sense; in consequence, are not adjudicable in terms of empirical evidence. Preliminary conceptual elucidations, preconditional for a dissolution of the ambiguity, and, hence, the entry of the issues into a fruitful, cumulative, line of empirical investigation, are undertaken.

Keywords Psychopathy · Construct validity · Conceptual clarification · Term · Referent

8.1 Psychopathy and the Issue of Existence

Does psychopathy exist? (a question which, to the ears of some, it would seem, is roughly synonymous with, "is psychopathy *real*", and, hereafter, will be designated, Question I). If it exists, what are its causes? (hereafter, Question II). The systematic pursuit of answers to such questions of existence and causality, is, one is told, a hallmark of properly functioning science. Gratifying, it is, then, perhaps, the fact of the literature devoted to the subject of psychopathy, abounding with indications of the psychopathy researcher's preoccupation with Questions I and II. In Skeem et al. (2011), for instance, a motivating impetus is the question of whether psychopathy exists in non-western cultures. Dolan (2004), on the other hand, aims to gain insight into whether it can be found in children. Gowlett (2014) entertains the apparently ticklish question as to whether exists, something called *secondary psychopathy*.

M. D. Maraun (✉)
Simon Fraser University, Burnaby, Canada
e-mail: maraun@sfu.ca

© Springer Nature Switzerland AG 2022   121
L. Malatesti et al. (eds.), *Psychopathy*, History, Philosophy and Theory of the
Life Sciences 27, https://doi.org/10.1007/978-3-030-82454-9_8

And there have, of course, been tendered no small number of theories bearing on the causal basis of psychopathy.[1] The nascent and somewhat rabid enthusiasm for enlisting neuroimaging technology in the service of investigating psychopathy surely derives, in large part, from a strongly held belief that these novel investigative tools are destined to make definitive contribution to the resolution of both Questions I and II.

Whatever might be their views respecting the investigative approach which will, in the end, yield resolutions of Questions I and II, one would wager that few, if any, psychopathy researchers, doubt the essential resolvability of these questions; moreover, that answers will be delivered in the form of empirical evidence, carefully culled in accordance with scientific precepts, and processed with the aid of scientific methodology. At peril of being dismissed, out of hand, as merely fatuously iconoclastic, I will suggest, herein, that all of the talk and labour that has, as its focus, Questions I and II, must come to nought, for the simple reason that, as matters stand, Questions I and II are unresolvable. They are unresolvable because the meaning of each has been allowed to remain indeterminate.

The blame for this state of affairs will be laid squarely at the feet of the commitment of the behavioral scientist, to a badly distorted conception of science- herein, called Generalized Construct Validity (GCV, hereafter)-, the propagation of which can be traced back to Cronbach and Meehl (1955). Under GCV, the conceptual facets of science, preconditional to a fruitful, intelligible, line of empirical investigation- and, notably, wherein is undertaken the *specification* of that which is to stand as the target of investigation-, are bypassed in favour of the making of vacuous references to constructs. To cut to the chase, under commitment to GCV, the identity of that to which Questions I and II make reference- that thing that goes by the appellation, psychopathy- has been allowed to remain unspecified; and this failure to specify, vitiates both the questions, themselves, and the empirical investigative efforts undertaken in their service. Evidence cannot bear on either of the issues of the existence, or causes, of some constituent of natural reality, the identity of which has been left unspecified.

In the service of sustaining these verdicts, the organization of the paper is as follows: in the first section, I review the logic relating to the conceptual and empirical facets of science, with special attention directed at the foundational relation between *term* and its *class of referents*; in the second, characterize the manner in which commitment to GCV precludes the scientific treatment of Questions I and II; finally, in the third section, take preliminary steps towards a reconciliation of Questions I and II with coherent science, by way of a presentation of some of the missing conceptual clarification, directly relevant to the antecedent task of specifying the phenomena to which, in these questions, reference is ostensibly made.

---

[1] For a critical review, see Jalava et al. (2015).

8.2 The Conceptual and Empirical Facets of Science

Let there be a term "φ" introduced within the frame of an empirical (scientific) investigation. Then: (a) being as it is a term (a concept), "φ" is a constituent of language, and because language is a human creation, so, too, is "φ"; (b) "φ" is employed by humans as a constituent part of linguistic expressions, and, more particularly, as a component of scientific expressions- those asserting of hypotheses, expressing of facts and results, making contribution to theories, etc.-, and any *particular* employment is either linguistically correct (i.e., *coherent*; the average height of Canadian bachelors is 174 cm, for example) or incorrect (*incoherent*; the shortest Canadian bachelors tend to be female)[2]; (c) any behavioural practice- linguistic or otherwise- wherein there is a normative basis for adjudicating behaviour as either correct or incorrect (e.g., the driving of motor vehicles on public roads, the playing of games such as chess and bridge and poker, the making of mathematical calculations), is a rule governed practice; (d) a rule is not an empirical assertion of any sort (for it expresses no claim as.to what, within some particular behavioural context, an agent *will*, in actuality, do). It is simply a standard of correctness, specifying of what an agent *must* do in order that his behaviour be correct (in respect whatever behavioural particular it is, to which the rule makes reference). A stop sign, for example, expresses no claim relating to the actual driving behaviour of extant vehicle drivers, but only that a driver must bring his vehicle, before it, to a complete stop (else, be guilty of having violated a rule of the road); (e) within certain rule governed behavioural practices, the rules definitive of correct (incorrect) behaviour are codified- those, e.g., that comprise motor vehicle law-, and, within others, not so (see Hacker, 1988); within certain practices, are of a *constitutive* nature, and, within others, a *modifying* nature[3] (see Duintjer, 1977; Ter Hark, 1990); (f) being as they are *linguistic* rules, the rules which fix what constitutes correct (incorrect) employment of term "φ" are of the constitutive sort, and are referred to, collectively, as the grammar of "φ"; (g) by the *meaning* of term "φ", one means, simply, "φ"'s range of correct employments in linguistic expressions. Because it is the grammar of "φ" which determines what constitutes correct (incorrect) employment of "φ", any meaningful term has a (extant, public) grammar (to wit, the web of rules which fix its range of correct employments), and to grasp some segment of "φ"'s grammar is to grasp part

---

[2] In the event that the expression is of propositional form, not to be confused with the correctness (truth value) of that which is asserted. A *precondition* for a propositional form to *have* a truth value is precisely that it be coherent (i.e., that the combination of words of which it is formed is legitimate, and, in so being, does, in fact, describe a possible state of affairs).

[3] In the case of a modifying rule- e.g., a culinary rule- there exists an independent criterion of identity for the goal on which it bears. Thus, one's having followed a recipe for cherry pie is not identical to, and does not guarantee, his having produced a cherry pie (Ter Hark, 1990). Whether one has succeeded in doing the latter is adjudicated with reference to an independent criterion (the look and taste of the pie, its chemical profile, etc.). A constitutive rule, in contrast, is *internally related* to that which satisfies it (in that it is impossible to both flout the rule, and achieve the aim on which it bears).

of its meaning (as when a student of physics is given credit for comprehending the meaning of the term *alpha particle*, in consequence of their being able to recite the constitutive rule, "an alpha particle is a positively charged nuclear particle, consisting of two protons bound to two neutrons"[4]); (h) to inquire as to "φ"'s meaning is, accordingly, to seek elucidation of a relevant segment of an extant, public, web of rules, and, more specifically, a segment of "φ"'s grammar; (i) a term can be usefully classified as either *signifying* or *non-signifying*, in accordance with whether or not one of its correct employments involves its ascription to constituents of natural reality (hereafter, CNR; material entities, substances, conditions, processes, phenomena, forces, etc.); (j) because a particular employment of term "φ" is correct just when it accords with "φ"'s grammar, "φ"'s grammar determines whether or not it is a signifying term; (k) if- by virtue of its grammar containing constitutive rules of ascription- "φ" happens to be a signifying term, then: ki) the particulars of its correct ascription to CNR (*inter alia*, to *which* CNR it is legitimately ascribed) is determined by its grammar; kii) the CNR to which it is correctly ascribed are called its *referents* (generically, say, φ-things, the full set of extant φ-things forming of the class C_φ of referents of "φ").[5]

Within the context of scientific inquiry, both conceptual and empirical issues feature prominently, and these classes of issue are, in fundamental ways, both logically related and disjoint. Contrary, it would seem, to the understanding of the behavioural scientist, conceptual issues are neither theoretical issues, nor hunches, nor speculations (lay or otherwise). They are, rather, issues pertaining to the web of constitutive, linguistic, rules, extant and public, which fixes the meanings of terms; and, so, depend, for their resolution, upon an elucidation of segments of this very web of rules. Thus, it is a conceptual issue, to inquire as to which subset of humans will, in an empirical inquiry into the tastes and behavioural tendencies of bachelors, stand as the targets of investigation. For to inquire as such, is to inquire as to which humans the term *bachelor* is correctly ascribed; the answer given in the form of an articulation of the term's rule of ascription; to wit, the term *bachelor* is correctly ascribed to just those humans who are unmarried, adult, males.[6]

So, too, is it a conceptual matter, the dismissal of a body of empirical results billed as pertaining to bachelors, on grounds that they were, in actuality, recorded

---

[4] This is *not* an empirical assertion; for an empirical assertion *about* alpha particles, is about those things, independently identifiable *as* alpha particles. And in the case of alpha particles, the normative criterion of identity, commonly presented in text books under the heading of *definition* of the term *alpha particle*, is precisely the constitutive rule, "an alpha particle is a positively charged nuclear particle, consisting of two protons bound to two neutrons."

[5] Of course, the correct manner of speaking of the elements of C_φ will depend upon what kinds of things they are. E.g., in the case of a liquid substance such as oil, it is conventional to speak of the *amount* of the substance; the elements of C_φ, volumetric partitionings expressed in some particular units.

[6] Though the rules of ascription of many signifying terms- notably those of the natural and technical sciences- refer to necessary and sufficient conditions, by no means do all. Grammar is term-specific, and rules of ascription are manifold, varying widely in form over the class of signifying terms.

on unmarried, adult, *females* (for the case rests on the grammatical point that the term *bachelor* is not correctly ascribed to such individuals). More generally, the adjudication of the veracity of a claim to the effect that *these* results pertain to some particular class C_φ of CNR, is a conceptual matter; as is to wrestle with the issue of whether there is (or should be) "overlap in the study of psychopathy and criminality." Empirical issues, on the other hand, are those that have to do with the properties and behaviours, broadly conceived, of CNR. It is to the resolution of empirical issues, that the manifold tools and practices most saliently identifiable with the scientific method, have proven to be so formidably well-suited.

The conceptual and empirical spheres of concern have, within scientific praxis, many logical points of contact. It is elementary logic, for example, that a precondition for a sentence to be a scientific proposition is that it have a truth value (the latter, potentially determinable with reference to empirical evidence)[7]; and a precondition that it have a truth value, is that it have a *sense*.[8,9] Even more directly relevant to the concerns of the current work, is the fact of the conceptual and empirical meeting in the requirement that, antecedently to the commencement of fruitful empirical investigation, the CNR which are to stand as the targets of investigation must be *specified*. To specify that the targets of investigation are some particular type of CNR, say, φ-things, is to specify that they are the elements of a particular class C_φ of CNR; and the composition of C_φ- which CNR qualify as its elements- is unambiguously settled through the articulation of a rule of inclusion into C_φ. In many instances, however, the elements of C_φ will be the referents of some particular signifying term "φ"; in which case, the rule of inclusion into C_φ is simply the rule of ascription of signifying term "φ." Because its rule of ascription is part of its grammar, the elucidation of a rule of ascription of a particular signifying term (equivalently, the corresponding rule of inclusion) is an antecedent conceptual undertaking.

When one specifies the targets of inquiry to be, for example, *planets*, one does so by citing the rule of ascription, to CNR, of the term *planet*; to wit, the term *planet* is ascribed to just such a CNR as is in orbit about the sun, has sufficient mass to assume hydrostatic equilibrium, and has cleared the neighbourhood around its orbit. Of all the CNR, planets are, uniquely, those to which the term *planet* is correctly ascribed; and the term *planet* is ascribed in accordance with its grammar, its rule of ascription, recently- in 2006- and after intense disputation, having been altered, and codified anew, by the International Astronomical Union. When the target of investigation is, on the other hand, cancer, then one is setting out to investigate those entities to which the term *cancer* is correctly ascribed; to wit, the elements forming of the class of diseases, each of which is characterized by abnormal cell growth in a

---

[7] Called the *principle of bivalence* (see, e.g., Audi, 1999).

[8] I.e., that the combination of words of which it is formed, accords with the rules of language; in so doing, is meaningful, and, in consequence, unambiguously descriptive of a state of affairs (which either does or does not hold).

[9] Though the term *sense*, in its philosophical employment, derives from the sense/reference distinction of Frege (1892), it is, herein, employed in a manner consonant with that of Wittgenstein; i.e., as a synonym for meaning.

bodily region *A*, along with the potential for the abnormal cell growth to spread within the body, beyond *A*.

As when I claim to be studying leopard sharks, when the ocean going CNR on which I am, in actuality, collecting data, do not have gills, a study billed as an investigation into φ, is, in actuality, *not so*, if it happens that the CNR which, in reality, are under investigation, are not those to which- by virtue of its grammar- term "φ" is correctly ascribed. The impetus for scientific investigation is, indeed, a state of uncertainty over- questions posed about- the characteristics of some particular aspect of reality; and it is nature (broadly conceived), and not language, that engenders the manifold characteristics of aspects of reality. The capacity to *identify* the particular aspect about which empirical questions have arisen, is, however, an *a priori* conceptual matter.

The two-tiered structure of inquiry, wherein linguistic, constitutive, rules- normative standards determining of the correct employments of the terms which organize and denote- settle the issue of *which* CNR will be studied in an investigation of φ-things (precisely those CNR to which the rules grant ascription of term "φ"), and contingent scientific investigation targeted at φ-things (those particular CNR antecedently singled out *as* the targets of investigation by virtue of their being referents of term "φ") yields knowledge of their characteristics, is, within scientific praxis, everywhere in evidence. Textbooks devoted to all manner of empirical domains, manifest this structure, for example, in so much as they present definitions of terms- these, specifying of classes of CNR-, followed by listings of facts relating to, and theory bearing on, the elements of the classes of CNR, thereby specified. Prior to turning the focus of his treatment on empirical questions such as the amount of work a car does against a frictional force, Bueche, in his introductory remarks to his chapter on work and energy, both acknowledges the logical dependency of fruitful science on conceptual clarity- "Does a baseball player work when he is playing baseball? Many people would say that since he is playing a game he is not working. But what if he were being paid to play baseball? Is the ground underneath a house doing work? It is holding the house. Is it, therefore, basically different in function from a pillar holding the roof over the porch of the house? Yet some would insist that the pillar was doing work. Clearly, if we are to use the term *work* in physics, we need to define it in a precise way" (1972, p.83)-, and provides the necessary clarity in the form of a rule of application of the technical term, *work*.

Imagine organizing a team of research assistants to aid in a new investigation into *greks*. You inform these assistants that, "theory suggests that, if they exist, greks should be found in the woods out back of this office." Perhaps, gesturing towards an array of technical recording and sensing equipment, you indicate that each assistant should take what they need, and get down to the business of investigating the scientific question of the existence of greks. Unless they had been brought to a particularly well entrenched state of deferential obsequiousness, one or more of these hypothetical assistants would stare quizzically at you, and exclaim, "but what is a grek?!" And in so doing, they would have inadvertently articulated the logical precept that, in absence of a rule which settles the basis of application of the term *grek* to CNR, putative empirical propositions bearing on *greks* (the putative referents of the term) will be absent a sense; so, too, then, the logical hallmark of the propositional form, *to wit*, the property of bivalence. Unless there can be cited, antecedently, a rule which settles admission of CNR into the category of *greks*, sentences such as "*greks* exist" and "*greks* weigh, on average, 68 grams", are not propositional forms; accordingly, cannot be adjudicated with respect to empirical evidence.

It is notable that, unless one is introducing into discourse, a novel term, the constitutive linguistic rules which settle the basis of ascription, to CNR, of signifying terms, are *extant, public, manifold*, and *term-specific. Elucidation* of the rules of terms focal to a line of scientific investigation is the antecedent, conceptual, part of the scientific enterprise. Terms deriving from the natural and technical sciences are, standardly, technical terms of trade, the grounds of ascription of which are given in terms of necessary and sufficient conditions. Elucidation, here, is often straightforward, for it is commonly effected through the citing of a definition. The terms appearing in the discourses of the psychologist are, by contrast, in the main, the non-technical terms of ordinary language.[10] The correct employments of *these* terms are standardly fixed, not by necessary and sufficient conditions, but by ramifying rule structures of astonishing diversity and complexity (for examples, see, e.g., Baker and Hacker (1982); Bennett and Hacker (2003)). This makes the task of elucidation an exceedingly challenging one.

Related in scientific praxis, though they are, the conceptual and empirical facets of investigation are, in fundamental ways, logically disjunct. In the first place, the linguistic rules which fix the grounds of ascription of a signifying term to its referents, are mute with respect to the characteristics of these referents. That the grammar of the term *bachelor* grants ascription of the term to some particular individual, Jim, in virtue of Jim's happening to be an adult, unmarried, male, implies nothing about- among countless other things- Jim's personality, employment history, or financial profile. It is precisely for this reason- that the *specification* of a class of CNR of scientific interest through the antecedent citing of the rule of ascription of a signifying term, says nothing about what these CNR are like-, that, if such issues are to be addressed, contingent empirical (scientific) investigation will have to be undertaken. The rule for the ascription of the term *bachelor* to humans, settles the matter as to *which* humans it is, about which one is expressing scientific interest, when one poses questions of the sort, "what are bachelors like", but is silent with respect the answers to such questions; settles the matter as to *which* empirical results pertain to bachelors- precisely those deriving from the investigation of elements belonging to the class of bachelors, *to wit*, those CNR to which is correctly ascribed, the term *bachelor*-, but is wholly uninformative as to the contents of these results.[11]

---

[10] The discipline of psychology did not arise as a technical science with its own novel vocabulary, but in response to interest in the phenomena denoted by ordinary language concepts.

[11] There are those (see, e.g., Jost & Gustafson, 1998) who believe that, if it were the case that, as argued, herein, the meaning of a term is fixed by constitutive linguistic rules, this would imply a "closing up of the epistemic space," under which nothing would be left for science to discover. But this is a misunderstanding, akin to believing that the existence of rules of chess closes up the epistemic space relating to chess playing: e.g., the chess moves that an individual player might make;

In the second place- and, as we will see, a point of logic of fundamental relevance to any consideration of the impact of GCV on social and behavioural research-, the characteristics of the CNR denoted by a particular signifying term, have no direct role in the determination of the linguistic rules that fix the term's correct employments (i.e., have no direct bearing on the meaning of the term). The fact that, say, the personality characteristics of bachelors alter in some respect between the years 2020 and 2021, would effect no alteration to the meaning of the term *bachelor*. In particular- and exemplifying of a logical precondition for cumulative empirical science-, a discovered empirical characteristic of bachelors has no direct bearing on- no power to alter- the rule in accordance to which the term *bachelor* is correctly ascribed to humans. It is essential to appreciate why this is so. Consider, then, a scenario in which a term "φ" is correctly ascribed to certain extant humans in accordance with a rule r_φ, these humans- the referents of term "φ"- forming of a class, C_φ, of φ-humans. Research undertaken on the elements of C_φ leads to the discovery that φ-humans have a characteristic f_φ. Now, consider what it would mean for f_φ to have the power to alter the meaning of term "φ" and, in particular, "φ"'s rule of ascription. Suppose, then, that f_φ *does* effect alteration of "φ"'s rule of ascription, so that "φ" is no longer correctly ascribed in accordance with r_φ but, instead, with new rule, r'_φ. Because "φ" is ascribed, now, to those humans who satisfy r'_φ, the class of φ-humans would undergo corresponding alteration to $C'_\varphi \neq C_\varphi$. Because characteristic f_φ is a characteristic of the elements of class C_φ, and *not* of the elements of class C'_φ- the *latter*, now, comprised of φ-humans-, instantly upon the discovery of such a rule-of-ascription-altering characteristic f_φ, f_φ would no longer be *about* φ-humans.

Ergo, the annihilation of the possibility of cumulative science. It is precisely because reference of term can be maintained through constancy of linguistic rules, that the network of facts pertaining to some particular class of CNR can be built up. For when a rule r_φ remains in force as the grounds for ascribing a term "φ" to CNR, then it remains, that an empirical result e, novel or otherwise, is *about* φ-things, just so long as it pertains to those CNR which, by virtue of their satisfying r_φ, are, in fact, elements of C_φ. It is *humans*, not facts, who have the power to alter the meaning of a term. When they do so, they do so- as per the case of the re-definition of the term *planet*- by laying down new rules. And when they do so within a scientific context, the status of empirical results signified by the term, prior to the re-engineering of its meaning, must be carefully reappraised in light of the term's re-engineered signification.[12]

---

to what *non*-chess uses the king playing piece can be put (e.g., as a door-stop); the personality characteristics of those who happen to be chess players; the neural correlates of legitimate movements of pawn playing pieces. To fix rules merely defines correct and incorrect, and, so, establishes the logical space within which behaviour occurs (Hacker, 1986). What is settled by the rules of chess are such antecedent conceptual issues, as the *grounds* for legitimately asserting that A and B are playing chess, and that one or the other has won (this latter, prerequisite to contingent empirical issues, such as the won/lost record of each player, his world ranking, etc.); what *constitutes* correct (and incorrect) movement of the king playing piece; etc.

[12] On the issue of conceptual engineering, see, e.g., Carnap (1950), Creath (1990), and Brun (2016).

Frequently, it seems, people take the (correct) claim that the range of correct employments (the meaning) of a term is fixed by linguistic, constitutive, rules, as implying that concept meaning is immutable and, accordingly, temporally invariant. This, however, is a misunderstanding. For the potential to revise concept meaning *follows from* the fact of the determination of concept meaning by linguistic, constitutive, rules. For rules are not immutable abstractions, but, rather, standards of correctness. They are created- in manifold ways- by humans, and, so too- for manifold reasons-, may be altered by humans. Though it was an empirical discovery that acids are proton donors, "…this proposition was transformed into a rule: a scientist no longer calls something "an acid" unless it is a proton donor, and if it is a proton donor, then it is to called "an acid", even if it has no effect on litmus paper". New rules were laid down when the standard metre in Sevres was replaced "…by a certain number of wavelengths of the light in a certain line in the spectrum of cadmium" (Baker & Hacker, 1982). Mach devoted forty pages to the analysis of *why* Newton's definition of *mass* was circular, explained why such a poorly created concept could not provide a reasonable conceptual basis for scientific investigation within physics, and offered an alternative definition- *new rules* and, hence, a new concept- unafflicted by circularity. Physics post the theory of relativity, "employs the expressions 'space' and 'time' differently from the ways in which ordinary speakers do and also from the ways Newtonian physics does" (Hacker, 1986, p.193). On the other hand, the fact that concept meaning is revisable in principle, does not imply that it can be altered willy-nilly, by individual language users. A term's meaning has altered when an old standard of correctness has been supplanted by a new one, and so has come to have- however it came about- a normative role.

Let us turn, now, to some commonly arising questions of existence. The most basic would seem to be something along the lines of, "does (do) there exist φ(s)" (neutrinos, aliens, humans exceeding nine feet in height), the particulars of the grammar, here, depending, once again, on what kind of thing(s) φ(s) happens to be.[13] Such a question of existence is empirical and *contingent*, in as much as that a precondition for it's being addressable in science is the capacity to *identify*, in nature, what is (and is not) (a) φ. And to be able to identify what is (and is not) (a) φ is nought but to grasp the rule by which signifying term "φ" is correctly ascribed to CNR. A question of existence is coherent, and, hence, addressable in science, just when the conceptual issue of specification has been unambiguously resolved. One must know, unambiguously, which sort of thing it is, the existence of which is in question. Questions of existence posed about the referents of a term, the grammar of which has *not* been adequately clarified- in particular, for which the rule of ascription has not been elucidated-, are not coherent; hence, not answerable.

Though scientists established, eventually, that the substance phlogiston does not exist, they could not have done so if the meaning of the term *phlogiston* had been in doubt. For if, surrounding the term's meaning, there had been unclarity, there would

---

[13] If, e.g., φ is a substance (phlogiston, say), one says, "does there exist", rather than do there exist.

have existed a consequent uncertainty over the grounds of the term's correct ascription. Such uncertainty, however, would have been equivalent to uncertainty over what the term *phlogiston* signifies; hence, to an inability to specify- *identify*- the CNR, the existence of which was in question. Clearly, one can neither verify, nor disprove, the existence of that which one cannot identify. The question, "do there exist any greks?", is not open to being resolved on the basis of empirical evidence, in consequence of there never having been laid down a rule of inclusion into the class C_{greks} of greks. Though its surface grammar confers upon it the ring of such, it is not, in actuality, a question of existence; for it has no meaning. In light of the power inherent to linguistic formulations relating to issues of existence, to engender science precluding metaphysics, it is useful to note that an equivalent re-phrasing of the question, "do there exist φs", is the question, "does there exist at least one CNR to which term "φ" can be correctly ascribed" (equivalently, is the class C_φ, containing of the referents of term "φ", empty or not). Finally, and with, in mind, the GCV perspective to which, next, we will turn, the reader is reminded that an issue of existence does not pertain to a signifying term "φ" (it not being a CNR), but, rather, to its referents (which *are* CNR).[14]

Let there be a class C_φ of CNR, the elements of which- say, φ-things (the specified targets of investigation)- are the referents of a signifying term "φ", and let it be that at least one φ-thing has been established as existing (i.e., that C_φ is non-empty). Then, many other empirical questions, contingent upon the existence of φ-things, can be posed, among these: a) what is, in actuality, the number of extant φ-things (i.e., what is the cardinality of C_φ); b) what are the characteristics of φ-things (what are they *like*, those elements of C_φ[15]). Depending upon the kinds of things φ-things happen to be, it may make sense to inquire as to their causes (factors that bring them into being), and/or possible material or physical preconditions (conditions necessary, but not sufficient) for their existence.[16] Note that: (a) only things that exist can have causes and material preconditions. Accordingly, the search for causes or material preconditions is, generally speaking, contingent upon the establishment of existence (which, the latter, has, as a conceptual precondition, the capacity to identify whatever it is, the existence of which is in question); (b) an extant CNR and its causes- or material preconditions of existence- are logically distinct. Causes- material preconditions- are always *of* something (some CNR *independently specifiable* through citation of the rule of ascription of a signifying term).

---

[14] E.g., it is the substance phlogiston, not the term *phlogiston*, the existence of which was in question, and eventually disproved.

[15] It should be noted that it would be viewed as tautological, and illegitimate, to cite, in answer to this question, properties mentioned in rule of inclusion r_φ into C_φ. For a characteristic of φ-things, is simply a characteristic of those CNR which, by virtue of their having satisfied r_φ, belong to the category C_φ. Thus, to reply, "they are unmarried", in answer to the question, "what are bachelors like", would be illegitimate. Unmarriedness is not a characteristic *of* bachelors.

[16] E.g., scientific investigation can legitimately be targeted at the causes of cancer or the physical preconditions under which neutrinos came into being.

8.3   The Distorting Influence of GCV

Because, to inquire as to whether "φ exists" (φ-things exist) is to inquire as to whether or not is non-empty, the class C_φ of referents of a signifying term, "φ", a precondition for the fruitful application of science to the resolution of such an issue, is the possession of a rule of inclusion for CNR into C_φ. For if one cannot say what is and is not legitimately assignable to C_φ- equivalently, cannot single out, or identify, those particular CNR, the existence of which is in question-, then one is absent the logical capacity to determine the cardinality of C_φ. Because the causes of- material preconditions for- φ-things, are the causes of the elements of C_φ, to enlist science in the service of investigating such, requires all of a: a) rule of inclusion for CNR into C_φ; b) knowledge that C_φ is, in fact, non-empty; and c) the capacity to draw a sample of elements from C_φ. To possess a rule of inclusion into C_φ, is, of course, nought but to grasp the linguistic rule of ascription of signifying term "φ".

Accordingly, a logical precondition for Questions I and II to be answerable by means of scientific investigation- to be intelligible-, is that there be an unambiguously articulable rule of ascription of the term *psychopathy* (which, ostensibly, is to be taken as signifying the CNR to which, in these questions, reference is made). We are, then, confronted immediately by a very grave difficulty. For what is absent from the voluminous literature devoted to the subject of psychopathy is just such a transparently articulated rule. In place of the preconditional conceptual work, bearing on the explication of the term *psychopathy*- its cognates, and terms, to it, related-, the end product of which would be clarity in respect the CNR that, in an empirical investigation of psychopathy, are to stand as the targets of investigation, one encounters interminable, and vacuous, references to "the construct of psychopathy" and intensive discussion and disputation over the psychometric properties of instruments purported to scale individuals with respect it.[17]

The blame for this state of affairs, can be laid squarely at the feet of the discipline's commitment to GCV. The manner in which the GCV distorts and undermines the scientific enterprise, can be described as follows. In place of *term* and *class of referents*, which marks the point of contact between the conceptual and empirical facets of science, the GCV installs the single notion of *construct*, which is portrayed as having both a meaning and empirical characteristics. Thus, for example, in an iconic quote from their 1955 exposition, Cronbach and Meehl, explain that "… the meaning of theoretical constructs is set forth by stating the laws in which they occur, our incomplete knowledge of the laws of nature produces a vagueness in our constructs…" (Cronbach & Meehl, 1955, p.69). The replacement

---

[17]A recent exception is Cooke (2018), wherein, in very much the same spirit as the current paper, it is urged that "It is only when conceptual clarity is achieved that valid operations and measures can be created" (p.15). The approach Cooke adopts with the aim of achieving conceptual clarity, i.e., the creation of concept maps, is somewhat different than that which, in the current work, is deemed appropriate.

of the *term/referent* pairing, with the single notion of *construct*, effectively eradicates the distinction between the conceptual and empirical facets of science.

Whereas in science, there is the conceptual task of specifying the class of CNR, the elements of which are the targets of investigation (effected by articulating the rule of ascription to CNR, of a focal term), and, *contingently*, the empirical task of studying those targets of investigation, under GCV, everything is re-cast as an empirical matter.[18] The consequence is endemic conflating of the conceptual and empirical, as when, e.g., Cronbach and Meehl (1955, p.69) state that, "We will be able to say "what anxiety is" when we know all of the laws involving it; meanwhile, since we are in the process of discovering these laws, we do not yet know precisely what anxiety is." The corruption of science thereby enacted is explained, in part, through the propagation of an assortment of mythologies, chief among these: (i) that through the doing of empirical research, little by little, clarity regarding the meaning of a focal term can be brought into focus; (ii) that the conceptual confusion which surrounds terms- the grammars of which have been left unelucidated, as they are under GCV-, is a commonplace of science; and (iii) that psychological constructs are unobservable, and conceptually problematic for this reason.

Respecting the first, they being CNR, it is the characteristics of the *referents* of a signifying term, which empirical investigation is capable of bringing into focus. It has no power to reveal the meaning of a term, because a term's meaning is set by linguistic rules, and rules are not CNR, at all, but, rather, standards of correctness. They are, accordingly, not *discoverable*, but, instead, are taught and learned and clarified. The second is, merely, a blatant misportrayal. Conceptual *clarity* is the hallmark of fruitful scientific enterprise, and in the natural sciences, paradigmatically, this clarity and transparency is achieved and assured for the technical terms of trade, through the laying down of definitions (chiefly of the necessary and sufficient sort), which experts and students, alike, are expected to master.[19] Finally, as to the third mythology: (i) not everything can be sensically classified as either observable

---

[18] There appears to exist a widely held misconception that, in his paper, *Two Dogmas of Empiricism*, Quine (1951) established the nonexistence of grounds for distinguishing between a term's *meaning* and *empirical properties*- facts and theories about, laws pertaining to, etc.- of its referents (so, too, then, the conceptual and empirical facets of science). In fact, the target of Quine's attack was Carnap's- entirely different- distinction between *analytic* and *synthetic* statements (sentences of propositional form). In light of Carnap's criterion of analyticity- i.e., a statement is analytic, if it is necessarily true by virtue of the meanings of its constituent terms and logical operators-, it is evident that both the distinction he set forth, and Quine's attack, *presuppose*, in fact, the existence of the very distinction- that between *meaning* of a term and contingent empirical facts relating to a term's referents- that folks, nowadays, misconstrue the latter as assailing against; for what Quine attacks is the assertion that there exists a special class of propositions which, in virtue of nought but the *meanings of their constituent terms* [italics added] and logical operators, are necessarily true.

[19] Once again, I am neither asserting that the concepts of psychology are technical terms of trade (for they are, in fact, in the main, the very psychological concepts of ordinary language), nor that their meanings are fixed in accordance with necessary and sufficient conditions (for, in fact, the meanings- range of correct employments- of the psychological concepts of ordinary language are fixed, not by necessary and sufficient conditions, but other rule structures, manifold and term-

or unobservable; (ii) in the paradigmatic cases which arise in the natural sciences, unobservability is a characteristic- relative to human perceptual capacities- of the *referents* of a term; (iii) the fact of the elements of a class of referents happening to be unobservable, engenders no *conceptual difficulties* respecting the signifying term in question, but, rather, difficulties in detecting the presence of the referents (the consequent need to engineer instruments of detection); (iv) the unobservability alluded to in GCV is not a true (perceptual) unobservability, at all, for it does not pertain to CNR. It is not potentially removable through the employment of tools of detection (and no attempts are made to develop any such tools). It is, rather, a platonic unobservability- an unobservability in principle-, conjured as part of a misdiagnosis of the source of the endemic conceptual confusion, engendered, in actuality, by commitment to GCV.

The literature on psychopathy is verily steeped in GCV. Hare and Neumann (2008), for example, offer a letter perfect accounting of psychopathy research in the terms of GCV.[20] The predictable upshot is that psychopathy research is founded on a focal term, attaching to which is an abject unclarity of meaning; consequently, that there is an indeterminacy relating to the identity of the targets of investigation; and, accordingly, an intractable indeterminacy of meaning attaching to empirical results generated by psychopathy researchers. Psychopathy researchers, themselves, though prone to explain them as empirical in nature, inadvertently attest to the presence of the conceptual confusions which vitiate their investigations, as when Skeem et al. (2011, p. 103) observe that, "There is a striking degree of continuing debate among contemporary scholars about the nature and scope of the psychopathy construct. Unresolved controversies include (a) what clinical features are, and are not, intrinsic to psychopathy (i.e., essential or core elements of psychopathy as opposed to concomitants or sequelae); and (b) whether psychopathy should be viewed as a single homogeneous entity or as encompassing subgroups of individuals with distinguishable characteristics."

Embedded in the empirical discourse bearing on the issue of psychopathy, are, most assuredly, conceptual fragments, some of which do bear on the issue of what, precisely, the term *psychopathy* denotes. The book by Jalava et al. (2015) offers an analytical consideration of the historical flux in the grounds of ascription of the term. As matters stand, however, it is a fact that this considerable history has not eventuated in the establishment of a *normative* basis of ascription; a state of affairs which, under influence of the GCV, researchers do not see as problematic. Of course, the fact that there is endemic conceptual unclarity within a domain of empirical inquiry, does not bring activity to a halt (any more than to occlude the lettering of a traffic sign would bring to a halt, driving behaviour). The historical record shows, in fact, that empirical investigation marches onwards, irrespective of the presence of conceptual confusion. In manifold ways, however, conceptual

---

specific). I am asserting, merely, that, in all but the social and behavioural sciences, antecedent conceptual clarity is a hallmark of scientific praxis.

[20] For a more general consideration of the GCV, and the ways in which it undermines fruitful science, see Maraun et al. (2008) and Maraun and Gabriel (2013).

confusion assails against the doing of fruitful science; e.g., by visiting uncertainty upon the *meaning* of empirical claims and questions, making it impossible to determine what constitutes evidence in support of scientific propositions, etc. This is why- with the aim of extirpating science undermining conceptual confusion-, in the natural and technical sciences, conceptual clarification and elucidation are undertaken as fundamental components of the scientific process. One of Einstein's many contributions to physics was to point out that, though the standard employment of the concept *simultaneity* was unproblematic for events occurring at close proximity, it broke down in the case of events occurring at great distance from each other (Waismann, 1965)... and to offer a novel concept applicable in these latter, empirically novel, settings.

8.4   Questions I and II Brought into Alignment with Science

Questions I and II will not be addressable in science until the science precluding incoherence of the GCV is discarded, and the requisite conceptual work, undertaken. Though little hope do we feel that the former will come to pass any time soon, we, nevertheless, conclude the current work by taking some preliminary steps in the direction of the latter.[21] We begin with Question I, which will not be addressable in science until psychopathy- the thing, the existence of which is in question- can be *identified*; equivalently, until has been elucidated, the rule of ascription, to CNR, of the term *psychopathy*. To begin, the expression "does psychopathy exist", has a surface structure, erected, deceptively, on the model of substance terms (e.g., does phlogiston exist, or does gold exist). As when one asks the analogous, "does haemophilia exist", it is evident that something is awry. As with haemophilia, psychopathy is a condition. And, just as haemophilia is the *condition* which a haemophiliac has, psychopathy is the condition which a psychopath has. Question I, therefore, is more transparently- less misleadingly- expressed as, "do there exist any psychopaths"; equivalently, is the cardinality of the class C_P, the elements of which are psychopaths, nonzero. As already noted, the problem is that, in absence of a rule of ascription, to humans, of the condition-term *psychopathy*, we are absent a rule of inclusion into C_P, and, so, are in no position to determine the cardinality of the latter.

Of those condition-terms which can be legitimately ascribed to humans, the grounds of ascription are, of course, manifold and term specific; for the grounds of ascription of a particular term are part of its grammar. There do, however, exist broad categories, the distinctions between which are both important, and useful, to draw. The class of condition-terms can be partitioned into subclasses of *nonpsychological* (of which, e.g., *haemophilia* is an element) and *psychological* (to which

---

[21] The author anticipates that, to many, what is, herein, offered, will strike the ear as theoretical speculations. His belief, however, is that, to the contrary, they are conceptual clarifications; i.e., elucidations of relevant segments of the web of extant, public, linguistic, constitutive, rules which fix the range of correct employments of the term *psychopathy* (and related terms).

psychopathy belongs) terms; each of these two, into *state* (which signify phenomena with a genuine duration, and an identifiable beginning and ending) and *disposition* (which signify enduring characteristics possessing of no true beginning or ending, and for which it makes no sense to check on with the aim of ascertaining whether or not they are ongoing) terms.[22] Both of the terms *haemophilia* and *psychopathy* are condition-terms of the dispositional sort; *drunkenness*, for example, of the state variety. All told, then, we can conclude, at this point, that *psychopathy* is a psychological condition-term of the dispositional variety.

Now, key to the elucidation of the meaning of any given psychological term- not merely condition-terms-, is the noting as to whether or not the term's grammar induces a first person/third person asymmetry of ascription (Baker & Hacker, 1982). For a term φ belonging to the latter category- paradigmatically, the *abilities* and *capacities*-, the justification for the ascription of φ to an individual A (either oneself or another; i.e., in both the first and third-person modes) is that A has performed behaviours which are criterial for φ.[23] For a term belonging to the former category- *hopes*, *desires*, *sensations* (e.g., *pain*), etc.-, first person ascriptions are not justified by behaviour, at all, but are, rather, *avowals*.[24] Now, psychological disposition terms belong to the latter category, from which we conclude that: (i) there does not exist a coherent basis for ascribing to A the dispositional condition-term *psychopathy*, in absence of A's having performed particular (criterial) behaviours, which, in accordance with the term's grammar, instantiate it; (ii) one cannot legitimately avow that one is a *psychopath*.[25]

It is essential to emphasize that, for *no* psychological term, does the correct ascription to an individual A, rest on material characteristics- neurobiological, or otherwise- of A, or putative hidden processes or mechanisms unfolding within A. This is not an empirical conjecture, but, rather, an observation about the manner in which psychological terms are, *in fact*, correctly employed; accordingly, about the grammars of such terms. As such, it is an antecedent, logical, fact. For, as a matter of fact, the grounds of ascription of psychological terms are learned- and such terms, ascribed correctly-, without reference to- and, standardly, in complete ignorance of- such material and putative hidden features of people. A leader is not someone carrying *within him* some particular set of material conditions, but, rather,

---

[22] Certain terms have a range of correct employments under which they appear in both sub-classes.

[23] See, e.g., Baker and Hacker (1982). Behaviours criterial for φ: i) are internally (grammatically) related to φ in the sense that there is no criterion of identity for that which φ signifies, independent of these behaviours; ii) constitute logical- rather than empirical- justification for the ascription of φ; iii) are part of the meaning of φ, and, accordingly, are taught and learned as part of the teaching and learning of φ's correct employments.

[24] One can coherently *avow* one's desire for a holiday, or ambition to climb Mount Everest, but not so, one's intellectual brilliance or superiority as a leader. The latter, both as first person and third person utterances, stand in need of behavioural support. Equally, it is incoherent to assert, for example, that "I am a leader, but have never behaved as such", but not so, "I love B, but have never behaved as such" (the latter, an avowal).

[25] An individual could, of course, coherently *claim* to have performed the criterial behaviours.

someone who, in relevant circumstances, has behaved in a particular fashion. The grounds for ascribing the dispositional condition-term *psychopathy* to another, is logically independent of brain structures, hypothesized material deficits, unobservable processes, and the like. This is not to say that the latter are not involved in the story of psychopathy (as, perhaps, characteristics *of* psychopaths, or material bases of a proclivity towards), but only that the grammars of psychological terms- of which *psychopathy* is an instance- make no reference to them; accordingly, that they are not constituent parts of the term's rule of ascription; and, so, play no role in the rule of inclusion into class C_P, the elements of which are psychopaths.

Psychopathy is a psychological condition-term of dispositional variety. It's rule of ascription must, then, make reference to criterial behaviours. What is the precise form of this rule? What *are* the criterial behaviours? An important distinction to draw, preliminary to attempting to answer such a question, is that between *technical* and *ordinary language* terms. A technical term- e.g., *neutrino, phlogiston, derivative, determinant of matrix, blastomere,* or *superego*- is a term which was introduced into the language by a technical discipline. In contradistinction to ordinary language terms, which were not, the rule of ascription of a technical term standardly takes the form of a set of necessary and sufficient conditions. The vast majority of psychological terms are of the ordinary language sort, their rules of ascription having what Baker and Hacker (1982) call an open-circumstance relativity.[26]

Whatever be its precise conceptual lineage, the term *psychopathy* was, it seems undeniable, introduced as part of technical discourse. However, as we have indicated, there is little point in looking to the technical disciplines wherein the term is employed, with the aim of ascertaining the criterial behaviours, the performance of which is necessary and sufficient to instantiate the term. It has been suggested that the widespread employment of Hare's Psychopathy Checklist-Revised (PCL-R), within the criminal justice system, has served to install the PCL-R as a *de facto* rule of ascription (its items, the required criterial behaviours); to wit, "by *psychopath* is meant a human who scores thirty or greater on the PCL-R." However, psychopathy researchers are anything but unanimous in holding the view that the PCL-R should be granted right to play this role (see, e.g., Edens, 2001; Martens, 2008; Skeem & Cooke, 2010). To the contrary, under sway of GCV, the received view- even that of Hare (see Hare & Neumann, 2008)- is that: (a) psychopathy is an (unobservable) construct; (b) it is not possible to lay down, or elucidate, a rule of ascription relating to an (unobservable) construct; and (c) this state of affairs poses no fundamental obstacle to the undertaking of empirical investigation into psychopathy. We draw to a close our preliminary observations on the conceptual foundations of Question I, by recording the view that until is settled upon, a normative set of behaviours, criterial for the term *psychopathy*- equivalently, a technical rule of inclusion into C_P-, Question I will remain unanswerable, and the empirical results generated under the heading of psychopathy research will do nought but engender confusion (certainly,

---

[26] Meaning that: (a) the criterial behaviours to which each rule makes reference form an unbounded set; (b) for no subset of criterial behaviours, does the performance of the elements constitute necessary and sufficient conditions for ascription of the term.

will not qualify as building blocks of a cumulative science). It is the scientific responsibility of psychopathy researchers to engineer a discipline-level settlement on a technical rule of inclusion into C_P.

Let us turn, now, to Question II. Because questions of causality are contingent upon demonstrations of existence, it might be thought that the sensicality of Question II rests wholly on the satisfactory resolution of Question I. That is to say, that once the discipline comes to agreement on a rule of inclusion into C_P, and, consequently, those who belong in C_P can be unambiguously identified, the issue of the causal story of psychopathy will be addressable empirically. Regrettably, the issue turns out to have hidden- or, at least, rarely acknowledged- complexities, an appreciation of which begins with the following observations: a) it is not the case- contrary to the dogma of the social sciences- that every type of phenomena in need of a scientific explanation, *has* causes (equivalently, "valid explanatory account" is not synonymous with "causal account"); b) what can validly feature in an explanatory account relating to elements φ of a class C_φ of CNR, depends upon what sort of a thing a φ is; or, as it is sometimes put, what qualifies as a candidate explanans, depends upon the nature of the explanandum.[27]

Let us imagine that the discipline has settled on a set of criterial behaviours, S_p, and, on the basis of S_p, formulated a rule of inclusion, r_p, into C_P. To offer up a valid explanatory account of psychopathy, is to offer up a valid explanation as to why individuals wind up in C_P. Now, psychopaths are humans with a dispositional condition; they are in C_P in virtue of their having performed the elements of S_p. Because an individual winds up in C_P in virtue of his having performed the elements of S_p, the form that a candidate explanation of *psychopathy* must take in order to be valid, is determined by the form of explanatory account, valid for the elements of S_p. And when the explanandum is a particular behaviour, what can, and cannot, validly serve as explanans, depends upon the logical *type* of behaviour the explanandum is.[28]

Behaviours can be exhaustively classified as *voluntary* or *nonvoluntary*. Among other things, a voluntary behaviour, γ: (a) involves the exercise, by a performing agent, A, of a two-way power to perform or refrain from performing (this being the chief point of contrast between they and behaviours that are not voluntary); (b)

---

[27] Psychopathy researchers are strongly inclined towards tying together the issues of existence and causality into a single Gordian knot of confusion. Krupp, for example, in his review of Jalava et al. (2015), does so when he asserts that, "…we cannot rely on statistical procedures to tell us whether psychopathy is real, because statistics tend to be incapable of detecting causality."

[28] Once again, the reader might be predisposed to see, in what follows, a theoretical account of certain types of behaviour, and wonder why the author has not addressed "other" theoretical perspectives, such as that of Donald Davidson (e.g., Davidson, 1963). The observations registered, herein, are not, however, of a theoretical nature; they do not serve the purpose of, say, "explaining voluntary behaviours." Note that to provide an explanation *of* voluntary behaviours, presupposes an independent criterion of identity of the thing for which the explanation is being offered (i.e., voluntary behaviours). What I provide, herein- with substantial assistance from P.M.S. Hacker-, are grammatical elucidations relating to the concept *voluntary behaviour* (and cognates), which settle the latter issue; i.e., the issue of what is *meant* by a voluntary (involuntary, intentional) behaviour, a motive, etc.

involves the control, by A, of the behaviour's inception, continuation, and termination; (c) can be engaged in at will; and d) is a behaviour that an agent can *try*, *intend*, or *decide* to perform (Bennett & Hacker, 2003). The class of voluntary behaviours can be partitioned into the sub-classes of *intentional, unintentional*, or *neither intentional nor unintentional* behaviours; the class of nonvoluntary behaviours, into subclasses of *intentional, involuntary*, or *neither intentional nor involuntary* behaviours (Bennett & Hacker, 2003).

Causes can validly appear as explanans in candidate explanations of nonvoluntary behaviours- paradigmatically, elements of the sub-class of involuntary behaviours-, but not so, voluntary behaviours. This is because in having performed an involuntary (voluntary) behaviour, A was not (was) exercising a two-way power to perform or refrain from performing. Causes are not valid explanans of voluntary behaviours, because such behaviours *have* no causes.[29] Among those which can validly appear in candidate explanations of voluntary behaviours, *reasons*- and, as will be suggested, of particular relevance to the explanation of psychopathy, the subclass of reasons known as *motives*- are the characteristic explanans.

If the doing in question is an involuntary behaviour ρ, then: (i) ρ can have one or more causes ρ_c; (ii) ρ_c is the thing that brought about ρ; (iii) A, who performed ρ, cannot have had reasons for having done so. Though it would make no sense to inquire as to the *reasons* A had for slipping and falling, it would make perfect sense to investigate the conditions under which the accident occurred, the aim being to discover its *causes*. One inquires as to *why* Agent A performed voluntary behaviour γ (and, if A had reasons for having performed, and these reasons are known, cites these reasons), but *what* is the cause of ρ, say, his involuntary twitch. Whereas A's reasons for performing the voluntary behaviours he performs are *his* reasons (are formulated by him out of his knowledge, beliefs, and desires), the causes ρ_c of ρ, are not A's causes. A has reasons, not causes. A may *avow* his reasons for performing a voluntary behaviour, but not the cause of his twitch. Causes are as they are. They are independent of what A knows, believes, and desires. It is incoherent to say that A has reasons for having performed a behaviour, but does not know what these reasons are, but not at all so, to say, for example, that A's twitch has a cause, but A does not know what it is.

Motives can validly appear as explanans in candidate explanations of voluntary, *intentional*, behaviours (Bennett & Hacker, 2003). Let it be the case that A has performed an intentional behaviour β. Then it is a *possibility* (allowable under the rules of language) that, in performing β, A was *acting out of* a motive ζ (Bennett &

---

[29] No doubt, many will scoff, and remonstrate that, "This is mere talk. *All of it* is caused!.. even the *feelings* of two-way empoweredness." To this, I reply, that the reach of every assertion is tied to the reach of the meanings of its constituent terms. Words are not *mere* anything. To "prove" that one has "climbed Mt. Everest", by walking the 29,029 pancake-flat feet, stretching out from one's house, is not to have revealed heretofore unappreciated senses of the terms *climb* and *Mt. Everest*, but, rather, one's incomprehension of the meanings of these terms. It is my contention that, along similar lines, what such remonstrators have in mind, rests on a *feeling* about an inarticulable sense of causality, the presence of which, in their remonstrances, renders the latter (unprovable) metaphysics.

Hacker, 2003). If, in performing β, A happened to have been acting out of a motive ζ, then ζ is an *appetite, emotion,* or *desire* of A's; more generally, is a psychological phenomenon which has a formal or formal and specific object. *Clarification*: Appetites such as *hunger, thirst,* and *lust* are blends of sensation and desire (Bennett & Hacker, 2003). Sensations do not have objects, but desires do (a formal, but not a specific, object). The formal object of *hunger* is food/nutritional sustenance, and of *lust*, sexual intercourse (Bennett & Hacker, 2003). Emotions such as *grief, love,* and *jealousy,* have both formal and specific objects (Bennett & Hacker, 2003).

It is the fact of appetites, emotions, and desires having (formal or formal *and* specific) *objects,* that singles them out as logically suitable for the role of motive relative to intentional acts; for it is characteristically human to address the objects of one's emotions, desires, and appetites, through the performance of sequences of intentional behaviours. If, in his performance of intentional behaviour β, A was acting out of a motive (appetite, emotion, or desire) ζ, then ζ is the (non-causal) origin of, or impetus for, a *pattern of behavior* (sequence of intentional behaviours, of which β was an element) oriented towards the object of ζ (Bennett & Hacker, 2003). *Clarification*: When, for example, an Agent A is in love: (a) the object of his love is some individual, say, B; (b) A must address, in some way, his passively acquired feelings of love for B; and (c) might well do so by performing intentional behaviours oriented towards B (e.g., by acting in a manner that he believes will please B, seeking out B's company, etc.) His love for B is the motivating emotion of a sequence of intentional behaviours.

As noted, there has not, to date, been fixed, as a normative standard within the discipline, a set of behaviours, jointly necessary and sufficient for the instantiation of the term *psychopathy.* However, the fact of the sorts of criteria bandied about by psychopathy researchers- notably, the categories comprising of the PCL-R- resting on voluntary, intentional, behaviours, gives indication that it is this class of behaviours that is the relevant one. Question II is not, then, coherent. For, even if a rule of inclusion into C_P were available, an individual winds up in C_P in virtue of his having performed voluntary, intentional, behaviours; and causes are not valid explanans in candidate explanations of such behaviours.[30] It is the *reasons* agents had for having

---

[30] Consider the following test scenario. Agent A is performing behaviour s, which happens to count among behaviours criterial for the term *psychopathy*, when the police arrive on the scene. If A has the capacity to terminate performance of s and attempt to flee, then A is exercising a two-way power to perform or not perform s. That is to say, s is a voluntary, intentional, behaviour. Imagine, now, that, as he is attempting to flee, A happens to be suffering from a cramp in his leg. Even if its presence greatly militated against his likelihood of escape, he would be absent, by way of contrast, the capacity to terminate *it*. Being, as it is, an involuntary act, over s, A does not exercise a two-way power. The cramp has causes, operative irrespective of A's needs and requirements. When Ted Bundy accosted Carol DaRonch in Utah, he *intended* to kidnap her, *attempted* to kidnap her, and, when the attempt "went wrong", and he perceived a threat of detection, controlled the *termination* of the attempt; all of which are characteristic marks of the voluntary, intentional, act. If the act were the sort of thing that had causes, then Bundy would not have been able to terminate it, for the operation of a cause is indifferent to an agent's psychological state (among those, his fear of being apprehended).

performed the- yet to be specified- intentional behaviours, criterial for the term *psychopathy*, which would constitute valid explanans in candidate explanations of psychopathy. What, then, of the role of neurobiology? What of the postulated neurobiological basis of psychopathy?

There are two ways in which material characteristics of an element of C_P, can validly enter into an explanatory account of his having qualified for admission into this category. Firstly, there may exist material *preconditions* for the capacity to perform the elements of S_p. In fact, for virtually any voluntary, intentional, behaviour, β, there will exist such preconditions. For example, preconditional for the performance of the voluntary, intentional, behaviour of dancing, is the possession of two functioning legs. The material precondition that is the possession of two functioning legs, neither *causes A*'s dancing (the latter, in fact, absent of causes), nor is a *reason* for *A*'s having danced. It takes little imagination to list off material characteristics, preconditional for an individual's capacity to perform the intentional acts, the performance of which would qualify him for inclusion into C_P.

The second way is, from the perspective of psychological science, far less trivial. To begin, those individuals- Ted Bundy, e.g.- who are cited as prototypical psychopaths are frequently described as acting out of certain- societally unacceptable-motivating appetites. According to Bennett and Hacker (2003), conceptually, an appetite: a) is a form of unease which disposes the agent to action, with the aim of satisfying it, and, in so doing, removing the unease; b) is characterized by an intensification which is progressively more unpleasant; c) is cyclical, and recurrent, in that, once satisfied there ensues a state of temporary satiation (equivalent to the disappearance of the appetites sensation component), out of which builds, eventually, and once again, the characteristic sensations of unease. Now, as already noted, an appetite is a blending of sensation and desire, and as these latter are passively received (the agent does not exert, over them, a two-way power), they are, logically, the right sorts of phenomena to have causes. So, then, were it to be established that elements of C_p were, as a rule, acting out of a motivating appetite when they performed the intentional acts, the performance of which is the logical basis for their having been assigned to C_p, it would be coherent to apply science to the task of investigating the causal basis- neurobiological and otherwise- of these motivating appetites. Discovered causes of such would stand as a material basis for a proclivity towards psychopathy.

Russell Williams was, apparently, able to stand out by a fence, for a good while, at the rear of Jessica Lloyd's residence, *bide his time*, and *decide upon* the correct moment to enter the residence and commit his crimes; all features characteristic of the performance of a voluntary, intentional, act. The motivating appetite out of which he was acting was, by contrast, passively received, and, over it, he did not exercise a two-way power. It- one may coherently speculate-, was engendered by causes, both unknown to him, and- at present- those who seek to understand his behaviour.

8.5   Conclusion

The methods of science, broadly conceived, can often be enlisted, with striking effectiveness, in the service of addressing inquiries posed- adjudicating the truth status of assertions made- about the empirical world. Both the proper scientific handling of a claim or inquiry, and what constitutes evidence relevant to its empirical consideration, depends upon what, precisely, the claim or inquiry expresses. And it is language, the mode of representation upon which science rests, which settles what, if anything, is expressed by a particular assemblage of words; in particular, a comparison of a putative declarative or interrogative sentence to the constitutive linguistic rules which fix the range of correct employments of its constituent terms. The matter is not empirical, but, rather, antecedent and logical. Is it true or false, that "25% of all Canadians feel hopeless"? What form of evidence is relevant to an adjudication of the claim? Prior to setting the wheels of science turning in the hope that it will settle such issues, it is prudent to consider the issue of what, precisely- if anything, at all- is being asserted; the issue of what- if any- state of affairs is specified by the assemblage of words in question. This antecedent, logical, issue comes down to component issues such as "what is meant by hopelessness" and "on what grounds is one justified in designating another as feeling hopeless", and all of these issues are linguistic issues. More particularly, they are settled through an elucidation of what *we* mean by *hopelessness* and *the condition of feeling hopeless*; consequently, with reference to a clarification of relevant segments of the logical grammar which determines how words are to be employed, when they are employed correctly.

Not every arrangement of words constitutes an inquiry or assertion. Though quite possibly striking the ear as an issue adjudicable with respect evidence, the statement, "teams working at the CERN accelerator, have been able to use Alice and LHC_b detection theory to isolate, for the first time, the long theorized Charpak particle" is mere nonsense. Because certain of its constituent terms are absent a normative place within language- i.e., there do not exist constitutive rules which fix their ranges of correct employments-, this particular assemblage does not specify a particular state of affairs (which, in turn, either does or does not hold). It is conceivable that, some day, it will be possible to form word assemblages expressing of scientific issues, wherein reference is made to "the existence of" or "the causes of" psychopathy. However, until the range of correct employments of the key constituent term *psychopathy* is settled- this, a matter of the clarification of extant, or the laying down of novel, constitutive linguistic rules, as the case may be-, the insidious ambiguity to which David Cooke (2018) alludes, will continue to visit itself upon such word assemblages, and hold them firmly outside of the purview of the scientific method.

References

Audi, R. (1999). *The Cambridge dictionary of philosophy*. Cambridge University Press.

Baker, G., & Hacker, P. (1982). The grammar of psychology: Wittgenstein's Bemerkungen über die philosophie der psychologie. *Language and Communication, 2*(3), 227–244.

Bennett, M., & Hacker, P. M. S. (2003). *Philosophical foundations of neuroscience*. Blackwell Publishing.

Brun, G. (2016). Explication as a method of conceptual re-engineering. *Erkenntnis, 81*(6), 1211–1241.

Carnap, R. (1950). *Logical foundations of probability*. University of Chicago Press.

Cooke, D. (2018). Psychopathic personality disorder: Capturing an elusive concept. *European Journal of Analytic Philosophy, 14*(1), 15–32.

Creath, R. (1990). *Dear Carnap, dear Van: The Quine-Carnap correspondence and related work: Edited and with an introduction by Richard Creath*. University of California Press.

Cronbach, L. J., & Meehl, P. E. (1955). Construct validity in psychological tests. *Psychological Bulletin, 52*, 281–302.

Davidson, D. (1963). Actions, reasons, and causes. *Journal of Philosophy, 60*(23), 685–700.

Dolan, M. (2004). Psychopathic personality in young people. *Advances in Psychiatric Treatment, 10*, 466–473.

Duintjer, O. (1977). *Rondon Regels*. Boom.

Edens, J. (2001). Misuses of the Hare psychopathy checklist-revised in CourtTwo case examples. *Journal of Interpersonal Violence, 16*(10), 1082–1093.

Frege, G. (1892). Über Sinn und Bedeutung. *Zeitschrift für Philosophie und philosophische Kritik, 100*, 25–50.

Gowlett, C. (2014). Does secondary psychopathy exist? Exploring conceptualizations of psychopathy and evidence for the existence of a secondary variant of psychopathy. *Internet Journal of Criminology*. https://958be75a-da42-4a45-aafa-549955018b18.filesusr.com/ugd/b93dd4_cea6276862c341ad9da93d2f57f79c71.pdf

Hacker, P. (1986). *Insight and illusion*. Oxford University Press.

Hacker, P. (1988). Language, rules and pseudo-rules. *Language and Communication, 8*, 159–172.

Hare, R., & Neumann, C. (2008). Psychopathy as a clinical and empirical construct. *Annual Review of Clinical Psychology, 4*, 217–246.

Jalava, J., Griffiths, S., & Maraun, M. (2015). *The myth of the born criminal: Psychopathy, neurobiology, and the creation of the modern degenerate*. University of Toronto Press.

Jost, J., & Gustafson, D. (1998). Wittgenstein's problem and the methods of psychology: How 'grammar depends on facts'. *Theory and Psychology, 8*(4), 463–479.

Maraun, M., & Gabriel, S. (2013). Illegitimate concept equating in the partial fusion of construct validation theory and latent variable modelling. *New Ideas in Psychology, 31*, 32–42.

Maraun, M., Gabriel, S., & Slaney, K. (2008). The Augustinian methodological family of psychology. *New Ideas in Psychology, Special Edition: Wittgenstein's Relevance to Psychology, 27*(2), 148–162.

Martens, W. (2008). The problem with Robert Hare's psychopathy checklist: Incorrect conclusions, high risk of misuse, and lack of reliability. *Medicine and Law, 27*(2), 449–462.

Quine, W. (1951). Two dogmas of empiricism. *The Philosophical Review, 60*(1), 20–43.

Skeem, J., & Cooke, D. (2010). Is criminal behavior a central component of psychopathy? Conceptual directions for resolving the debate. *Psychological Assessment, 22*(2), 433–445.

Skeem, J., Polaschek, D., Patrick, C., & Lilienfeld, S. (2011). Psychopathic personality; bridging the gap between scientific evidence and public policy. *Psychological Science in the Public Interest, 12*(3), 95–162.

Ter Hark, M. (1990). *Beyond the inner and the outer: Wittgenstein's philosophy of psychology*. Kluwer.

Waismann, F. (1965). In R. Harré (Ed.), *Principles of linguistic philosophy*. Macmillan.

Chapter 9
Assessment of Psychopathy: Addressing Myths, Misconceptions, and Fallacies

Martin Sellbom, Scott O. Lilienfeld, Robert D. Latzman, and Dustin B. Wygant

Abstract Psychopathy is a contentious construct with respect to both theory and operationalization. A number of myths, misconceptions, and fallacies regarding the assessment of psychopathy hinder scientific progress and impede this construct's applied consideration across settings and contexts. We review these widespread erroneous ideas with an eye to guiding the reader towards an improved understanding of scientifically-informed means of conceptualizing and applying psychopathy assessment. More specifically, we consider whether "gold standards" exist for psychopathy measurement, differing psychopathy measures capture the same underlying subdimensions, and the DSM-5 diagnosis of antisocial personality disorder affords an adequate operationalization of psychopathy. We also consider the relevance of the controversial construct of boldness to psychopathy. Moreover, we discuss whether it is scientifically acceptable to rely on total psychopathy scores or use cut scores for a "psychopathy diagnosis." In light of psychopathy being an important criminal justice construct, we ask whether the Psychopathy Checklist – Revised is an unparalleled measure of violent recidivism, as numerous scholars have asserted. We conclude by considering the appropriateness for various assessment modalities (self-report, informant-reports, and brief forms) for psychopathy measurement.

Keywords Psychopathy · Measurement · Self-report · Informant report · Psychopathy checklist – revised · Boldness · Antisocial personality disorder

M. Sellbom (✉)
University of Otago, Dunedin, New Zealand
e-mail: martin.sellbom@otago.ac.nz

S. O. Lilienfeld
Emory University, Atlanta, GA, USA

University of Melbourne, Melbourne, Australia

R. D. Latzman
Georgia State University, Atlanta, GA, USA

D. B. Wygant
Eastern Kentucky University, Richmond, KY, USA

© Springer Nature Switzerland AG 2022
L. Malatesti et al. (eds.), *Psychopathy*, History, Philosophy and Theory of the Life Sciences 27, https://doi.org/10.1007/978-3-030-82454-9_9

Psychopathy fascinates people. It is often depicted in characters encountered in books, movies, and other stories. There are many lay views of what constitutes psychopathy, some more accurate than others (Berg et al., 2013), but as scientists we tend to shrug off erroneous beliefs regarding this condition as uninformed and unimportant. This cavalier attitude is unfortunate. Myths, misconceptions, and fallacies are not specific to lay people's understanding of psychopathy, and a number of beliefs often encountered among mental health professionals are irreconcilable with scientific evidence. In the current chapter, we focus specifically on the operationalization and assessment of the psychopathy construct. We describe the various myths, misconceptions and fallacies that we believe to be the most important and prevalent, and address them in reference to the empirical literature.

We believe that it will help readers to have a sense of how we view psychopathy, which is a contentious area with respect to definition. We regard it as a severe personality pathology that is reflected in a maladaptive constellation of personality traits. There is unlikely to be a single "psychopath" but rather a mélange of individuals who present with varying degrees of these traits. We view the triarchic model of psychopathy (Patrick et al., 2009) as a helpful descriptive framework in that individuals with high levels of psychopathic traits are coldhearted, unemotional, at time even callous and exploitative, have reduced capacity for empathy and remorse, and are deceitful and manipulative ("meanness"); these traits are the least controversial. We also believe that some of these individuals are highly impulsive, sensation seeking, irresponsible, nonplanful, and are disinclined to delay gratification ("disinhibition"), which is also reflected in most psychopathy theories. Finally, we believe that many individuals with psychopathic traits are fearless, risk taking, resilient in light of stressful situations, and socially dominant, self-assured, and even grandiose ("boldness"). We discuss this last domain with respect to psychopathy relevance later.

9.1 Prologue: Constructs and Measurement

"Whatever exists at all exists in some amount. To know it thoroughly involves knowing its quantity as well as its quality" – E. L. Thorndike (1918)

We start with a discussion of the basic tenets of psychological assessment. The distinction between theoretical constructs and their measurement is critical. Constructs are the central 'building blocks' of a theoretical entity; they are theoretical abstractions given that they are hypothetical concepts that are derived from theory. Constructs are built by theorists. Moreover, they are latent in that they need to be inferred through overt operationalization, which includes manifest (observable) quantitative indicators that are measured in some manner (Thorndike, 1918). Thus, measurement is invariably fallible in the sense that it represents the best effort at approximation of the (latent) construct. Almost always, this measurement will contain both unsystematic and systematic sources of error along with some component of the true score (Cronbach & Meehl, 1955; Loevinger, 1957). Measurement scholars consider both reliability[1] (a function of unsystematic error) and validity[2] (a function of systematic error) in evaluating the degree to which overt measurement is able to sufficiently approximate the theoretical construct of interest. Of course, as indicated by our citations, these concepts are far from new, but in the psychopathy assessment literature, they seem at times to have been largely forgotten (Cooke, 2018; Skeem & Cooke, 2010).

For example, researchers and clinicians often erroneously regard the widely used Psychopathy Checklist – Revised (PCL-R; Hare, 1991/2003) as isomorphic with the psychopathy construct itself. Instead, this measure is merely one useful but imperfect approximation of the (underlying) construct of psychopathy (Cooke, 2018). Indeed, we (the authors) have all at times fallen victim to reviewers and editors who objected to our use of well-validated psychopathy measures because they were not the PCL-R, with anonymous reviewers going as far as to say "the PCL-R *is* psychopathy" or "the PCL-R is the gold standard of psychopathy measurement" (see "Are Certain Measures "Gold Standards" for Assessing Psychopathy?"). Such perspectives are not only misguided, but they defy the basic principles of measurement theory.

Good measurement hinges on well-defined and articulated constructs (e.g., Cook & Campbell, 1979) – a requirement that is particularly problematic in the psychopathy field (Cooke, 2018; Patrick, 2018). There are numerous theoretical perspectives on psychopathy, often with little consensus on the actual definition. The evaluation of our operationalizations, therefore, depends on the theoretical perspective that underpinned their development. To say that the Psychopathic Personality Inventory's (PPI; Lilienfeld & Andrews, 1996) Fearless-Dominance factor is not a valid indicator of one key feature of psychopathy (e.g., Miller & Lynam, 2012) cannot be made outside of a theoretical context; it depends on the theoretical perspective to which one adheres (Lilienfeld et al., 2012). For instance, from Cleckley's (1941) perspective, the PCL-R might have experienced "construct drift" (i.e., measurement moving away from its actual target construct), but evaluation of our measurement tools with respect to construct validity can also reshape how we see a construct (Cooke, 2018; Hare & Neumann, 2008). Those who adhere to sociologist Lee Robins's (1966) conceptualization of psychopathy (or what she termed "sociopathic personality") might view the DSM-5 (American Psychiatric Association, 2013) criteria for Antisocial Personality Disorder as an adequate operationalization of this construct, although most psychopathy experts disagree (e.g., Hare, 1996; Lilienfeld, 1994; Lykken, 1995; Sellbom & Boer, 2019).

---

[1] Reliability refers to the degree to which a test score is measuring an underlying theoretical construct with consistency.

[2] Validity refers to the degree to which a test score is measuring what it is purported to measure per the theory underlying the test.

9.2   Part 1: General Issues in Psychopathy Assessment

First, we confront 7 largely conceptual myths, misconceptions, and fallacies that have been perpetuated in the literature. These concern myths regarding the existence of "gold standards" in assessment, misconceptions regarding a number of theoretical issues that pertain to operationalization, and fallacies that pertain to interpretation of psychopathy scores.

9.2.1   Are Certain Measures "Gold Standards" for Assessing Psychopathy?

A frequently voiced assertion in the psychopathy literature is that certain extensively validated measures are "gold standards" for detecting this condition. This claim appears to be especially prevalent for the PCL-R and its variants (see Baskin-Sommers et al., 2011; Ermer et al., 2012, for examples). As of this writing (May, 2019), the phrase "PCL-R is the gold standard" (or "golden" standard) yields 24 hits in *Google Scholar* alone, not to mention over 2000 hits in *Google*. Another author team referred to the "PCL-R throne as the sole tool of choice for psychopathy measurement" (Evans & Tully, 2016; p. 79). Anecdotally, we have sometimes heard scholars maintain that "psychopathy consists of 20 criteria," in clear reference to the 20 items of the PCL-R. In fairness, the PCL-R is not alone in this regard. At least one set of authors referred to the Psychopathic Personality Inventory (PPI; Lilienfeld & Andrews, 1996) as "the gold standard self-report psychopathy measure" (Witt et al., 2009, p. 1007).

Nevertheless, there are no genuine gold standards – largely or entirely infallible criteria – for detecting psychopathy or arguably, any mental disorder for that matter (Faraone & Tsuang, 1994). Even the best-validated measures of psychological conditions are necessarily fallible indicators of their respective constructs. As Cronbach and Meehl (1955; see also Loevinger, 1957; Westen & Rosenthal, 2005) observed in their classic article on construct validation, and as stated earlier, psychological constructs are by definition latent entities that cannot be directly observed. As a consequence, no psychological measure perfectly or exhaustively captures the construct of interest. Furthermore, construct validation, like the testing of scientific theories, is inherently an ongoing and self-correcting process, never to be regarded as final.

By using the phrase "gold standard" to apply to the PCL-R, perhaps some authors instead merely intend to advance the less controversial proposition that this measure possesses substantially greater construct validity[3] compared with other psychopathy

---

[3] Construct validity refers to whether a test score converges with theoretically expected extra-test variables (convergent validity) as well as diverges from theoretically unrelated extra-test variables (discriminant validity).

measures. Even here, however, the evidence is at best equivocal. We are unaware of any meta-analytic evidence that the PCL-R consistently outperforms other widely used psychopathy measures, such as the PPI, Levenson Self-Report Psychopathy Scale (LSRP; Levenson et al., 1995), or Triarchic Psychopathy Measure (TriPM; Patrick, 2010), in its convergent and discriminant validities with relevant external criteria,[3] such as laboratory indicators, relations with personality dispositions, brain imaging findings, course and outcome, or family history (see Kendler et al., 2009; Robins & Barrett, 1989; Robins & Guze, 1970, for frameworks for the external validation of measures of psychopathology). To the contrary, there is even some evidence pointing in the opposite direction. For example, although many authors maintain that the PCL-R is distinctly useful in the prediction of violence (see section entitled "Are the PCL-R and Allied Measures Unparalleled Predictors of Violence?" for more information), Camp et al. (2013) reported that, in a sample of 158 offenders, the PPI – especially its Self-Centered Impulsivity subdimension – displayed incremental validity above and beyond the PCL-R total score and its subdimensions for predicting future violence. In any case, the assumption that the PCL-R possesses inherently superior construct validity to other well-validated psychopathy measures is largely or entirely unsupported.

9.2.2 Are Psychopathy Subdimensions Isomorphic Across Measures?

Some authors appear to presume that the key subdimensions across psychopathy measures are largely interchangeable (e.g., Poythress et al., 2010). This misunderstanding may derive in part from two influential articles (Harpur et al., 1988, 1989) demonstrating that the PCL, the precursor to the PCL-R, displayed a correlated two-factor structure. Factor I comprises the core affective (e.g., lack of guilt and empathy) and interpersonal (e.g., grandiose sense of self-worth) features of psychopathy, whereas Factor II comprises poor impulse control and antisocial lifestyle features often associated with psychopathy (a number of later factor analyses suggested that these two factors can be further deconstructed into two correlated facets, yielding a four facet structure; Hare, 1991/2003; Hare & Neumann, 2005). Perhaps because several other psychopathy measures, especially the PPI (Lilienfeld & Andrews, 1996) and PPI-R (Lilienfeld & Widows, 2005), also often yield a two factor higher order structure, a number of authors appear to have assumed that the corresponding subdimensions of these measures are largely isomorphic (e.g., Poythress et al., 2010).

Nevertheless, PCL and PCL-R Factors I and II are by no means synonymous empirically or conceptually with PPI and PPI-R Factors I and II, with the latter two subdimensions commonly dubbed Fearless Dominance and Self-Centered Impulsivity, respectively. In particular, the associations between the first subdimensions (Factor I) of these measures tend to be only modest, with correlations typically falling between $r = .20$ and $.25$ (Malterer et al., 2010; Marcus et al., 2013; Miller & Lynam, 2012). The correlations between the Factor II scores on these measures tend to be considerably higher (e.g., $r = .56$ in Miller & Lynam's, 2012, meta-analytic review), although again far from unity even after controlling for measurement error.

What accounts for the striking discrepancy in the Factor I scales of these two widely used measures? The primary source appears to be the trait of boldness (Patrick et al., 2009), which is heavily represented in PPI Factor I (Fearless Dominance) but at best only modestly represented in PCL-R Factor I, and even then, only its interpersonal facet (Lilienfeld et al., 2016; Wall et al., 2015). Boldness appears to be even less well-represented in certain other measures derived from the PCL-R, especially the LSRP (Sellbom & Phillips, 2013), at least in part accounting for the low correlation between PPI/PPI-R Factor I and LSRP Factor I (Marcus et al., 2013; but see Sellbom, 2011, for evidence that the standard two-factor structure of the LSRP is scientifically suboptimal).

9.2.3 Is Antisocial Personality Disorder an Adequate Operationalization of Psychopathy?

"It is about time that we cease making psychopathy, criminal and antisocial behavior, identical and interchangeable" – B. Karpman (1946)

As illustrated in this quote, early prominent psychopathy scholar Benjamin Karpman recognized the problematic union of psychopathy and antisocial behavior. Still, over 70 years later, we continue to see a problematic conflation of these two interrelated but separable constructs. Our current diagnostic manuals (DSM-5, ICD-10) characterizes Antisocial (or Dyssocial) Personality Disorder (ASPD), ostensibly the formal diagnostic operationalization of psychopathy, as a pattern of behavior involving the "disregard for, and violation of, the rights of others that begins in childhood or early adolescence and continues into adulthood." (p. 659). The manual further states "this pattern has also been referred to as *psychopathy, sociopathy,* or *dyssocial personality disorder.*" (p. 659, emphasis in original). But using ASPD as an operationalization of psychopathy has not been supported by empirical evidence. Hare et al. (1991) observed that whereas 50–80% of incarcerated offenders meet criteria for DSM-III-R ASPD, only 15–20% of such individuals would be classified as psychopaths based on the PCL-R (although see later discussion of the fallacy of using cut scores to indicate psychopathy). Moreover, recent studies have found important neurobiological differences between psychopathy and ASPD (e.g., Hyde et al., 2014), particularly with respect to the functioning of the amygdala and other brain regions involved in the processing of fear and other emotions. In addition, evidence has also shown that the construct of boldness (see subsequent section) particularly separate the two (e.g., Venables et al., 2015; Wall et al., 2015).

Although ASPD is aimed at measuring psychopathic personality traits that manifest in externalizing and antagonistic behavior, substantial construct drift has been observed since the first two versions of the DSM (Crego & Widiger, 2014; Widiger & Crego, 2018). The first edition of the DSM (APA, 1952) characterized an antisocial reaction as one manifestation of "sociopathic personality disturbance (p. 38)" that captured cases previously classified as "psychopathic personality (p. 38)." DSM-II (APA, 1968) characterized those with Antisocial Personality as selfish, callous, irresponsible, and impulsive (among other traits). Conceptually, these early DSM characterizations were aligned with classic (e.g. Cleckley, 1976; Karpman, 1941; McCord & McCord, 1964) and contemporary models (e.g., Patrick et al., 2009) and measures (Hare, 1991/2003; Lilienfeld & Widows, 2005) of psychopathy. Interestingly, the first two versions of the DSM noted that criminal history alone was insufficient for fulfilling the criteria for the condition. Subsequent versions of the DSM utilized a criterion-based approach to diagnosing ASPD highly influenced by the work of Robins (1966), who identified the behaviors of adolescents in St. Louis child guidance clinics that were predictive of later antisocial and criminal careers. Although perhaps improving diagnostic reliability, DSM-III (APA, 1980), DSM-III-R (APA, 1987), and DSM-IV (APA, 1994) ASPD, which focused largely on discernable behaviors rather than underlying personality traits, drifted dramatically from most theoretical perspectives of psychopathy, particularly in what many characterized as the affective-interpersonal domain (e.g., Harpur et al., 1988). At around the same time that DSM-III was released, Robert Hare was in the process of developing the Psychopathy Checklist (Hare, 1980), which was based in part on Cleckley's (1976) influential list of psychopathic traits and as noted earlier, became one of the most influential advances in our field's understanding of psychopathy. As the construct of psychopathy became increasingly influenced by the PCL in the 1980's and 1990's, the gap between it and ASPD as a psychopathy operationalization has become substantial.

Although the DSM-5 (APA, 2013) retained the categorical model of personality disorders from the DSM-IV, it also included an alternative model of personality disorders (AMPD) that incorporates evidence of personality impairment and the presence of maladaptive personality traits based on a five-factor model. The AMPD operationalizes ASPD as manifested in disinhibition (risk-taking, impulsivity, irresponsibility) and antagonism (manipulativeness, deceitfulness, callousness, hostility). Additionally, a "psychopathy-specifier" was included to capture the construct of boldness, which is one of the three primary domains of psychopathy in the triarchic psychopathy model (Patrick et al., 2009). Studies have examined the AMPD operationalization of ASPD in relation to psychopathy in both community (Anderson et al., 2014; Few et al., 2015) and correctional (Sleep et al., 2018; Wygant et al., 2016) samples. Each of these studies revealed that the AMPD operationalization of ASPD exhibited stronger associations with facets of psychopathy than the categorical version of ASPD. This finding was particularly evident for the interpersonal and affective facets of psychopathy. The psychopathy specifier was strongly associated with markers of boldness and fearless-dominance. Research suggests that the AMPD offers a renewed alignment of diagnostic operationalization with its target construct, similar to earlier versions of the DSM, but with a more sophisticated measurement model.

9.2.4 Is Boldness Irrelevant to Psychopathy?

The role of Boldness in psychopathy, however, has been quite contentious (see e.g., Lilienfeld et al., 2018, and Sellbom, 2019, for reviews). Boldness, also termed fearless dominance in the PPI family of instruments, is a higher-order dispositional dimension associated with the capacity to remain calm under pressure coupled with interpersonal effectiveness and decreased avoidance of risky situations, both social situations and physical (Patrick et al., 2009; Lilienfeld et al., 2018). Scholars in the field disagree regarding whether boldness is an intrinsic feature of psychopathy as opposed to being peripheral or even irrelevant. Although space constraints preclude a full discussion of this debate, a number of overviews are available for interested readers (e.g., Lilienfeld et al., 2018; Miller & Lynam, 2012; Sellbom, 2019). In short, among those that argue that boldness is of limited importance to psychopathy, arguments often point to the generally weak to negligible association of boldness indicators with nonspecific antisocial behavior, aggression, and broad externalizing features (Miller & Lynam, 2012; Sleep et al., 2019). These authors tend to consider such behaviors to be the central and perhaps even exclusive outcomes of interest. Furthermore, boldness tends to correlate negatively with conditions marked by emotional distress (e.g., anxiety and depression) and positively with healthy adjustment (Benning et al., 2005; Latzman et al., 2019a, b, 2020). These findings have led some scholars (e.g., Lynam & Miller, 2012; Miller & Lynam, 2012; Sleep et al., 2019) to conclude boldness has little or no place in conceptualizations of psychopathy given that emotional stability and relative immunity to internalizing forms of psychopathology seem inconsistent with the notion that psychopathy is a mental disorder, which in turn should be marked by distress, impairment, or both (Spitzer & Endicott, 1978).

We believe this perspective is fallacious or at best incomplete, however, because it depends on the theoretical lens through which boldness is considered. Indeed, not only is the boldness construct clearly represented within prominent historic conceptions of psychopathy (i.e., Cleckley, 1976; Karpman, 1941; Kraepelin, 1904; Lykken, 1957, 1982), boldness has been reliably found to be robustly associated with numerous non-PCL-measures of psychopathy. These even include psychopathy assessment instruments developed by outspoken skeptics of the boldness construct, including the Elemental Psychopathy Assessment (Lynam et al., 2011) and the Psychopathy Resemblance Index (Lynam & Widiger, 2007) (e.g., Lilienfeld et al., 2016). Controversy aside, such findings are not especially surprising as boldness has been found to be deemed central to psychopathy by psychopathy scholars broadly, including both academicians and clinicians (Berg et al., 2017; Sörman et al., 2016).

It is further possible that dimensions of psychopathy function not only additively but configurally. Indeed, it is likely that the impact of psychopathy is most pronounced when considering configural patterns of trait dimensions (Lilienfeld et al., 2019). From this perspective, the presence of boldness, in conjunction with other trait dimensions may serve as a protective or risk factors for psychopathological outcomes (e.g., Latzman et al., 2020; Sellbom, 2015; Venables et al., 2015). For example, although the empirical evidence to date is still mixed, it is likely that whereas boldness is rarely pathological by itself, it may be that it is problematic in presence of certain other traits, such as disinhibition (poor impulse control). Along these lines, the inclusion of boldness within models of psychopathy is important when considering the distinction between psychopathy and antisocial personality disorder. The sole focus on antisocial outcomes restricts the nomological network to those already identified within the broader antisocial personality disorder diagnosis. It is only when one considers the traits that comprise boldness that one is able to differentiate the two conditions (Lilienfeld et al., 2015; Venables et al., 2015; Lilienfeld et al., 2016, 2018; Wall et al., 2015).

9.2.5 Is It Scientifically Acceptable to Rely Exclusively on Psychopathy Total Scores?

Numerous researchers rely exclusively on global (total) measures of psychopathy (e.g., Kosson, 1996). In doing so, they appear to assume that they gain valuable psychometric information by summing scores across psychopathy subdimensions. For example, in a recent meta-analysis, Poeppl et al. (2019) examined the functional brain imaging correlates of global psychopathy without reporting separate effect size data on psychopathy subdimensions (see Latzman et al., 2020, for a critique). Other authors commonly use brief measures of psychopathy that do not even provide information on subdimensions, as discussed in a later section. The exclusive reliance on psychopathy global scores is problematic for at least two reasons.

First, as Smith et al. (2009) observed in a useful analysis, the use of total scores on measures of personality and psychopathology can often obscure or dilute important associations that emerge when fine-grained associations at the lower-order level are examined. This appears to be the case for most if not all psychopathy measures (e.g., Watts et al., 2017). For example, PCL (and PCL-R) Factor I is largely unassociated or even slightly negatively associated with trait anxiety, whereas PCL (and PCL-R) Factor II is robustly positively associated with trait anxiety (e.g., Harpur et al., 1989). In some cases, the subdimensions of psychopathy measures fractionate in opposing directions, often resulting in near-zero associations when these subdimensions are collapsed into a total score (see Lilienfeld et al., 2015). For example, suicidal ideation and behavior tends to be negatively associated with PPI-R Factor I (Fearless Dominance) but positively associated with PPI-R Factor II (Self-centered Impulsivity; see Lilienfeld et al., 2019). This important finding is lost when PPI total scores, which effectively cancel out these diverging correlations, are used.

Second, numerous authors have reported cooperative (reciprocal) suppressor effects for psychopathy subdimensions. Such effects are especially pronounced for psychopathy measures whose subdimensions are robustly correlated, such as the PCL-R and LSRP (see Lilienfeld et al., 2019, for a review), thereby allowing sizeable suppressor effects to emerge. For example, the negative association between PCL-R Factor I and internalizing (e.g., anxiety, depression) symptoms tends to be become more pronounced following statistical control for Factor II scores, and the positive association between PCL-R Factor II and internalizing symptoms tends to become more pronounced following statistical control for Factor I scores (Blonigen et al., 2010). Such suppressor effects are potentially important theoretically, as they often point to the presence of conceptually distinctive processes underlying observed measures (Watson et al., 2013). Needless to say, the discovery of suppressor effects is precluded by sole reliance on total psychopathy scores.

9.2.6 Is it Scientifically Acceptable to Rely on Psychopathy Cut-Off Scores?

The manuals of several widely used psychopathy measures come with recommended cut-off scores for psychopathy. For example, the PCL-R manual recommends a cut-off score of 30 (out of 40) for psychopathy (Hare, 1991/2003), although some authors use lower cut-off scores for women or for European participants. Consistent with this recommendation, numerous researchers distinguish "psychopaths" and "non-psychopaths" by comparing high PCL-R scorers with low scorers or by comparing those above and below the standard PCL-R cut-off. Anecdotally, one of the authors on this chapter, who was a co-developer of the PPI-R (Lilienfeld & Widows, 2005), frequently receives inquiries from researchers and clinicians regarding what cut-off he suggests to distinguish psychopaths from nonpsychopaths. He routinely informs them that he does not use or recommend cut-off scores on his measure, a response that is commonly met with puzzlement, if not bewilderment.

Nevertheless, there is surprisingly little empirical foundation for the prevalent practice of using cut-off scores on psychopathy measures, let alone for the specific cut-offs recommended in psychopathy manuals. Taxometric analyses, which permit researchers to determine whether an observed distribution of scores is underpinned by two or more discrete distributions, have with few exceptions indicated that psychopathy is not a taxon, that is, a class (category) in nature. Instead, the latent structure of psychopathy appears to be dimensional (e.g., Edens et al., 2006; Edens et al., 2011; Guay et al., 2007; Marcus et al., 2004), meaning that the differences between psychopathic and nonpsychopathic individuals are of degree, not of kind.

The use of cut-off scores on psychopathy measures is psychometrically problematic as well. Dichotomizing (psychopaths vs. nonpsychopaths) or trichotimizing (high, medium, or low psychopathic individuals) scores on continuous measures generally results in a marked diminution of statistical power, especially when distributions are normal or quasi-normal (DeCoster et al., 2009; MacCallum et al., 2002). Consistent with this observation, a meta-analysis of the broader psychopathology literature revealed robust increases in reliability (15% on average) and construct validity (37% on average) when continuous as opposed to discrete measures are used (Markon et al., 2011). Because of the loss of statistical power that typically results when dichotomized or trichotomized scores are used, statistically significant results detected in such studies may be especially likely to be Type I errors (spurious findings) or genuine findings that are overestimated in magnitude (see Button et al., 2013, for an accessible discussion of this "winner's curse" phenomenon). At the same time, this loss of statistical power is likely to result in an abundance of Type II errors as well. Alternatively, researchers who compare high versus low psychopathy scorers are likely to introduce psychometric problems of their own. Extreme-groups designs tend to inflate variance, resulting in artificially large effect sizes. At the same time, if scores are normally distributed, the most reliable variance is typically observed around average scores per test information functions (Preacher et al., 2005). Furthermore, such designs presume, typically without sufficient evidence, a linear rather than curvilinear relation between the variable in question and other criteria (Preacher, 2014). Conceptually, extreme group designs are often difficult to interpret. If one finds that extremely high PCL-R or PPI-R scorers, for example, differ on a given laboratory variable from extremely low PCL-R or PPI-R scorers, does this necessarily mean that the "psychopathic' group is abnormal? Perhaps the abnormality lies with the low psychopathy group.

In sum, it seems likely that the widespread use of cut-off scores in the psychopathy literature has distorted the results and conclusions of meta-analyses in largely unknown ways. Hence, authors who use cut-off scores for the PCL-R and other psychopathy measures should always supplement their analyses with dimensional (continuous) analyses using the full range of psychopathy scores. Furthermore, reviewers and editors should insist that authors supplement categorical analyses of psychopathy measures with dimensional analyses.

9.2.7 Are the PCL-R and Allied Measures Unparalleled Predictors of Violence?

One of the reasons that the psychopathy construct and measures of it, the PCL-R in particular, has become so popular in forensic risk assessment is the assumption that these measures are excellent predictors of future violence. Indeed, many authors appear to assume that the PCL-R and its variants are distinctively, if not uniquely, suited for the prediction of violence. For example, one group of authors argued that the PCL-R appears to be an "unparalleled …measure for making risk assessments with white male inmates" (Salekin et al., 1996, p. 211) and a number of authors have advanced comparable assertions (see Gendreau et al., 2002, for a discussion). For example, one writer recently described the PCL-R's ability to predict violence as "unprecedented and unparalleled" (Helfgott, 2019, p. 8).

Nevertheless, although the PCL-R is a consistent and robust predictor of criminal violence and recidivism (Leistico et al., 2008; Salekin et al., 1996), the evidence that it is uniquely "talented" in this regard is slim to none. In a meta-analytic review of multiple risk assessment devices across 68 studies (N = 25,980), Singh et al. (2011) ranked the PCL-R among the lowest at predicting risk for general criminal offending, falling below a number of other widely used measures that incorporate past criminal history and allied risk variables. Similarly, Walters (2012) reported that the PCL-R facets did not display statistically significant incremental validity for forecasting general or violent criminal recidivism above and beyond age and prior criminal history. These findings remind us of the wisdom of Meehl's Maxim (named after influential clinical psychologist Paul Meehl), namely, that the best predictor of future behavior tends to be past behavior (Ruscio, 2004). More specifically, to the extent to which the PCL family of measures is helpful for predicting crime and reoffending, much of its validity and clinical utility may stem from its incorporation of criminal indicators (see also Walters et al., 2008), and in particular, on fourth antisocial facet. Indeed, the explicitly dispositional facets rarely add incrementally to the fourth "antisocial" facet in violence risk prediction (Walters et al., 2008), which follows Meehl's Maxim given its heavy saturation with criminality (see e.g., Cooke & Sellbom, 2019).

9.3   Part 2: Assessment Methods

Our second section considers issues that pertain to both clinical and research assessment methodology. In particular, we dispel myths and misconceptions concerning self- and informant-report measurement of psychopathy and consider the dangers of relying on brief measures of psychopathy and traditional scales embedded within clinical assessment instruments.

9.3.1   Are Self-Report Measures Inherently Problematic for Psychopathy Assessment?

For many years, scholars have argued that the self-report assessment[4] of psychopathy is highly problematic and more problematic than other modes of assessment (e.g., Hare, 1985; Harpur et al., 1989), citing a variety of concerns about dishonest responding that goes undetected, poor insight, and inability to self-reflect on emotions (e.g., guilt, shame, fear) with which psychopathic individuals have little experience. In fact, scant research supports these beliefs. Sellbom et al. (2018) addressed

---

[4] Self-report assessment refers to the practice by which individuals respond to questions about themselves, usually via questionnaires or interviews. In the current chapter, self-report refers exclusively to questionnaire-based assessment.

many of these misconceptions in detail; we summarize some of the most pertinent issues here.

A major misconception is that the validity of psychopathy scale scores assumes veridical responding (e.g., Lilienfeld, 1994), in that responses to test items need to be factually accurate and that lack of insight exacerbates this problem. For instance, Sellbom et al. (2018) noted the example of the PPI-R item "I often get blamed for things that aren't my fault." This item is hardly factually accurate for most individuals with high levels of psychopathy and also probably reflects poor insight in most (as they likely should get blamed for even more things that go wrong in their lives), but it is nevertheless a valid indicator of psychopathic individuals' failure to accept responsibility. Indeed, the extent to which item responses reflect a *distortion of* reality may in some cases afford valuable information regarding the skewed interpretations of individuals with certain personality (e.g., narcissistic, psychopathic) and psychotic (e.g., delusional) disorders. The insight that responses to self-report items need not be veridical to provide valid psychometric information dates back to a classic article by Meehl (1945), but appears to have been largely overlooked in the psychopathy literature. Furthermore, research has clearly shown strong convergence in both mean scores and predictive validity of self- and informant report scores on psychopathy measures (Miller et al., 2011; Jones & Miller, 2012), indicating that "lack of insight" may not necessarily be a problem in the self-report assessment of psychopathy. Still, it is entirely plausible that individuals with psychopathic features lack insight into the impact of their behavior on others, suggesting that researchers may want to supplement self-report indices with informant-based measures (see "Are Informant Reports Not Useful in Adult Psychopathy Assessment"?).

Another common misconception is that because individuals with psychopathic traits tend to lie, and that they are prone to positive impression management in their responding to self-report measures of psychopathy. This conjecture, however, is not compellingly supported by scientific evidence. Psychopathy scale scores tend to be *negatively* correlated with scores on social desirability and positive impression management scales (Marion et al., 2013; Ray et al., 2013). We do not view such findings as surprising, however. Most psychopathic personality traits, especially those tied to disinhibition and callousness, tend to be socially undesirable, and those high in them seem willing to report them, which is likely a reflection of their different conceptions of what is "normal" and "desirable." Furthermore, there is relatively little evidence that, at least in non-incentivized samples, such as those studied in research settings in which responses are confidential or anonymous, scores on social desirability measures diminish the validity of self-reported psychopathy scales (Watts et al., 2016).

A final misconception discussed here is that those with high scores on psychopathy measures are more skilled at manipulation and therefore more prone to getting away with dishonest responding on self-report questionnaires. As already noted, research has shown that psychopathy measures tend to be positively correlated with negative impression management in forensic psychological evaluations (Marion et al., 2013; see Ray et al., 2013, for a meta-analysis). There is no evidence, however, that individuals with high scores on psychopathy measures are *better* at it.

Marion et al. (2013) found that psychopathy scores did not moderate individuals' ability to feign good or bad compared with those responding honestly. If anything, those high in meanness traits were *worse* at avoiding detection when feigning good compared with those low in such traits.

9.3.2 Should Informant Reports Be Ignored in Adult Psychopathy Assessment?

Although it is clear from the existing literature that self-report methods can provide valid and reliable assessments of psychopathy, a slowly growing literature is examining the validity of informant-report-based assessments[5] of adult psychopathy. Indeed, whereas informant-report has been widely-used in the child and adolescent psychopathy literatures, informant reports have been insufficiently emphasized and rarely used in the assessment of adult psychopathy, which could give rise to the impression (and misconception) that they are not helpful in the assessment of psychopathy. Moreover, given recent research that has begun to consider the utility of informants in the assessment of psychopathy (for data on the incremental validity of informant reports beyond self-reports for personality pathology more broadly, see Fiefler et al., 2004), it is important to dispel potential misconceptions that might be partly impeding their more widespread use.

There exist a number of good reasons to incorporate informant-report-based methodologies in the assessment of psychopathy, including the possibility that informant data may help to circumvent "blind spots" in psychopathic individuals' self-reporting (e.g., Grove & Tellegen, 1991; Sellbom et al., 2018). Within this growing literature, however, the question of whether informant-report data provides an incremental contribution over self-report data is not clear. For example, in a student sample, Fowler and Lilienfeld (2007) found no incremental contribution of informant-reported psychopathy over self-reported psychopathy in the statistical prediction of self-reported antisocial behavior, although Jones and Miller (2012) found in a community sample that informant-reported psychopathy afforded incremental validity in the statistical prediction of a variety of externalizing behaviors. Weiss et al. (2018) found similar evidence of incremental validity in the prediction of relationship outcomes among newlyweds: partner-rated psychopathic traits provided substantial incremental validity beyond self-ratings in the prediction of marital functioning outcomes.

In addition, recent research has also confirmed a parallel factor-analytic structure in self-report and informant-report methodologies. Extending factor analytic work by Somma et al. (2019) on the structure of the Triarchic Psychopathy Measure

---

[5] Informant-based assessment refers to a measurement modality in which someone other than the person being evaluated, typically a friend, romantic partner, family member, or co-worker, reports on the person's psychological characteristics.

(TriPM; Patrick et al., 2009), Latzman and colleagues (2019a) examined model fit of an informant-report version of the TriPM. Model fit comparisons between the model fitted with informant-report data on "an individual they know well" and self-reported data in the original Somma et al. Italian-speaking sample and the Latzman et al. English-speaking sample revealed statistically comparable fit across all three samples. This finding demonstrates evidence of a common latent structure of psychopathy across informant modalities, at least as assessed by the TriPM.

9.3.3   Are Brief Measures Appropriate for the Assessment of Psychopathy?

According to William Shakespeare's ill-fated character Polonius in *Hamlet*, "brevity is the soul of wit." Brevity is not, however, necessarily the best way to operationalize a complex, multifactorial psychological construct, such as psychopathy. Although the underlying factor structure of psychopathy has been vigorously debated with respect to matters such as the view of antisocial behavior as either a component or consequence of the construct (Hare & Neumann, 2010; Skeem & Cooke, 2010) or the relevance of boldness (Miller & Lynam, 2012, 2015; Lilienfeld et al., 2012) as mentioned earlier, there is little debate that psychopathy is complex and multifaceted.

Self-report measures of psychopathy have gained increasing traction in recent years for a variety of reasons (see Sellbom et al., 2018 for a review), their ease of administration and time efficiency being only two advantages over more labor-intensive interview and file-based clinical ratings, such as the PCL-R. In this vein, some have developed very brief measures to further expediate and ease the collection of data for research purposes. Given the content and factorial complexity of the psychopathy, it is unlikely that brief scales can accomplish the broad coverage needed to assess the construct. To cover the expansive theoretical terrain of psychopathy in a reliable and content-valid manner, a sufficient number of items is required, along with multiple subscales assessing the diverse features of psychopathy in adequate depth (see also earlier discussion of using unitary or "total" scores for psychopathy). The most prominent examples come from the measurement of the so called "dark triad" (i.e., narcissism, Machiavellianism, and psychopathy; see Paulhus & Williams, 2002), which rely upon brief, unidimensional measurement of complex constructs. Two common examples are the Short Dark Triad (Jones & Paulhus, 2014) and the Dirty Dozen (Jonason & Webster, 2010). The latter is a particularly popular 12-item questionnaire, which assesses the psychopathy factor using only 4 items: *I tend to lack remorse. I tend to be unconcerned with the morality of my actions. I tend to be callous or insensitive. I tend to be cynical.* These 4 items clearly capture some manifestations of psychopathy, particularly antagonism, which is allied with meanness in the triarchic model (Patrick et al., 2009). However, it is not clear that they assess any of the other aforementioned aspects of psychopathy (e.g., boldness, disinhibition), a point demonstrated empirically by Miller et al. (2012). Although Jonason and Webster (2010) and Miller et al. (2012) reported good internal consistency for the Dirty Dozen Psychopathy scale, such reliability comes at the expense of construct validity, including content validity,[6] when viewed from the lens of the broader nomological network of the psychopathy construct. Shortcuts rarely pay off in science, and psychopathy assessment is no exception.

9.3.4 Is It Acceptable to Rely on Embedded Psychopathy Scales and Indices in Clinical Practice?

There are several highly popular clinical assessment tools that contain standard scales or indices that are purported to detect psychopathy with reasonably high fidelity. We focus on four commonly used instruments: the Minnesota Multiphasic Personality Inventory – 2 (MMPI-2; Butcher et al., 2001), Personality Assessment Inventory (PAI; Morey, 1991/2007), Millon Clinical Multiaxial Inventory – IV (MCMI-IV; Millon et al., 2015), and the Rorschach Inkblot Method. The first three are structured instruments that rely on a self-report format; the last is a performance-based instrument.

Since its inception, the MMPI has included a scale, Psychopathic deviate (Pd; Scale 4), that ostensibly measures psychopathy. Indeed, many early studies of psychopathy relied on this scale to detect this condition (e.g., Astin, 1961; Blanchard et al., 1977). Respected MMPI scholar Alex Caldwell, author of a widely used interpretative report, wrote relatively recently that *"[i]f the scale 4 (Pd) T-score is moderately to highly elevated (a bit over T-70), all of the components originally identified by McKinley and Hathaway (1944) as well as with the PCL–R and PCL–SV typically are clinically apparent, if not severely so"* (Caldwell, 2006, p. 195). Nevertheless, the empirical evidence does not support this inference. As early as 60 years ago, Lykken (1957) questioned the validity of this scale as an indicator of psychopathy, as it did not adequately differentiate offenders who were rated as psychopaths (based on Cleckley's, 1941, criteria) from nonpsychopathic offenders.

Hawk and Peterson (1974) observed that the Pd scale is a measure of general social deviance rather than psychopathic traits specifically. Moreover, Hare (1985) and Harpur et al. (1989) found only weak or at best modest associations between Scale 4 and the PCL. More recent research showed that Pd does not correlate meaningfully with affective and interpersonal psychopathy traits derived from the PPI (Lilienfeld, 1996; Sellbom et al., 2005) or the PCL:SV (Sellbom et al., 2007). Hence, we caution clinicians or researchers from equating elevated scores on MMPI

---

[6] Content validity refers to whether a test or scale provides sufficient content coverage to adequately canvas the universe of the theoretical construct being measured.

or MMPI-2 Pd with psychopathy; at most, these scores reflect a reasonably stable disposition toward antisocial behavior.

Morey (1991/2007) developed the Antisocial Features (ANT) scale for the PAI, and in particular, an ANT-Egocentricity subscale designed to capture the "pathological egocentricity and narcissism often thought to lie at the core of this disorder." (p. 112). Despite its conceptual promise, the ANT scale, including the ANT-E subscale, correlates primarily with the behavioral, but not affective or interpersonal, features of psychopathy on both the PCL-R (e.g., Douglas et al., 2007; Edens et al., 2000) and PPI (Patrick et al., 2006). Similarly, Blackburn (2007) found no evidence that any MCMI-II scale correlated significantly with PCL-R Factor 1 (which assesses the affective and interpersonal features of psychopathy) in a sample of male mentally disordered offenders, though the Antisocial scale evinced a moderate association with PCL-R Factor 2 (which assesses the impulsive and antisocial lifestyle features sometimes associated with psychopathy). Hart et al. (1991) reported similar findings in a sample of Canadian prison inmates, although the MCMI Sadistic scale displayed a moderate and significant correlation ($r = .28$) with PCL-R Factor 1. By and large, and similar to the MMPI-2 Pd and PAI ANT scales, there is little evidence that the MCMI-IV adequately detects the affective-interpersonal features of psychopathy, which many scholars deem to be "core" to the construct (e.g., Verschuere & te Kaat, 2019).

Finally, the Rorschach Inkblot Method (RIM) has been proposed as an alternative to other methods of psychopathy assessment that rely on self-reported information (e.g., Gacono & Meloy, 1994; Hartmann et al., 2006). The main premise is that the RIM can be particularly useful because its scores tap more unconscious or implicit processes relevant to psychopathy (e.g., self-concept, unconscious drives). Gacono and Meloy (2009, p. 571) went so far as to conclude that the RIM is "ideally suited" to the assessment of psychopathy, contrary to previous critiques of their research. In response, Wood et al. (2010) conducted a meta-analysis of 37 RIM scores, such as low number of texture responses, purported to differentiate psychopathy from non-psychopathy. The results were not supportive of the RIM for this purpose, with Wood et al. concluding, *"the relationship of Rorschach scores to psychopathy appears to be at best weak in both comparative and absolute terms. Overall, the present findings contradict the view that the Rorschach is a clinically useful instrument for discriminating psychopaths from nonpsychopaths in forensic settings"* (p. 346).

In summary, there is little evidence that classic scales and indices on common clinical assessment instruments can capture the full range of psychopathic personality traits. We suspect that this shortcoming is due largely to the multifaceted nature of psychopathy, which is challenging to capture using standalone and relatively brief measures (see earlier section). The classic trade-off between bandwidth and fidelity (Cronbach, 1960) probably applies here; high bandwidth (omnibus) measures such as the MMPI-2 and PAI are useful for capturing a broad array of psychopathological and personality traits in a relatively brief amount of time, but they may come at the cost of detecting complex disorders, such as psychopathy, with relatively low fidelity. We do note that the MMPI-2 Restructured Form (MMPI-2-RF;

Ben-Porath & Tellegen, 2008) can capture the full range of psychopathic traits, but it relies on numerous conceptually relevant scales to do so rather than one single measure (e.g., Sellbom et al., 2012; Wygant & Sellbom, 2012). Other clinical assessment instruments would be wise to follow this lead.

9.4   Summary and Conclusions

There are numerous myths, misconceptions, and fallacies with respect to psychopathy and its assessment. There are many more (e.g., Arkowitz & Lilienfeld, 2017; Berg et al., 2013; Furnham et al., 2009) but we confined our discussion to those errors that were most pertinent to its assessment. As we have indicated, there are long-held beliefs that the PCL-R is a "gold standard" assessment tool, perhaps even equivalent to the psychopathy construct itself, but such thinking runs afoul of basic principles of psychometrics. Moreover, the psychopathy literature is theoretically diverse, with various measurement modalities reflecting this diversity, and it is therefore important to be explicit regarding one's nomological network when evaluating assessment tools. Moreover, psychopathy scores are dimensional and multifaceted; as such, reliance on total scores (including on brief measures) is often ill-advised; so are cut scores, which are largely or entirely arbitrary and can lead to problematic statistical artifacts. Furthermore, both well-conceptualized self-report and informant report measures boast adequate construct validity, running counter to many myths and misconceptions that are not reflected in scientific observations. Finally, the PCL-R can yield useful information in violence risk assessment, but clinicians need to be aware that other standard risk assessment tools, as well as general violence history, are likely to be more potent risk factor indices than are psychopathy measures alone.

We encourage readers to rely on psychopathy assessments that exhibit robust construct validity for the purposes and populations with which they should be used; this recommendation is no different than it would be for any psychological test. We (as a field) need to move beyond the idea that one measurement modality is inherently superior to others, as there is scant evidence that clinician ratings, self-report, or informant reports consistently outperform one another, and instead focus on the measure that best maps onto the theoretical perspective of psychopathy to which one prescribes. (Of course, it is possible that certain modalities may be better suited than others for detecting certain psychopathic traits). We need to consider measurement in dimensional (trait severity) rather than categorical terms, and pay particular attention to constellations of psychopathic personality trait dimensions and richly describe individuals accordingly as opposed to discussing psychopathy in unitary terms. Overcoming myths, misconceptions, and fallacies in psychopathy assessment will only serve to improve research and clinical practice.

References

American Psychiatric Association. (1952). *Diagnostic and statistical manual of mental disorders* (1st ed.). Author.

American Psychiatric Association. (1968). *Diagnostic and statistical manual of mental disorders* (2nd ed.). Author.

American Psychiatric Association. (1980). *Diagnostic and statistical manual of mental disorders* (3rd ed.). Author.

American Psychiatric Association. (1987). *Diagnostic and statistical manual of mental disorders, revised* (3rd ed. Rev ed.) Author.

American Psychiatric Association. (1994). *Diagnostic and statistical manual of mental disorders* (4th ed.). Author.

American Psychiatric Association. (2013). *Diagnostic and statistical manual of mental disorders* (5th ed.). Author.

Anderson, J. L., Sellbom, M., Wygant, D. B., Salekin, R. T., & Krueger, R. F. (2014). Examining the associations between DSM-5 section III antisocial personality disorder traits and psychopathy in community and university samples. *Journal of Personality Disorders, 28*, 675–697.

Arkowitz, H., & Lilienfeld, S. O. (2017). *Facts and fictions in mental health*. Wiley.

Astin, A. W. (1961). A note on the MMPI psychopathic deviate scale. *Educational and Psychological Measurement, 21*, 895–897.

Baskin-Sommers, A. R., Newman, J. P., Sathasivam, N., & Curtin, J. J. (2011). Evaluating the generalizability of a fear deficit in psychopathic African American offenders. *Journal of Abnormal Psychology, 120*, 71–78.

Benning, S. D., Patrick, C. J., Blonigen, D. M., Hicks, B. M., & Iacono, W. G. (2005). Estimating facets of psychopathy from normal personality traits: A step toward community epidemiological investigations. *Assessment, 12*, 3–18.

Ben-Porath, Y. S., & Tellegen, A. (2008). *MMPI-2-RF: Manual for administration, scoring, and interpretation*. University of Minnesota Press.

Berg, J. M., Lilienfeld, S. O., & Sellbom, M. (2017). The role of boldness in psychopathy: A study of academic and clinical perceptions. *Personality Disorders: Theory, Research, and Treatment, 8*, 319–328.

Berg, J. M., Smith, S. F., Watts, A. L., Ammirati, R., Green, S. E., & Lilienfeld, S. O. (2013). Misconceptions regarding psychopathic personality: Implications for clinical practice and research. *Neuropsychiatry, 3*, 63–74.

Blonigen, D. M., Patrick, C. J., Douglas, K. S., Poythress, N. G., Skeem, J. L., Lilienfeld, S. O., ... Krueger, R. F. (2010). Multimethod assessment of psychopathy in relation to factors of internalizing and externalizing from the Personality Assessment Inventory: The impact of method variance and suppressor effects. *Psychological Assessment, 22*, 96–107.

Blackburn, R. (2007). Personality disorder and psychopathy: Conceptual and empirical integration. *Psychology, Crime & Law, 13*(1), 7–18.

Blanchard, E. B., Bassett, J. E., & Koshland, E. (1977). Psychopathy and delay of gratification. *Correctional Psychologist, 4*(3), 265–271.

Butcher, J. N., Graham, J. R., Ben-Porath, Y. S., Tellegen, A., Dahlstrom, W. G., & Kaemmer, B. (2001). *MMPI-2: Manual for administration, scoring, and interpretation* (Rev. ed.). University of Minnesota.

Button, K. S., Ioannidis, J. P., Mokrysz, C., Nosek, B. A., Flint, J., Robinson, E. S., & Munafò, M. R. (2013). Power failure: Why small sample size undermines the reliability of neuroscience. *Nature Reviews Neuroscience, 14*, 365–376.

Caldwell, A. B. (2006). Maximal measurement or meaningful measurement: The interpretive challenges of the MMPI-2 Restructured Clinical (RC) scales. *Journal of Personality Assessment, 87*(2), 193–201.

Camp, J. P., Skeem, J. L., Barchard, K., Lilienfeld, S. O., & Poythress, N. G. (2013). Psychopathic predators? Getting specific about the relation between psychopathy and violence. *Journal of Consulting and Clinical Psychology, 81*, 467–480.

Cleckley, H. (1976). *The mask of sanity* (5th ed.). Mosby. (Original work published 1941).

Cook, T. D., & Campbell, D. T. (1979). *Quasi-experimentation, design and analysis issues for field settings*. Rand McNally College Publishing Company.

Cooke, D. J. (2018). Psychopathic Personality Disorder: Capturing an elusive concept. *European Journal of Analytic Philosophy, 14*(1), 1–15.

Cooke, D. J., & Sellbom, M. (2019). An examination of Psychopathy Checklist-Revised latent factor structure via exploratory structural equation modeling. *Psychological Assessment, 31*(5), 581–591.

Crego, C., & Widiger, T. A. (2014). Psychopathy and the DSM. *Journal of Personality, 83*, 665–677.

Cronbach, L. J. (1960). *Essentials of psychological testing* (2nd ed.). Harper & Row.

Cronbach, L. J., & Meehl, P. E. (1955). Construct validity in psychological tests. *Psychological Bulletin, 52*, 281–302.

DeCoster, J., Iselin, A. M. R., & Gallucci, M. (2009). A conceptual and empirical examination of justifications for dichotomization. *Psychological Methods, 14*, 349–366.

Douglas, K. S., Guy, L. S., Edens, J. F., Boer, D. P., & Hamilton, J. (2007). The Personality Assessment Inventory as a proxy for the Psychopathy Checklist–Revised: Testing the incremental validity and cross-sample robustness of the Antisocial Features Scale. *Assessment, 14*(3), 255–269.

Edens, J. F., Hart, S. D., Johnson, D. W., Johnson, J. K., & Olver, M. E. (2000). Use of the Personality Assessment Inventory to assess psychopathy in offender populations. *Psychological Assessment, 12*(2), 132–139.

Edens, J. F., Marcus, D. K., Lilienfeld, S. O., & Poythress, N. G., Jr. (2006). Psychopathic, not psychopath: Taxometric evidence for the dimensional structure of psychopathy. *Journal of Abnormal Psychology, 115*, 131–144.

Edens, J. F., Marcus, D. K., & Vaughn, M. G. (2011). Exploring the taxometric status of psychopathy among youthful offenders: Is there a juvenile psychopath taxon? *Law and Human Behavior, 35*, 13–24.

Ermer, E., Kahn, R. E., Salovey, P., & Kiehl, K. A. (2012). Emotional intelligence in incarcerated men with psychopathic traits. *Journal of Personality and Social Psychology, 103*, 194–204.

Evans, L., & Tully, R. J. (2016). The triarchic psychopathy measure (TriPM): Alternative to the PCL-R? *Aggression and Violent Behavior, 27*, 79–86.

Faraone, S. V., & Tsuang, M. T. (1994). Measuring diagnostic accuracy in the absence of a "gold standard". *American Journal of Psychiatry, 151*, 650–657.

Few, L. R., Lynam, D. R., Maples, J. L., MacKillop, J., & Miller, J. D. (2015). Comparing the utility of DSM-5 Section II and III antisocial personality disorder diagnostic approaches for capturing psychopathic traits. *Personality Disorders: Theory, Research, and Treatment, 6*, 64–74.

Fiefler, E. R., Oltmanns, T. F., & Turkheimer, E. (2004). Traits associated with personality disorders and adjustment to military life: Predictive validity of self and peer reports. *Military Medicine, 169*, 207–211.

Fowler, K. A., & Lilienfeld, S. O. (2007). The psychopathy Q-sort: Construct validity evidence in a nonclinical sample. *Assessment, 14*, 75–79.

Furnham, A., Daoud, Y., & Swami, V. (2009). How to spot a psychopath. *Social Psychiatry and Psychiatric Epidemiology, 44*, 464–472.

Gacono, C. B., & Meloy, J. R. (1994). *The Rorschach assessment of aggressive and psychopathic personalities*. Erlbaum.

Gacono, C. B., & Meloy, J. R. (2009). Assessing antisocial and psychopathic personalities. In J. N. Butcher (Ed.), *Oxford handbook of personality assessment* (pp. 567–581). Oxford University Press.

Gendreau, P., Goggin, C., & Smith, P. (2002). Is the PCL-R really the "unparalleled" measure of offender risk? A lesson in knowledge cumulation. *Criminal Justice and Behavior, 29*, 397–426.

Grove, W. M., & Tellegen, A. (1991). Problems in the classification of personality disorders. *Journal of Personality Disorders, 5*, 31–41.

Guay, J.-P., Ruscio, J., Knight, R. A., & Hare, R. D. (2007). A taxometric analysis of the latent structure of psychopathy: Evidence for dimensionality. *Journal of Abnormal Psychology, 116*, 701–716.

Hare, R. D. (1980). A research scale for the assessment of psychopathy in criminal populations. *Personality and Individual Differences, 1*, 111–119.

Hare, R. D. (1985). Comparison of procedures for the assessment of psychopathy. *Journal of Consulting and Clinical Psychology, 53*(1), 7–16.

Hare, R. D. (1991/2003). *Manual for the Revised Psychopathy Checklist* (2nd ed.). Multi-Health Systems.

Hare, R. D. (1996). Psychopathy: A clinical construct whose time has come. *Criminal Justice and Behavior, 23*(1), 25–54.

Hare, R. D., Hart, S. D., & Harpur, T. J. (1991). Psychopathy and the DSM-IV criteria for antisocial personality disorder. *Journal of Abnormal Psychology, 100*, 391–398.

Hare, R. D., & Neumann, C. S. (2005). Structural models of psychopathy. *Current Psychiatry Reports, 7*, 57–64.

Hare, R. D., & Neumann, C. S. (2008). Psychopathy as a clinical and empirical construct. *Annual Review of Clinical Psychology, 4*, 217–246.

Hare, R. D., & Neumann, C. S. (2010). The role of antisociality in the psychopathy construct: Comment on Skeem and Cooke (2010). *Psychological Assessment, 22*, 446–454.

Harpur, T. J., Hakstian, A. R., & Hare, R. D. (1988). Factor structure of the Psychopathy Checklist. *Journal of Consulting and Clinical Psychology, 56*, 741–747.

Harpur, T. J., Hare, R. D., & Hakstian, A. R. (1989). Two-factor conceptualization of psychopathy: Construct validity and assessment implications. *Psychological Assessment: A Journal of Consulting and Clinical Psychology, 1*, 6–17.

Hart, S. D., Forth, A. E., & Hare, R. D. (1991). The MCMI-II and psychopathy. *Journal of Personality Disorders, 5*(4), 318–327.

Hartmann, E., Nørbech, P. B., & Grønnerød, C. (2006). Psychopathic and nonpsychopathic violent offenders on the Rorschach: Discriminative features and comparisons with schizophrenic inpatient and university student samples. *Journal of Personality Assessment, 86*(3), 291–305.

Hawk, S. S., & Peterson, R. A. (1974). Do MMPI psychopathic deviancy scores reflect psychopathic deviancy or just deviancy? *Journal of Personality Assessment, 38*(4), 362–368.

Helfgott, J. B. (2019). *No remorse: Psychopathy and criminal justice*. Praeger.

Hyde, L. W., Byrd, A. L., Votruba-Drzal, E., Hariri, A. R., & Manuck, S. B. (2014). Amygdala reactivity and negative emotionality: Divergent correlates of antisocial personality and psychopathy traits in a community sample. *Journal of Abnormal Psychology, 123*, 214–224.

Jonason, P. K., & Webster, G. D. (2010). The Dirty Dozen: A concise measure of the Dark Triad. *Psychological Assessment, 22*, 420–432.

Jones, D. N., & Paulhus, D. L. (2014). Introducing the short dark triad (SD3) a brief measure of dark personality traits. *Assessment, 21*(1), 28–41.

Jones, S., & Miller, J. D. (2012). Psychopathic traits and externalizing behaviors: A comparison of self- and informant reports in the statistical prediction of externalizing behaviors. *Psychological Assessment, 24*, 255–260.

Karpman, B. (1941). On the need of separating psychopathy into two distinct clinical types: The symptomatic and the idiopathic. *Journal of Criminal Psychopathology, 3*, 112–137.

Karpman, B. (1946). A yardstick for measuring psychopathy. *Federal Probation, 10*, 26–31.

Kendler, K. S., Muñoz, R. A., & Murphy, G. (2009). The development of the Feighner criteria: A historical perspective. *American Journal of Psychiatry, 167*, 134–142.

Kosson, D. S. (1996). Psychopathy and dual-task performance under focusing conditions. *Journal of Abnormal Psychology, 105*, 391–400.

Kraepelin E. (1904). *Psychiatrie. Ein Lehrbuch fur Studierende und Artzte. 7. Auflage. II. Band. Klinische Psychiatrie.* Leipzig, Germany

Latzman, R. D., Palumbo, I. M., Krueger, R. F., Drislane, L. E., & Patrick, C. J. (2020). Modeling relations between triarchic biobehavioral traits and DSM internalizing disorder dimensions. *Assessment, 27*(6), 1100–1115.

Latzman, R. D., Palumbo, I. M., Sauvigné, K. C., Hecht, L. K., Patrick, C. J., & Lilienfeld, S. O. (2019a). Psychopathy and internalizing psychopathology: A triarchic model perspective. *Journal of Personality Disorders, 33*, 262–287.

Latzman, R. D., Patrick, C. J., & Lilienfeld, S. O. (2019b). Heterogeneity matters: implications for Poeppl et al.'s (2019) meta-analysis and future neuroimaging research on psychopathy. *Molecular Psychiatry.*

Leistico, A. M. R., Salekin, R. T., DeCoster, J., & Rogers, R. (2008). A large-scale meta-analysis relating the Hare measures of psychopathy to antisocial conduct. *Law and Human Behavior, 32*, 28–45.

Levenson, M. R., Kiehl, K. A., & Fitzpatrick, C. M. (1995). Assessing psychopathic attributes in a noninstitutionalized population. *Journal of Personality and Social Psychology, 68*(1), 151–158.

Lilienfeld, S. O. (1994). Conceptual problems in the assessment of psychopathy. *Clinical Psychology Review, 14*, 17–38.

Lilienfeld, S. O. (1996). The MMPI—2 Antisocial Practices Content Scale: Construct validity and comparison with the Psychopathic Deviate Scale. *Psychological Assessment, 8*(3), 281–293.

Lilienfeld, S. O., & Andrews, B. P. (1996). Development and preliminary validation of a self-report measure of psychopathic personality traits in noncriminal population. *Journal of Personality Assessment, 66*, 488–524.

Lilienfeld, S. O., Patrick, C. J., Benning, S. D., Berg, J., Sellbom, M., & Edens, J. F. (2012). The role of fearless dominance in psychopathy: Confusions, controversies, and clarifications. *Personality Disorders: Theory, Research, and Treatment, 3*, 327–340.

Lilienfeld, S. O., Smith, S. F., Sauvigné, K. C., Patrick, C. J., Drislane, L. E., Latzman, R. D., & Krueger, R. F. (2016). Is boldness relevant to psychopathic personality? Meta-analytic relations with non-Psychopathy Checklist-based measures of psychopathy. *Psychological Assessment, 28*, 1172–1184.

Lilienfeld, S. O., Watts, A. L., Murphy, B., Costello, T. H., Bowes, S. M., Smith, S. F., … & Tabb, K. (2019). Personality disorders as emergent interpersonal syndromes: Psychopathic personality as a case example. *Journal of Personality Disorders, 33*(5), 577–622.

Lilienfeld, S. O., Watts, A. L., Smith, S. F., Berg, J. M., & Latzman, R. D. (2015). Psychopathy deconstructed and reconstructed: Identifying and assembling the personality building blocks of Cleckley's chimera. *Journal of Personality, 83*, 593–610.

Lilienfeld, S. O., Watts, A. L., Smith, S. F., & Latzman, R. D. (2018). Boldness: Conceptual and methodological issues. In C. J. Patrick (Ed.), *Handbook of psychopathy* (2nd ed., pp. 165–188). The Guilford Press.

Lilienfeld, S. O., & Widows, M. R. (2005). *Manual for the psychopathic personality inventory-revised* (PPI-R). Psychological Assessment Resources.

Loevinger, J. (1957). Objective tests as instruments of psychological theory. *Psychological Reports, 3*, 635–694.

Lykken, D. T. (1957). A study of anxiety in the sociopathic personality. *Journal of Abnormal and Social Psychology, 55*, 610.

Lykken, D. T. (1982). Fearlessness: Its carefree charm and deadly risks. *Psychology Today, 16*(9), 20–28.

Lykken, D. T. (1995). *The antisocial personalities.* Erlbaum.

Lynam, D. R., Gaughan, E. T., Miller, J. D., Miller, D. J., Mullins-Sweatt, S., & Widiger, T. A. (2011). Assessing the basic traits associated with psychopathy: Development and validation of the Elemental Psychopathy Assessment. *Psychological Assessment, 23*, 108–124.

Lynam, D. R., & Miller, J. D. (2012). Fearless dominance and psychopathy: Response to Lilienfeld et al. *Personality Disorders: Theory, Research, and Treatment, 3*, 341–353.

Lynam, D. R., & Widiger, T. A. (2007). Using a general model of personality to identify the basic elements of psychopathy. *Journal of Personality Disorders, 21*, 160–178.

MacCallum, R. C., Zhang, S., Preacher, K. J., & Rucker, D. D. (2002). On the practice of dichotomization of quantitative variables. *Psychological Methods, 7*, 19–40.

Malterer, M. B., Lilienfeld, S. O., Neumann, C. S., & Newman, J. P. (2010). Concurrent validity of the Psychopathic Personality Inventory with offender and community samples. *Assessment, 17*, 3–15.

Marcus, D. K., Fulton, J. J., & Edens, J. F. (2013). The two-factor model of psychopathic personality: Evidence from the Psychopathic Personality Inventory. *Personality Disorders: Theory, Research, and Treatment, 4*, 67–78.

Marcus, D. K., John, S. L., & Edens, J. F. (2004). A taxometric analysis of psychopathic personality. *Journal of Abnormal Psychology, 113*, 626–635.

Marion, B. E., Sellbom, M., Salekin, R. T., Toomey, J. A., Kucharski, L. T., & Duncan, S. (2013). An examination of the association between psychopathy and dissimulation using the MMPI-2-RF validity scales. *Law and Human Behavior, 37*(4), 219.

Markon, K. E., Chmielewski, M., & Miller, C. J. (2011). The reliability and validity of discrete and continuous measures of psychopathology: A quantitative review. *Psychological Bulletin, 137*, 856–879.

McCord, W., & McCord, J. (1964). *The psychopath: An essay on the criminal mind*. D. Van Nostrand.

Meehl, P. E. (1945). The dynamics of "structured" personality tests. *Journal of Clinical Psychology, 1*, 296–303.

Miller, J. D., Few, L. R., Seibert, L. A., Watts, A., Zeichner, A., & Lynam, D. R. (2012). An examination of the Dirty Dozen measure of psychopathy: A cautionary tale about the costs of brief measures. *Psychological Assessment, 24*, 1048–1053.

Miller, J. D., Jones, S. E., & Lynam, D. R. (2011). Psychopathic traits from the perspective of self and informant reports: Is there evidence for a lack of insight? *Journal of Abnormal Psychology, 120*(3), 758–764.

Miller, J. D., & Lynam, D. R. (2012). An examination of the Psychopathic Personality Inventory's nomological network: A meta-analytic review. *Personality Disorders: Theory, Research, and Treatment, 3*, 305–326.

Miller, J. D., & Lynam, D. R. (2015). Psychopathy and personality: Advances and debates. *Journal of Personality, 83*, 585–592.

Millon, T., Grossman, S., & Millon, C. (2015). *Millon clinical multiaxial inventory–IV manual*. Pearson Assessments.

Morey, L. C. (1991/2007). *Personality Assessment Inventory professional manual*. Psychological Assessment Resources.

Patrick, C. J. (2010). Triarchic psychopathy measure (TriPM). PhenX toolkit online assessment catalog. Retrieved from https://www.phenxtoolkit.org/index.php?pageLink_browse.protocold etails&id_121601.

Patrick, C. J. (Ed.). (2018). *Handbook of psychopathy* (2nd ed.). Guilford Press.

Patrick, C. J., Edens, J. F., Poythress, N. G., Lilienfeld, S. O., & Benning, S. D. (2006). Construct validity of the Psychopathic Personality Inventory two-factor model with offenders. *Psychological Assessment, 18*(2), 204.

Patrick, C. J., Fowles, D. C., & Krueger, R. F. (2009). Triarchic conceptualization of psychopathy: Developmental origins of disinhibition, boldness, and meanness. *Development and Psychopathology, 21*, 913–938.

Paulhus, D. L., & Williams, K. M. (2002). The dark triad of personality: Narcissism, Machiavellianism, and psychopathy. *Journal of Research in Personality, 36*, 556–563.

Poeppl, T. B., Donges, M. R., Mokros, A., Rupprecht, R., Fox, P. T., Laird, A. R., … Eickhoff, S. B. (2019). A view behind the mask of sanity: Meta-analysis of aberrant brain activity in psychopaths. *Molecular Psychiatry, 24*, 463–470.

Poythress, N. G., Lilienfeld, S. O., Skeem, J. L., Douglas, K. S., Edens, J. F., Epstein, M., & Patrick, C. J. (2010). Using the PCL-R to help estimate the validity of two self-report measures of psychopathy with offenders. *Assessment, 17*, 206–219.

Preacher, K. J. (2014). Extreme groups designs. In R. L. Cautin & S. O. Lilienfeld (Eds.), *The encyclopedia of clinical psychology* (pp. 1189–1192). Wiley.

Preacher, K. J., Rucker, D. D., MacCallum, R. C., & Nicewander, W. A. (2005). Use of the extreme groups approach: A critical reexamination and new recommendations. *Psychological Methods, 10*(2), 178–192.

Ray, J. V., Hall, J., Rivera-Hudson, N., Poythress, N. G., Lilienfeld, S. O., & Morano, M. (2013). The relation between self-reported psychopathic traits and distorted response styles: A meta-analytic review. *Personality Disorders: Theory, Research, and Treatment, 4*(1), 1–14.

Robins, E., & Guze, S. B. (1970). Establishment of diagnostic validity in psychiatric illness: Its application to schizophrenia. *American Journal of Psychiatry, 126*, 983–987.

Robins, L. N. (1966). *Deviant children grown up: A sociological and psychiatric study of sociopathic personality*. Williams & Wilkins.

Robins, L. N., & Barrett, J. E. (Eds.) (1989). *The validity of psychiatric diagnosis*. Raven Press.

Ruscio, J. (2004). Diagnoses and the behaviors they denote. *The Scientific Review of Mental Health Practice, 3*, 1–5.

Salekin, R. T., Rogers, R., & Sewell, K. W. (1996). A review and meta-analysis of the psychopathy checklist and psychopathy checklist-revised: Predictive validity of dangerousness. *Clinical Psychology: Science and Practice, 3*, 203–215.

Sellbom, M. (2011). Elaborating on the construct validity of the Levenson Self-Report Psychopathy Scale in incarcerated and non-incarcerated samples. *Law and Human Behavior, 35*, 440–451.

Sellbom, M. (2015). Elucidating the complex associations between psychopathy and post-traumatic stress disorder from the perspective of trait negative affectivity. *International Journal of Forensic Mental Health, 14*, 85–92.

Sellbom, M. (2019). The Triarchic psychopathy model: Theory and measurement. In M. DeLisi (Ed.), *Routledge international handbook of psychopathy and crime* (pp. 241–264). Routledge.

Sellbom, M., Ben-Porath, Y. S., Lilienfeld, S. O., Patrick, C. J., & Graham, J. R. (2005). Assessing psychopathic personality traits with the MMPI-2. *Journal of Personality Assessment, 85*(3), 334–343.

Sellbom, M., Ben-Porath, Y. S., & Stafford, K. P. (2007). A comparison of MMPI—2 measures of psychopathic deviance in a forensic setting. *Psychological Assessment, 19*(4), 430–436.

Sellbom, M., Ben-Porath, Y. S., Patrick, C. J., Wygant, D. B., Gartland, D. M., & Stafford, K. P. (2012). Development and construct validation of MMPI-2-RF indices of global psychopathy, fearless-dominance, and impulsive-antisociality. *Personality Disorders: Theory, Research, and Treatment, 3*(1), 17–38.

Sellbom, M., & Boer, D. (2019). Psychopathy vs antisocial personality disorder. In R. D. Morgan (Ed.), *The SAGE encyclopedia of criminal psychology*. Sage.

Sellbom, M., Lilienfeld, S. O., Fowler, K. A., & McCrary, K. L. (2018). The self-report assessment of psychopathy: Challenges, pitfalls, and promises. In C. J. Patrick (Ed.), *Handbook of psychopathy* (2nd ed., pp. 211–258). Guilford Press.

Sellbom, M., & Phillips, T. R. (2013). An examination of the triarchic conceptualization of psychopathy in incarcerated and nonincarcerated samples. *Journal of Abnormal Psychology, 122*, 208–214.

Singh, J. P., Grann, M., & Fazel, S. (2011). A comparative study of violence risk assessment tools: A systematic review and metaregression analysis of 68 studies involving 25, 980 participants. *Clinical Psychology Review, 31*, 499–513.

Skeem, J. L., & Cooke, D. J. (2010). Is criminal behavior a central component of psychopathy? Conceptual directions for resolving the debate. *Psychological Assessment, 22*, 433–445.

Sleep, C. E., Weiss, B., Lynam, D. R., & Miller, J. D. (2019). An examination of the Triarchic Model of psychopathy's nomological network: A meta-analytic review. *Clinical Psychology Review, 71*, 1–26.

Sleep, C. E., Wygant, D. B., & Miller, J. D. (2018). Examining the incremental utility of DSM-5 section III traits and impairment in relation to traditional personality disorder scores in a female correctional sample. *Journal of Personality Disorders, 32*, 738–752.

Smith, G. T., McCarthy, D. M., & Zapolski, T. C. B. (2009). On the value of homogeneous constructs for construct validation, theory testing, and the description of psychopathology. *Psychological Assessment, 21*, 272–284.

Somma, A., Borroni, S., Drislane, L. E., Patrick, C. J., & Fossati, A. (2019). Modeling the structure of the triarchic psychopathy measure: Conceptual, empirical, and analytic considerations. *Journal of Personality Disorders, 33*(4), 470–496.

Spitzer, R. L., & Endicott, J. (1978). Medical and mental disorder: Proposed definition and criteria. In R. L. Spitzer & D. F. Klein (Eds.), *Critical issues in psychiatric diagnosis* (pp. 15–39). Raven Press.

Sörman, K., Edens, J. F., Smith, S. T., Clark, J. W., Kristiansson, M., & Svensson, O. (2016). Boldness and its relation to psychopathic personality: Prototypicality analyses among forensic mental health, criminal justice, and layperson raters. *Law and Human Behavior, 40*, 337–349.

Thorndike, E. L. (1918). The nature, purposes, and general methods of measurements of educational products. In G. M. Whipple (Ed.), *Seventeenth yearbook of the national society for the study of education* (pp. 16–24). Bloomington Public School Publishing.

Venables, N. C., Sellbom, M., Sourander, A., Kendler, K., Joiner, T. E., Drislane, L. E., Sillanmäki, L., Elonheimo, H., Parkkola, K., Multimaki, P., & Patrick, C. J. (2015). RDoC separate and interactive contributions of weak inhibitory control and threat sensitivity to prediction of suicide risk. *Psychiatry Research, 226*, 461–466.

Verschuere, B., & te Kaat, L. (2019). What are the Core features of psychopathy? A prototypicality analysis using the psychopathy checklist-revised (PCL-R). *Journal of Personality Disorders, 33*, 1–9.

Wall, T. D., Wygant, D. B., & Sellbom, M. (2015). Boldness explains a key difference between psychopathy and antisocial personality disorder. *Psychiatry, Psychology and Law, 22*, 94–105.

Walters, G. D. (2012). Psychopathy and crime: Testing the incremental validity of PCL-R-measured psychopathy as a predictor of general and violent recidivism. *Law and Human Behavior, 36*, 404–412.

Walters, G. D., Knight, R. A., Grann, M., & Dahle, K.-P. (2008). Incremental validity of the Psychopathy Checklist facet scores: Predicting release outcome in six samples. *Journal of Abnormal Psychology, 117*, 396–405.

Watson, D., Clark, L. A., Chmielewski, M., & Kotov, R. (2013). The value of suppressor effects in explicating the construct validity of symptom measures. *Psychological Assessment, 25*(3), 929–941.

Watts, A. L., Lilienfeld, S. O., Edens, J. F., Douglas, K. S., Skeem, J. L., Verschuere, B., & LoPilato, A. C. (2016). Does response distortion statistically affect the relations between self-report psychopathy measures and external criteria? *Psychological Assessment, 28*, 294–306.

Watts, A. L., Waldman, I. D., Smith, S. F., Poore, H. E., & Lilienfeld, S. O. (2017). The nature and correlates of the dark triad: The answers depend on the questions. *Journal of Abnormal Psychology, 126*, 951–968.

Weiss, B., Lavner, J. A., & Miller, J. D. (2018). Self- and partner-reported psychopathic traits' relations with couples' communication, marital satisfaction trajectories, and divorce in a longitudinal sample. *Personality Disorders: Theory, Research, and Treatment, 9*(3), 239–249.

Westen, D., & Rosenthal, R. (2005). Improving construct validity: Cronbach, Meehl, and Neurath's ship: Comment. *Psychological Assessment, 17*, 409–412.

Widiger, T. A., & Crego, C. (2018). Psychopathy and DSM-5 psychopathology. In C. J. Patrick (Ed.), *Handbook of psychopathy* (pp. 281–296). The Guilford Press.

Witt, E. A., Donnellan, M. B., & Blonigen, D. M. (2009). Using existing self-report inventories to measure the psychopathic personality traits of fearless dominance and impulsive antisociality. *Journal of Research in Personality, 43*, 1006–1016.

Wood, J. M., Lilienfeld, S. O., Nezworski, M. T., Garb, H. N., Allen, K. H., & Wildermuth, J. L. (2010). Validity of Rorschach Inkblot scores for discriminating psychopaths from nonpsychopaths in forensic populations: A meta-analysis. *Psychological Assessment, 22*(2), 336–349.

Wygant, D. B., & Sellbom, M. (2012). Viewing psychopathy from the perspective of the personality psychopathology five model: Implications for DSM-5. *Journal of Personality Disorders, 26*, 717–726.

Wygant, D. B., Sellbom, M., Sleep, C. E., Wall, T. D., Applegate, K. C., Krueger, R. F., & Patrick, C. J. (2016). Examining the DSM-5 section III operationalization of antisocial personality disorder and psychopathy in a male correctional sample. *Personality Disorders: Theory, Research, and Treatment, 7*, 229–239.

Chapter 10
Psychopathy as a Scientific Kind: On Usefulness and Underpinnings

Thomas A. C. Reydon

Abstract This chapter examines the status of psychopathy as a scientific kind. I argue that the debate on the question whether psychopathy is a scientific kind as it is conducted at present (i.e., by asking whether psychopathy is a natural kind), is misguided. It relies too much on traditional philosophical views of what natural kinds (or: legitimate scientific kinds) are and how such kinds perform epistemic roles in the sciences. The paper introduces an alternative approach to the question what scientific (or: natural) kinds are. On this alternative approach, the *Grounded Functionality Account* of natural kinds, psychopathy emerges as a "good" scientific kind that is best understood as a region on a multidimensional space of behaviors rather than as a traditional natural kind.

Keywords Grounded functionality account · Natural kinds · Psychiatric kinds · Psychopathy · Scientific kinds

10.1 Introduction

Diagnosing a person as suffering from psychopathy – that is, as "being a psychopath" – seems to amount to placing that person into a particular category of people that supposedly share certain symptoms (i.e., a set of behavioral traits) as well as a common set of causes that underlie those symptoms. In the same way as for classifications in other domains of science and everyday practice, it seems that the grouping of people in the category of psychopaths is intended to serve a variety of epistemic functions in the biomedical sciences and in clinical practice. Presumably, it is used to serve certain functions in juridical settings and everyday contexts too

T. A. C. Reydon (✉)
Institute of Philosophy & Centre for Ethics and Law in the Life Sciences (CELLS),
Leibniz University Hannover, Hannover, Germany
e-mail: reydon@ww.uni-hannover.de; https://www.reydon.info/

(see, e.g., Malatesti & McMillan, 2014), but for reasons of space I shall ignore those here.

Important (but certainly not the only) epistemic functions of scientific classifications are the making of inferences (knowing the properties that the observed members of a kind share allows us to make reliable inferences about as yet unobserved members of the kind) and the construction of explanations (knowing that an entity is a member of a particular kind explains its behavior). Accordingly, the grouping of persons in the category of psychopaths is supposed to allow researchers, clinicians, aid workers, and so on to make predictions about how a person falling into the category will behave under particular circumstances, as well as how the person's behavior might be changed by means of therapies or other measures (Brazil et al., 2018). Thus, the category of psychopathy is supposed to serve researchers, clinicians, aid workers, and so on as a basis for investigations into the causes of psychopathy and possible ways of intervention.

A set of questions that now arises, is the following: Does the category of psychopathy indeed successfully serve the purposes that it is intended to serve? And if it does, *how* does it succeed? More generally, what makes classifications and the kinds that feature in them suitable to perform their epistemic functions, that is, what makes some classifications and kind suitable to stand at the focus of scientific investigations, and others much less so (Brazil et al., 2018)?[1] These are questions that have a long history of discussion in the philosophy of science, as well as analytic philosophy more broadly. A traditional way of approaching these questions is by using the philosophical theory of natural kinds as a tool for the analysis of individual cases and distinguishing "good" scientific kinds – i.e., kinds of things, phenomena, etc. that represent groups in nature – and thus can stand at the focus of scientific research – from groupings that are unsuitable for use in scientific research (MacLeod & Reydon, 2013; Reydon, 2010a, b). Thus, authors have asked whether kinds of entities in physics, chemistry, biology, and the social sciences can be conceived of as natural kinds – and, indeed, whether psychiatric kinds such as psychopathy *are* natural kinds (Beebee & Sabbarton-Leary, 2010; Brzović et al., 2017; Haslam, 2002a, b; Held, 2017; Kendler et al., 2011; Samuels, 2009; Tsou, 2013, 2016, 2019; Varga, 2018; Zachar, 2000, 2015). A guiding assumption in these debates is that natural kinds are "good", "legitimate" or "valid" kinds for the purposes of scientific research, while other kinds (such as arbitrarily constructed groupings) are not useful for scientific use. Whether or not psychopathy and other psychiatric kinds can be counted as natural kinds thus could have profound ramifications for psychiatric research and clinical practice.

In what follows, I will explore the question whether psychopathy is a "good" scientific kind and – answering this question affirmatively – I will try to clarify how psychopathy as a scientific kind is best conceived of. This will entail a perspective

---

[1] Presumably, classifications and kinds also perform non-epistemic roles, such as practical and social roles (e.g., the medical treatment of patients with certain symptoms, or implementing affirmative action measures for individuals of particular groups). For reasons of space, however, I shall ignore these in the present paper and only look at the most prominent epistemic roles.

of psychopathy as a kind that is quite different from traditional views of natural kinds. I will begin by briefly reviewing the state of the art in the philosophy of kinds and classification, and then examine the debate on the question whether kinds in psychiatric research and clinical practice, such as psychopathy, can be conceived of as natural kinds. I will argue that this debate as conducted at present is misguided, and introduce an alternative approach to the question what natural (or, scientific) kinds are.[2] By way of conclusion, I will suggest an alternative view of how psychopathy – if it cannot be understood as a natural or scientific kind – may still play a useful role in research and clinical practice.

10.2   The Philosophy of Kinds and Classification

The philosophical literature on (natural) kinds and classification is vast. Within the scope of the present chapter I can only provide a brief sketch of what (for the purposes of this chapter) I consider to be the most relevant aspects of the debate.

Traditionally, natural kinds are thought of as kinds of substances (gold, water, etc.) or of entities (material objects, processes, properties, events, behaviors, phenomena, etc.) that exist in nature independently of human classificatory activities, or – better formulated – as kinds of things that represent aspects of nature and as such have a firm foundation in nature. This natural foundation is supposed to hold the kind's member entities together, keep them separate from members of other kinds, explain the characteristic traits of the members of a kind, underpin the possibility of making generalizations about the members of a natural kind, and so on. Because of their natural foundation, natural kinds can stand at the focus of scientific investigations, and can be used to perform crucial epistemic roles in the sciences such as the ones mentioned in the preceding section. On this view, one of the principal aims of science is producing new knowledge about the various natural kinds of substances and/or things that exist in the world. As Bird and Tobin (2018) put it in their authoritative entry on natural kinds in the *Stanford Encyclopedia of Philosophy*:

> Scientific disciplines frequently divide the particulars they study into *kinds* and theorize about those kinds. To say that a kind is *natural* is to say that it corresponds to a grouping that reflects the structure of the natural world rather than the interests and actions of human beings. We tend to assume that science is often successful in revealing these kinds [...]. The existence of these real and independent kinds of things is held to justify our scientific inferences and practices.

Unfortunately, the situation is much less simple than the above quotation suggests. On the picture suggested by the quotation from Bird & Tobin – which is a widespread picture in the philosophical literature on natural kinds – there is something

---

[2] I will use the terms 'natural kind' and 'scientific kind' interchangeably. Whether one chooses to talk about scientific kinds or natural kinds, is merely a matter of terminological preference, I think. When it comes to this choice, though, nothing hinges on the usage of terms.

in nature that natural kinds latch onto. That is, natural kinds represent "the structure of the natural world" adequately and because science is interested in various aspects of this structure, natural kinds are suitable groupings for scientists to investigate. But what, exactly, *is* this presumed "structure of the natural world" that natural kinds are supposed to represent? And what about that structure should they represent – the core elements of that structure, those aspects that we happen to find interesting, or what? Does the world *have* a structure at all, or does it have no structure, or does it have many structures that all could be represented by natural kinds? How can we know whether the world has a definite structure, and what that structure consists in? These are long-standing metaphysical as well as epistemological questions that problematize the traditional picture of natural kinds and so far have not been conclusively answered.

Available accounts of natural kinds tend to assume that there is *one* specific aspect of the structure of the natural world that natural kinds should represent, but authors disagree on what this aspect is. For the longest time, the notion of natural kinds was connected to essentialism, i.e., the claim that for every natural kind there is a set of intrinsic, deep-level traits – the essence of the kind – that all and only members of that kind share, and that explain the observable traits that members of the kind typically exhibit. The idea thus was that essences are parts of the world's structure, and natural kinds latch onto those parts. For the various kinds of elementary particles (the kinds featuring in the Standard Model of particle physics) and atoms (the chemical elements featuring in the Periodic System) this picture seems to make sense. Consider one of the well-worn examples usually mentioned in the literature on natural kinds: gold. All gold atoms have 79 nuclear protons, and only gold atoms do, so the intrinsic property of having a nucleus that contains 79 protons is both necessary and sufficient for being a gold atom. The property of having a nucleus that contains 79 protons performs an explanatory role with respect to the typical interactions of gold atoms with other atoms. In this sense, having a nucleus that contains 79 protons can be thought of as the essence of the kind gold.

Essentialism has, however, come under fire in the philosophy of science as a view that does not fit many of the kinds and classifications that actually feature in scientific practice. Consider the biological classification of organisms into species. Once a paradigmatic example of a classification into natural kinds, the view that biological species are natural kinds characterized by essences in the aforementioned sense is very difficult to uphold (Reydon, 2010b, 2012, 2013: 207–208).[3] Most importantly, there usually is considerable variation among the members of any biological species, both synchronically and diachronically. Diachronically, due to evolutionary change, the early members of a species will not necessarily resemble late members of the same species. Synchronically, there must be variation among the members of a species to make future evolution by means of natural selection

---

[3] For the story of the demise of essentialism in biology, see Ereshefsky (2001: 95–102). For an extensive discussion of the notion of natural kinds and its connection to essentialism, as well as an argument against essentialism that has strongly influenced the contemporary discussion on natural kinds, see Dupré (1993, Chapters 1–3).

possible. In both cases, variation can occur with respect to any of the organisms' traits, such that no trait or set of traits can be singled out as the species' essence.

Notwithstanding the demise of essentialism, present-day accounts of natural kinds still tend to focus on one aspect of nature that all natural kinds are thought to represent. Consider the following examples. One account that is widely applied to kinds in the various sciences, Boyd's *Homeostatic Property Cluster* account (e.g., Boyd, 1991, 1999, 2000), treats all natural kinds as representing homeostatic mechanisms in nature that cause stable patterns of similarity between a kind's members to exist and in this way underwrite inferential statements about those kind-typical similarities.[4] In a similar way, a recent account proposed by Khalidi (2013, 2018) treats all natural kinds as representing nodes in the causal network structure of the world. And on an account proposed by Slater (2013, 2015), all natural kinds represent stable patterns in nature that should simply be taken as brute facts of nature. This is not to say that there are no causes underlying such patterns, but rather that the nature of these causes is irrelevant for accepting a pattern as a natural kind. Slater's point is that if we find a stable pattern of properties that regularly co-occur, we can highlight this pattern as a natural kind without knowing more (or anything) about what causes the co-occurrence. Moreover, there can be a plurality of such underlying causes on Slater's account. Notwithstanding the differences between Boyd's, Khalidi's and Slater's accounts, all treat natural kinds as representing *one* particular aspect of the world. In the contemporary debate, essences came to be replaced by homeostatic mechanisms, or by nodes in the world's causal network, or by stable patterns, or by other factors, depending on which account one prefers.

As argued elsewhere (Ereshefsky & Reydon, 2015, forthcoming), this focus on one single aspect of nature that is supposed to underwrite all natural kinds (be it homeostatic mechanisms, nodes in nature's causal nexus, stable patterns, or something else) is problematic. When considering the diverse sciences, we see that scientists can have a variety of aspects of nature in mind when they construct classifications. Some classifications may be aimed at supporting inferential statements, other at obtaining stable groups in which every entity under study can be uniquely located, others at obtaining groups of organisms that can stand at the focus of nature conservation efforts, others at mapping out all the causal factors that play a role in a particular domain of phenomena, still others at obtaining groups of people that are useful as the basis for therapeutic and social interventions, and so on. Sometimes homeostatic mechanisms might be in focus, while in other context causal nodes are highlighted, still other areas focus on brute stable patterns, and so on. Any adequate account of natural kinds should therefore recognize a plurality of

---

[4] Boyd's account has become popular among scientists in various areas of work as well as philosophers of science, and is often seen (but also criticized) as a promising account of kinds in psychology and psychiatry (Beebee & Sabbarton-Leary, 2010; Brzović et al., 2017; Held, 2017; Kendler et al., 2011; Samuels, 2009). However, Boyd's account is problematic as an account of natural kinds or kinds that successfully feature in the sciences – for reasons of space, I am unable to elaborate this matter here, but for detailed criticism see Ereshefsky and Reydon (2015).

aspects of nature that natural kinds might latch onto. Available accounts, thus, are too monistic and insufficiently attuned to what scientific practice is actually like.

Moreover, lacking direct access to what the natural world is like, we lack foundations for the assumption *that* the world has any clear-cut general structure (Waters, 2017, 2019). This makes it doubtful that the kinds that feature in the sciences can be interpreted as simply representing unique aspects of *the* world's structure. There might be many – even innumerably many – structures, or there might be no structure at all. As we don't have any direct access to the world's inner workings (i.e., we cannot straightforwardly see what the world is really like, but have to use mediated observations, experiments, inferences, and so on), the assumption that the world has a unique structure that natural kinds should latch onto is unwarranted.

For the reasons discussed above, the view according to which "[t]o say that a kind is *natural* is to say that it corresponds to a grouping that reflects the structure of the natural world rather than the interests and actions of human beings" (Bird & Tobin, 2018) does not seem tenable. We cannot say much about *the* structure of the natural world independently of the interests and actions of human beings, after all. Any account of natural kinds should account for those kinds that we use to describe, explain, and intervene in the world – i.e., the kinds that feature in the investigations of the world that we conduct because of the particular interests that we have.

By way of an alternative account (called the *Grounded Functionality Account* of natural kinds) that takes up the preceding considerations, Ereshefsky and Reydon (forthcoming) have suggested that "good" scientific kinds are kinds that further the specific epistemic and non-epistemic aims of the particular context in which they are used. The "goodness" of scientific kinds thus is a matter of functionality, i.e., of how well kinds perform as judged by the specific aims of the research context in which they feature. In addition, a kind's functionality must be explained in terms of what it is in the world that the kind represents. That is, a kind must not only be functional to count as a "good" scientific kind, but its functionality must also be explained by how it is grounded in specific aspects of nature. Both the functionality and the grounding of kinds are local matters on this account, and not global matters: functionality is assessed only with respect to the specific aims of a particular research context, and the explanation of how kinds achieve their functionality is given only in terms of the aspects of nature that are in focus of the corresponding research context. On the *Grounded Functionality Account*, natural kinds depend on the world as well as on human interests and classificatory activities.

This account follows the view of classification that was formulated more than three centuries ago in Locke's *Essay Concerning Human Understanding*. Locke observed that there always are two sides to every classification: as Locke put it, nature makes entities, phenomena, etc. similar and different, while we group entities, phenomena, etc. into kinds on the basis of the similarities and differences that we are interested in.[5] While many available accounts of natural kinds only look at

---

[5] For discussions of Locke's view and how it relates to the contemporary debate on kinds, see (Reydon, 2010a, 2014, 2016).

one side (what it is in nature that kinds represent) and as such can be said to think of natural kinds as "zooming in" onto aspects of nature, adequate accounts should look at both sides and think of kinds as equally being the result of nature and of human interests and classificatory activities (what I elsewhere called "co-creation" of kinds by nature and by us – Reydon, 2016). More generally (and that is the gist of the *Grounded Functionality Account*), when assessing the question at the beginning of this chapter – what makes some classifications suitable to stand at the focus of scientific investigations and others much less so – one has to look at both *what a classification is for* (i.e., the specific epistemic and non-epistemic aims in focus) and *how the classification achieves what it is for*.

10.3   Is Psychopathy a Natural Kind?

In the past 1–2 decades a body of literature has come into existence in which kinds of mental disorders are discussed as putative natural kinds. Several authors have noted that we have a – possibly innate – tendency to think of diseases and mental disorders in essentialist terms and thus as there being a set of deep-level, explanatory traits that, must necessarily be exhibited by a person to be classified as suffering from a particular disease or disorder, and is sufficient to classify the person as a sufferer from that disease or disorder (for discussion, see Haslam & Ernst, 2002; Adriaens & De Block, 2013). Because of such an essentialist understanding of natural kinds, a widespread consensus emerged that psychiatric disorders cannot be thought of as natural kinds.

For example, Zachar argues that "that it is a mistake to think of psychiatric syndromes as natural kinds, meaning bounded categories that have necessary and sufficient internal conditions for their diagnosis" (Zachar, 2000: 168) and that "this kind of essentialistic thinking is scientifically malignant" (Zachar, 2000: 169). Zachar – as well as other participants in the debate, such as Haslam (2002a, b) – think of natural kinds as grounded in essences (in the sense specified above) and reject psychiatric kinds as such natural kinds. Instead, Zachar suggests that they are practical kinds, that is, "stable patterns that can be identified with varying levels of reliability and validity" (Zachar, 2000: 167). But, as was briefly discussed in Sect. 10.2, there are many other accounts of what natural kinds are, some of which might fit psychiatric kinds better than essentialism. In particular, it is noteworthy that Zachar's notion of practical kinds seems the same as Slater's notion of natural kinds, mentioned above. While Zachar rejects mere stable patterns as natural kinds, Slater accepts them and argues that this is precisely what natural kinds *are*: stable patterns in the world. Zachar seems to object to the assumption that there always must be a clear-cut set of intrinsic traits underlying a pattern of behavior for it to count as a psychiatric natural kind. But the brief discussion in Sect. 10.2 showed that this is too strict a view of what natural kinds – or "good" scientific kinds – are. In this sense, the debate on psychiatric kinds is misguided.

In later work, Zachar (2015; Kendler et al. 2011) is more positive about Boyd's *Homeostatic Property Cluster* account of natural kinds, as he conceives it as less essentialistic than traditional views.[6] But he criticizes Boyd's account for not acknowledging the role of social factors in the construction of psychiatric kinds. As he put it:

> Natural kind concepts are supposed to represent what exists independent of our classifications, but in application, concepts for disorders become subject to our goals and interests. The clinical goals of practitioners and patients, the various scientific goals of researchers, philosophical theories about the nature of disorders, the priorities of health service administrators and social policy analysts, and commercial interests, for better or worse, have all played a role in how constructs for psychiatric disorders are developed. [...] The homeostatic property cluster model [...] says little about the role of background assumptions and goals in selecting "good" classifications. [...] Such is the inspiration behind the claim that psychiatric disorders are practical kinds. (Zachar, 2015: 289).

I agree with Zachar that the *Homeostatic Property Cluster* does not have social factors, such as clinical goals and priorities of health service administrators, in view when it comes to accounting for how kinds and classifications are constructed. The aim in view in Boyd's account is purely epistemic (i.e., the making of inferences). But I have quoted Zachar at length, because this quotation is illustrative of a further important aspect of the debate. Besides essentialism Zachar also assumes that natural kinds represent "what exists independent of our classifications" (Zachar, 2015: 289), thus expressing something similar to Bird & Tobin's view of natural kinds representing "the structure of the world", quoted above.

Zachar argues – correctly, I think – that psychiatric kinds do not necessarily represent nature as it is independently of human interests and classificatory activities, but to an important extent represent the interests of researchers, clinicians, health service administrators, and other parties with a vested interest in mental health. Indeed, as Hacking famously argued (e.g., Hacking, 1993, 1995, 1999, 2007) kinds in the human sciences (including psychiatric and psychological kinds and categories) to a considerable extent are products of social construction. What Hacking called "human kinds" are kinds of people, who are being grouped together by scientists into a kind because they exhibit the same scientifically interesting behaviors, characteristics, dispositions, etc. (Hacking, 1995: 351–352) – kinds like child abuser, genius, obese person, or unemployed person. Human kinds thus are intended as "good" scientific kinds, i.e., as kinds "about which we would like to have systematic, general, and accurate knowledge; generalizations sufficiently strong that they seem like laws about people" (Hacking, 1995: 352).

Hacking pointed to two important differences between human kinds and natural kinds. First, in contrast to the entities studied in the natural sciences the entities that the human/social sciences study (i.e., individual humans) may become aware that

---

[6] Zachar also proposed an account of psychiatric kinds as "moral-medical kinds" (Zachar & Potter, 2010) in response to Charland's (2004) suggestion that many personality disorders are moral disorders rather than medical disorders, that is, moral categories that are to be conceived of in a different way than the (natural) kinds that stand at the focus of biomedical science.

scientists classify them in a particular manner. Consequently, they may become motivated to alter their behavior or their characteristics and, in so doing, may bring it about that what scientists thought they knew to generally hold about members of the kind no longer holds. Thus, in contrast to the natural sciences, classifications in the human/social sciences may induce changes in the properties of the classified entities in such ways that useful scientific generalizations over the kinds will not be possible. Hacking calls such feedback effects between classifications and the classified subject matter "looping effects". Second, "[t]he chief difference between natural and human kinds is that the human kinds often make sense only within a certain social context" (Hacking, 1995: 362). Hacking argued that the kinds of people studied in the human/social sciences are not simply "given", but are kinds that are constructed at some point in human history in response to changes in the cultural context. The kind homosexual, for example, "as a kind of person came into being only late in the nineteenth century as homosexual behavior became an object of scientific scrutiny" (Hacking, 1995: 354). Thus, it is not the case that this kind was *discovered* by social scientists in the late-nineteenth century – as, for example, sodium was discovered earlier that century. Rather, according to Hacking, the kind was *created* at a particular time and in a specific cultural context as a result of a particular human behavior becoming socially and scientifically interesting. Moreover, the criteria for being a member of the kind may change with the interests of society and science, leading to changes in kind membership and even to kinds ceasing to be recognized.[7]

Such dependence of kinds on human interests makes it questionable, on Hacking's view, whether the natural grounding of such kinds is sufficiently strong and the kinds themselves sufficiently stable to perform its epistemic roles. At least, for Hacking it was sufficient reason to think that kinds in the human sciences do not live up to the expectations of "good" scientific kinds (and Zachar seems to agree on this point). But the assumption that natural kinds represent the world as it is independently of human interests, is much too strict. As I pointed out in Sect. 10.2, accounts of natural kinds exist that explicitly take natural kinds to represent human interests *as well as* aspects of the world (Ereshefsky & Reydon, 2015, forthcoming; Reydon, 2014, 2016). Moreover, this way of thinking about natural kinds has been present in philosophy for over 300 years. Zachar's conclusion that psychiatric kinds are not natural kinds because they are affected by our goals and interests, thus, is mistaken.[8]

But I believe that there is another reason why claims like the ones made by Zachar – and the debate on whether psychiatric kinds are natural kinds more

---

[7] Hysteria is a prominent example of a kind that made sense within a particular context, but ceased to be recognized at a later time.

[8] I also believe that Zachar's claim, that Boyd's account does not acknowledge the role of human interests and goals in the construction of classifications, is mistaken. In his work, Boyd explicitly refers to Locke's views on kinds and presents his work as a continuation of Locke's ideas and philosophers tend to understand Boyd's account as allowing human interests to play a part in classifications (e.g., Beebee & Sabbarton-Leary, 2010).

generally – are mistaken. The debate focuses on available philosophical accounts of what natural kinds are, and then asks whether psychiatric kinds fit those accounts. (And, as I argued, the debate does not include the whole spectrum of available accounts, but considers only essentialism and the *Homeostatic Property Cluster* account, and finds that psychiatric kinds do not fit those accounts.) Participants in the debate often come to a negative conclusion that psychiatric kinds in general, or at least one or several important psychiatric kinds, cannot be thought of as natural kinds (e.g., Brzović et al., 2017; Haslam, 2002a, b; Varga, 2018; Zachar, 2000, 2015).[9] Some authors, such as Zachar (discussed above) introduce other ways of thinking about psychiatric kinds. But what is gained by such arguments? The original question that at the beginning of this chapter I highlighted as standing at the heart of the philosophical debate on natural kinds – what makes some classifications suitable to stand at the focus of scientific investigations (and others much less so) – remains unanswered. Zachar's claim that psychiatric kinds are practical kinds takes psychiatric kinds out of the natural kinds fold, but does not explain why psychiatric kinds are successfully used in scientific research, clinical practice, and other contexts. The focus on what philosophers have written about the concept of natural kinds thus has led the debate astray, away from the question that the debate should have been about.

Taking available philosophical accounts and asking whether the case of psychopathy fits any of these accounts amounts to putting the cart before the horse, I suggest. So, in what follows I want to refocus the debate on its proper question: how can the successful use of psychiatric kinds be explained?[10] In so doing, I will also focus on the psychiatric kind of psychopathy, as so far I have said little about that kind.

Note first that traditional views of natural kinds, such as voiced by Bird and Tobin (2018), indeed do not fit the case of psychopathy. The kind psychopathy is not simply found in the world. We find individual human beings, each with a specific set of traits and behaviors and with a considerable variety between individuals, and some of those individuals we group together into the category of psychopaths on the basis of certain traits that are considered to be characteristic. In his influential account of case studies, *The Mask of Sanity*, Cleckley ((1976); see also Thomas-Peter, 1992: Table 1) listed 16 traits as characteristic of psychopaths, and contemporary definitions typically are variations on Cleckley's list. An article aimed at a general audience, for example, says:

---

[9] But see Tsou (2013, 2016, 2019) for a more positive view of psychiatric kinds as natural kinds.

[10] Assuming *that* psychopathy and other psychiatric kinds that feature in classificatory systems such as ICD-10 and DSM-5 are successfully used in the sciences, that is. I will not question that assumption at this point in the paper, but will discuss it further below. Here, it should be noted that several authors have shed doubt on the usefulness assumption, highlighting the problem that such kinds might not useful precisely because they do not capture unified sets of causal factors, mechanisms, etc., that underwrite re-occurring clusters of phenomena. For discussion, see Insel and Cuthbert (2015), Brazil et al. (2018) or Jurjako et al. (2019).

psychopathy consists of a specific set of personality traits and behaviors. Superficially charming, [...] self-centered, dishonest and undependable, and at times they engage in irresponsible behavior for no apparent reason other than the sheer fun of it. Largely devoid of guilt, empathy and love [...]. Psychopaths routinely offer excuses for their reckless and often outrageous actions, placing blame on others instead. They rarely learn from their mistakes or benefit from negative feedback, and they have difficulty inhibiting their impulses (Lilienfeld & Arkowitz, 2007/2008: 90).

At present, we have a situation in which different definitions encompass different lists of characteristic traits (see, e.g., MacKenzie, 2014; Brzović et al., 2017: 192ff.). At the phenomenological level, that is, the level of observable behavioral traits thought characteristic of members of the kind psychopath, there is profound disagreement with respect to the question which set of traits can be taken to delimit the kind.

This disagreement is deepened by a long-standing debate on the causal basis of the set of behavioral traits characteristic of psychopaths. In the "Hare vs. Blackburn" debate (discussed in detail by Thomas-Peter, 1992), for example, one side proposed a classification of psychopaths into "primary psychopaths" (characterized by low levels of anxiety and thought to be strongly genetically based) and "secondary psychopaths" (characterized by high levels of anxiety and thought to be strongly environmentally based) on the basis of clear differences in behavior,[11] while the other side in the debate argued that secondary psychopaths should not be classified as psychopaths at all, because the two kinds of behavior were due to different underlying causes. The debate ensued in different conceptualizations of psychopathy, which Thomas-Peter (1992: 339) identifies as a North American and a European conceptualization. But the debates are not limited to these two conceptualizations. In a recent inventory of the debates, Lilienfeld et al. (2015) highlighted that authors disagree on the questions which behavioral traits should be included in the set of characteristic traits, whether psychopathy is a unidimensional or multidimensional condition, whether adaptive traits (traits that generally benefit their bearers, such as boldness) are part of the condition, whether antisocial behaviors constitute an integral part of the condition or merely are consequences of it, and whether the correlations between behavioral symptoms is sufficient to understand psychopathy as a syndrome. As the authors point out, it is not clear whether psychopathy is a unified category or encompasses a plurality of quite distinct behavioral phenomena, how sharply delimited the category is or can be made, and what exactly psychopathy *is* (Lilienfeld et al., 2015). The authors' diagnosis is severe: "If researchers cannot agree on whether psychopathy is one condition or several, or on whether the traits that some researchers view as essential to the condition are even relevant to it, the field is bound to be in intellectual disarray." (Lilienfeld et al., 2015: 594). If this

---

[11] "[T]he inappropriate behavior of the primary psychopath is presumed to be a consequence of some intrinsic deficit that hampers self-regulation and normal adjustment, whereas secondary psychopathy is viewed as an indirect consequence of inadequate intelligence, psychotic thinking, excessive neurotic anxiety, unusual sex drive, or other attributes that increase a person's vulnerability to chronic misbehavior" (Newman et al., 2005: 319).

diagnosis is correct, it may be doubted whether psychopathy is a "good" scientific kind at all, i.e., a kind that can be used fruitfully in research and clinical contexts. Under such circumstances, asking whether psychopathy can be thought of as a natural kind is moot.

The problem with respect to the kind 'psychopath' as a "good" scientific kind (or, even, a natural kind), then, can be summarized as follows. First, there exists persistent unclarity regarding the question which behavioral traits are characteristic of the members of the kind 'psychopath'. At the level of the phenomena, thus, there is a debate whether 'psychopathy' denotes a stable pattern of behavioral traits at all. Second, there exists persistent unclarity regarding the question how the kind is grounded, i.e., which deep-level traits, causal factors, mechanisms, or other factors are responsible for the regular co-occurrence of the aforementioned characteristic traits. Some participants in the debate hold that, while 'psychopathy' does denote a stable pattern of behavioral traits, there is no common grounding for all instances, such that psychopathy cannot be seen as a natural kind but should be thought of as a mere practical kind (Zachar, discussed above). Third, and connected to the second point, there exists persistent unclarity regarding the question to what extent the kind 'psychopath' is grounded in nature (i.e., to what extent factors in nature play a role in supporting the kind, be it one factor for all instances or a diversity of factors) and to what extent it is grounded in the interests of researchers, clinicians, health administrators, and others. As we have seen in the above discussion, for authors such as Zachar, the partial grounding of psychiatric kinds in human interests reinforces the view of such kinds as practical – but not natural – kinds.

But if these considerations are right, what does this imply for the use of 'psychopathy' as a technical term in mental health research and clinical practice? The term continues to be widely used and even if the debate ends in the determination that psychopathy cannot be conceived of as a natural kind, its roles in research and clinical practice still must be explained. To conclude, I will now turn to this issue.

10.4 Psychopathy as a Behavioral Variant Rather Than a Kind

To what extent can we judge whether psychiatric kinds, such as psychopathy, are "good" scientific kinds? The philosophy of kinds and classification, briefly discussed above, can provide us with tools to accommodate this debate. Recall from Sect. 10.2 that a diverse spectrum of accounts of natural kinds is available in the philosophical literature. Most of these focus on the first and/or second issues mentioned in the previous paragraph, i.e., the properties that are thought to be characteristic of a kind's member entities at the phenomenal level and/or the deep-level properties or causal factors that cause the regular co-occurrence of those characteristic properties. That is, most available accounts of natural kinds focus on what it is *in nature* that kinds represent, either at the level of empirically accessible phenomena, or at the level of what underlies these phenomena, or both.[12] What is much less in focus is the question what a classification *is for* – in what ways kinds are successfully used in scientific practice and other contexts of practice, and (subsequently) what underwrites their success.

I want to suggest that it is not a fruitful approach to ask (1) what aspects of nature a kind represents without first considering (2) what it was *intended* to represent and why researchers, clinicians, and so on, in the first place chose to focus on those aspects of the world rather than on other aspects. The *Grounded Functionality Account* of kinds, mentioned above, inverts the order of questions (1) and (2) and begins with question (2). It begins by asking how well a kind or classification performs in light of the specific aims for which it was developed, and only then moves on to asking how the kind's functionality can be explained in terms of what it is in the world that the kind represents. Taking this perspective on the case of psychopathy or psychiatric kinds more generally leads us to focus on the questions for what aims the kind psychopathy was devised, whether it performs successfully in research, clinical practice and elsewhere, and if it does perform successfully, what underpins its success. I suggest that in this way the *Grounded Functionality Account* of kinds provides tools with which the debate can be reoriented in a more fruitful direction.

When asking whether psychopathy is a "good" scientific kind, then, the first question that presents itself is whether the kinds furthers the aims of *that* classificatory context within which it is used. An interesting aspect of the debate that was examined in the previous section is that it is conducted against the background of the assumption that psychopathy *is* a functional kind. That is, it is a kind that performs at least *some* role in scientific research and clinical practice. This assumption seems right: a quick search in Google Scholar yields a plethora of research publications that center around the kind psychopathy. But statements like the one from Lilienfeld et al. (2015), quoted above, suggest otherwise. The disagreement among researchers and clinicians about questions whether 'psychopathy' denotes one condition or multiple, which behavioral traits should be counted as part of the set of characteristic behaviors of psychopaths, and what psychopathy ultimately is, suggests that it may be doubted whether psychopathy is a useful scientific kind at all. It is odd that the debate that was reviewed briefly in the preceding section largely passes by the question *for what* the kind term 'psychopathy' is used, and lets itself be bogged down by the dichotomy between kinds that represent the world as it is independently of us (natural kinds, on traditional views of what natural kinds are)

---

[12] In a recent paper, for example, Brzović et al. (2017) argued that there is insufficient evidence to assume that psychopathy is associated with a stable behavioral pattern (i.e., a stable set of behavioral traits that is seen in most psychopaths) and that there is insufficient evidence to assume that the kind psychopathy is supported by a stable set of underlying mechanisms or causal factors. Thus, the authors conclude, psychopathy cannot be understood as a natural kind – neither on Boyd's *Homeostatic Property Cluster* account nor on Slater's account, which (alluding to Boyd) he called the *Stable Property Cluster* account.

and kinds that in part represent human interests (kinds that on traditional views cannot be thought of as natural kinds).

On the *Grounded Functionality Account* of scientific kinds this is the wrong way to approach the issue – first one has to clarify *what a kind does* or *what it is used* for within a particular context of practice, and then one can ask *in what way it is connected to nature* such that its grounding in nature supports what it does or what it is used for. The first questions to answer, then are what, exactly, the function(s) of the kind psychopathy is/are in the particular contexts of research and clinical practice in which it is used, how the kind relates to other kinds that feature in the same contexts, whether the kind's functions can be realized in a better way by subdividing the kind into multiple kinds or removing a subgroup from the kind, and so on. The metaphysical question regarding the grounding of the kind's functions in aspects of nature comes up second in line in the course of exploring these questions.

Consider Zachar's claim that psychopathy and other psychiatric kinds are not natural kinds but practical kinds. Zachar argues that psychiatric kinds are not generally well-grounded in deeper features of the world but nonetheless perform well in practice *as* stable patterns of behavior that we find in the world. Hence, such kinds are practical kinds. As Zachar writes: "Concepts for psychiatric disorders are constituted by discoveries *and* decisions. There is an interaction between what the world produces and what we find useful to notice." (Zachar, 2015: 289). That is surely right, but *this holds for all kinds in the sciences* (Reydon, 2016). There are no kinds in the sciences that represent the world as it is independently of human interests!

Kinds are always embedded in contexts of practice – they are always embedded in a context that is affected by decisions about which theoretical perspective should be used to understand the world and on which aspects of the world focus should be placed. Kinds always are both theory- and practice-laden.[13] Thus, the question whether psychiatric kinds are natural kinds or practical kinds is a red herring. Zachar's (2015) question whether psychiatric disorders are natural kinds made by the world or practical kinds made by us, is a non-starter. *All kinds* that are successfully used in research and clinical practice are practical kinds in Zachar's sense – they are stable patterns of recurrent properties (as in the case of kinds of elementary particles and the chemical elements) or behaviors (as in the case of kinds in animal ethology and kinds the sciences of human behavior), patterns of descent (such as biological species), patterns of stages (for example, kinds of physical, chemical, and biological processes), and so on, that we highlight because they are of interest in the

---

[13] A minority of contemporary accounts of natural kinds acknowledge this, including Boyd's *Homeostatic Property Cluster* account and the *Grounded Functionality Account*. A crucial difference between the *Homeostatic Property Cluster* account and the *Grounded Functionality Account*, though, is that the former only acknowledges epistemic interests as guiding classification, while the latter acknowledges both epistemic and non-epistemic interests as guiding classification. Another crucial difference is that the *Homeostatic Property Cluster* account makes a priori assumptions regarding how kinds are grounded in the world (namely, by homeostatic mechanisms), whereas the *Grounded Functionality Account* makes no such assumptions and allows for a plurality of ways of grounding.

context of our efforts to understand the world and intervene in it. But they cannot be merely stable patterns, because their successful use in scientific research and practices of application must be a matter of how the kinds are connected to the world (or, of what aspects of the world they represent).

To give a quick example, the grouping of fruits and vegetables in supermarkets (where fruits such as tomatoes and cucumbers are usually grouped with the vegetables) is a stable pattern in the context of a particular practice. But the pattern is exclusively grounded in our decisions regarding what goes well with what on your plate – it is wholly ungrounded in nature, as there is nothing in nature that makes tomatoes belong to the vegetables. This is where theories of natural kinds come into play – they are supposed to enable us to distinguish between stable patterns that are grounded in the world and those that are not grounded in the world. This means that nothing is gained by Zachar's proposal to think of psychiatric kinds as practical kinds rather than natural kinds: if they indeed are useful kinds in research and clinical practice, one cannot be content by the observation that they are stable patterns, but must move beyond that observation to explain what might underwrite their usefulness. Practical kinds, too, must have a metaphysical basis in the world – if they don't, they are as arbitrary as the common grouping of tomatoes and cucumbers with the vegetables. But while for the latter grouping its arbitrariness does not matter much, psychiatric kinds are aimed at understanding the causality behind certain patterns of behavior, and at interventions that change those behaviors. For psychiatric kinds, then, arbitrariness is not acceptable.

Given the debate about the functionality of 'psychopathy' as a kind term in psychiatric and behavioral research, as well as clinical practice, at present it is not possible to answer the question whether psychopathy is a "good" scientific kind. But we might achieve a little more clarity on this issue by looking at how the kind might be grounded in the world. Note that, while defending a view of psychiatric kinds as stable patterns (i.e., practical kinds), Zachar (2000: 173) leaves open the option that psychopathy is a maladaptive variant in the human population, rather than a dysfunction of the brain. Interestingly, in a recent paper Jurjako et al. (2019) argued that psychopathy might be an *adaptive* variant in the human population.[14] In the same way as Zachar, Jurjako observes that "psychopathic traits present a constant and stabile variation in human personality" (Jurjako, 2019: 12), or, in Zachar's terms, constitute a stable pattern. But while Zachar suspects that the pattern constitutes a maladaptive constellation of behavioral traits (and thus a constellation that does not need to be explained), Jurjako conjectures that it can be explained as a consequence of evolution by means of natural selection, that is, as a set of traits that is adaptive in relation to a particular social niche. As he writes,

the peculiar activation patterns of amygdala and related neural circuitry in psychopaths can be seen as adaptations to an environment where it pays off to engage in the antisocial lifestyle that is sustained by the balancing selection [i.e., selection where the fitness of a phenotype depends on the frequencies of other phenotypes in the population – clarification added]. [...] this life strategy can be beneficial in environments where life expectancy is

---

[14] For authors who made this suggestion earlier, see Mealey (1995) and Lalumière et al. (2008).

lower and thus it pays more to invest in reproductive efforts [...]. Since the amygdala's role in psychopathy might be to enable such a life strategy, we do not have grounds for claiming that it is malfunctioning, even if it is correlated with reduced longevity. [...] psychopathic traits instantiate an adaptive life strategy that is maintained by frequency-dependent selection. I maintain that until this hypothesis is proven false we should be reluctant in judging that psychopathic traits present harmful dysfunctions (Jurjako, 2019: 19).

In both Zachar's and Jurjako's account, then, psychopathy represents a behavioral variant, i.e., a particular area in the space of possible human behaviors. Contrary to how behaviors are often defined in evolutionary game theoretic models (in which, for example, two discrete behaviors such as "hawk" and "dove" are present in a population and after a number of rounds a stable pattern of behaviors emerges – Maynard Smith, 1982: 10ff.; Jurjako, 2019), human behaviors constitute a continuous many-dimensional space. In either case – whether psychopathy is an adaptive variant that is kept in the population due to natural selection, or a maladaptive variant that remains in the population due to other causes –, 'psychopathy' can be conceived of as denoting a non-strictly delimited area in the behavioral state space, where some people may occupy a position squarely within that area while others are located somewhere in the diffuse (and probably quite extensive) boundary area. This view would do justice to the fact that psychopathy is not a yes-or-no matter, but comes in degrees.

What does this mean for the question that we started out with, namely whether psychopathy is a "good" scientific kind (or a natural kind, even)? I want to suggest that the view of psychopathy as a behavioral variant can provide a clue to its function in contexts of research and clinical practice without it being relevant whether or not it can be counted as a natural kind on any of the traditional philosophical accounts of natural kinds. What the term 'psychopathy' does, I suggest, is to locate individual persons somewhere on a continuum of behaviors. Saying that someone is a psychopath does not put that person into a scientific kind or category, but rather locates that person somewhere in the space of distribution of behavioral traits. This is an important function for purposes of scientific research as well as clinical intervention (as well as in other contexts, such as attributions of moral responsibility in criminal trials – see Malatesti & McMillan, 2014), and we have seen two suggestions as to how that function may be supported by the world (as an adaptive variant kept in the population by natural selection, or as a maladaptive variant that may remain in the population as an evolutionary stable strategy).

What this leads to is a view of psychopathy not as a kind of people or a kind of behavior, characterized by a set of behaviors that are typical for psychopaths, but of psychopathy as a region on a multidimensional space of behaviors. Saying that a person is a psychopath, on this view, amounts to locating that person at a particular point within that region.

10.5   Conclusion

The conclusion of the considerations presented in this chapter must be: yes, psychopathy is a good scientific kind. Proponents of the *Grounded Functionality Account* would acknowledge it as a natural kind (but, as highlighted in footnote 2, it is merely a terminological matter whether one uses 'natural kind' or 'scientific kind'). But on (by far) most accounts of natural kinds, psychopathy would not count as a natural kind, as psychopathy does not fit their views of what natural kinds *are*.

Indeed, psychopathy as a grouping does not seem particularly useful as the basis of inferential statements (i.e., generalizations about the traits that all, or at least the large majority of, psychopaths exhibit) or as the basis of scientific explanations and as such it does not seem to perform the epistemic function that is commonly attributed to natural kinds in the philosophical literature. But there are many other functions besides supporting inferences, epistemic ones as well as non-epistemic ones, that kind terms perform in the sciences (Ereshefsky & Reydon, 2015, forthcoming). Focusing too much on just one epistemic role entails missing much of how kind terms function in actual scientific practice, clinical practice, and elsewhere.

In this chapter, I hope to have shown why asking whether psychopathy is a natural kind is the wrong question, at least when the question is approached using most of the available philosophical theories of natural kinds. We should ask what the grouping that the term denotes is good for, and whether its role is grounded in the world. Approaching the matter in this way avoids getting stuck on the question whether psychopathy represents a grouping in the world as it is independently of human interests and classificatory activities and allows us to focus on the questions that really matter.

References

Adriaens, P. R., & De Block, A. (2013). Why we essentialize medical disorders. *Journal of Medicine and Philosophy, 38*, 107–127.

Beebee, H., & Sabbarton-Leary, N. (2010). Are psychiatric kinds "real"? *European Journal of Analytic Philosophy, 6*, 11–27.

Bird, A., & Tobin, E. (2018). Natural kinds. In E. N. Zalta (Ed.), *The Stanford encyclopedia of philosophy (Spring 2018 edition)*. https://plato.stanford.edu/archives/spr2018/entries/natural-kinds/

Boyd, R. N. (1991). Realism, anti-foundationalism and the enthusiasm for natural kinds. *Philosophical Studies, 61*, 127–148.

Boyd, R. N. (1999). Kinds, complexity and multiple realization. *Philosophical Studies, 95*, 67–98.

Boyd, R. N. (2000). Kinds as the "workmanship of men": Realism, constructivism, and natural kinds. In J. Nida-Rümelin (Ed.), *Rationalität, Realismus, Revision: Vorträge des 3. Internationalen Kongresses der Gesellschaft für Analytische Philosophie* (pp. 52–89). De Gruyter.

Brazil, I. A., van Dongen, J. D. M., Maes, J. H. R., Mars, R. B., & Baskin-Sommers, A. R. (2018). Classification and treatment of antisocial individuals: From behavior to biocognition. *Neuroscience & Biobehavioral Reviews, 91*, 259–277.

Brzović, Z., Jurjako, M., & Šustar, P. (2017). The kindness of psychopaths. *International Studies in the Philosophy of Science, 31*, 189–211.

Charland, L. C. (2004). Character: Moral treatment and the personality disorders. In J. Radden (Ed.), *The philosophy of psychiatry: A companion* (pp. 64–77). Oxford University Press.

Cleckley, H. (1976). *The mask of sanity* (5th ed.). Mosby.

Dupré, J. (1993). *The disorder of things: Metaphysical foundations for the disunity of science.* Harvard University Press.

Ereshefsky, M. (2001). *The poverty of the Linnaean hierarchy: A philosophical study of biological taxonomy.* Cambridge University Press.

Ereshefsky, M., & Reydon, T. A. C. (2015). Scientific kinds. *Philosophical Studies, 172*, 969–986.

Ereshefsky, M., & Reydon, T. A. C. (forthcoming). The grounded functionality account of natural kinds. In W. Bausman, J. Baxter, O. Lean, A. Love, & C. K. Waters (Eds.), *From biological practice to scientific metaphysics.* University of Minnesota Press.

Hacking, I. (1993). World-making by kind-making: Child abuse for an example. In M. Douglas & D. L. Hull (Eds.), *How classification works: Nelson Goodman among the social sciences* (pp. 180–231). Edinburgh University Press.

Hacking, I. (1995). The looping effects of human kinds. In D. Sperber, D. Premack, & A. J. Premack (Eds.), *Causal cognition: A multidisciplinary debate* (pp. 351–383). Clarendon Press.

Hacking, I. (1999). *The social construction of what?* Harvard University Press.

Hacking, I. (2007). Kinds of people: Moving targets. *Proceedings of the British Academy, 151*, 285–318.

Haslam, N. (2002a). Kinds of kinds: A conceptual taxonomy of psychiatric categories. *Philosophy, Psychiatry & Psychology, 9*, 203–217.

Haslam, N. (2002b). Practical, functional, and natural kinds. *Philosophy, Psychiatry & Psychology, 9*, 237–241.

Haslam, N., & Ernst, D. (2002). Essentialist beliefs about mental disorders. *Journal of Social and Clinical Psychology, 21*, 628–644.

Held, B. (2017). The distinction between psychological kinds and natural kinds: Can updates natural-kind theory help clinical psychological science and beyond meet psychology's philosophical challenges? *Review of General Psychology, 21*, 82–94.

Insel, T. R., & Cuthbert, B. N. (2015). Brain disorders? Precisely: Precision medicine comes to psychiatry. *Science, 348*, 499–500.

Jurjako, M. (2019). Is psychopathy a harmful dysfunction? *Biology and Philosophy, 34*, 5.

Jurjako, M., Malatesti, L., & Brazil, I. A. (2019). Some ethical considerations about the use of biomarkers for the classification of adult antisocial individuals. *International Journal of Forensic Mental Health, 18*, 228–242.

Kendler, K. S., Zachar, P., & Craver, C. (2011). What kinds of things are psychiatric disorders? *Psychological Medicine, 41*, 1143–1150.

Khalidi, M. A. (2013). *Natural categories and human kinds: Classification in the natural and social sciences.* Cambridge University Press.

Khalidi, M. A. (2018). Natural kinds as nodes in causal networks. *Synthese, 195*, 1379–1396.

Lalumière, M. L., Mishra, S., & Harris, G. T. (2008). In cold blood: The evolution of psychopathy. In J. D. Duntley & T. K. Shackelford (Eds.), *Evolutionary forensic psychology: Darwinian foundations of crime and law* (pp. 176–197). Oxford University Press.

Lilienfeld, S. O., & Arkowitz, H. (2007/2008). What "psychopath" means. *Scientific American MIND*, December 2007/January 2008: 90–91.

Lilienfeld, S. O., Watts, A. L., Francis Smith, S., Berg, J. M., & Latzmann, R. D. (2015). Psychopathy deconstructed and reconstructed: Identifying and assembling the personality building blocks of Cleckley's chimera. *Journal of Personality, 83*, 593–610.

MacKenzie, P. M. (2014). Psychopathy, antisocial personality & sociopathy: The Basics. A history review. *The Forensic Examiner*, December 2014, http://www.theforensicexaminer.com/2014/pdf/MacKenzie_714.pdf

MacLeod, M., & Reydon, T. A. C. (2013). Natural kinds in the life sciences: Scholastic twilight or new dawn? *Biological Theory, 7*, 89–99.

Malatesti, L., & McMillan, J. (2014). Defending psychopathy: An argument from values and moral responsibility. *Theoretical Medicine and Bioethics, 35*, 7–17.

Maynard Smith, J. (1982). *Evolution and the theory of games*. Cambridge University Press.

Mealey, L. (1995). The sociobiology of sociopathy: An integrated evolutionary model. *Behavioral and Brain Sciences, 18*, 523–541.

Newman, J. P., MacCoon, D. G., Vaughn, L. J., & Sadeh, N. (2005). Validating a distinction between primary and secondary psychopathy with measures of Gray's BIS and BAS constructs. *Journal of Abnormal Psychology, 114*, 319–323.

Reydon, T. A. C. (2010a). Natural kind theory as a tool for philosophers of science. In M. Suárez, M. Dorato, & M. Rédei (Eds.), *EPSA – Epistemology and methodology of science: Launch of the European philosophy of science association* (pp. 245–254). Springer.

Reydon, T. A. C. (2010b). How special are the life sciences? A view from the natural kinds debate. In F. Stadler (Ed.), *The present situation in the philosophy of science* (pp. 173–188). Springer.

Reydon, T. A. C. (2012). Essentialism about kinds: An undead issue in the philosophies of physics and biology? In D. Dieks, W. J. Gonzalez, S. Hartmann, M. Stöltzner, & M. Weber (Eds.), *Probabilities, Laws, and structures* (pp. 227–240). Springer.

Reydon, T. A. C. (2013). Classifying life, reconstructing history and teaching diversity: Philosophical issues in the teaching of biological systematics and biodiversity. *Science & Education, 22*, 189–220.

Reydon, T. A. C. (2014). Metaphysical and epistemological approaches to developing a theory of artifact kinds. In M. P. M. Franssen, P. Kroes, T. A. C. Reydon, & P. E. Vermaas (Eds.), *Artefact kinds: Ontology and the human-made world* (pp. 125–144). Springer.

Reydon, T. A. C. (2016). From a zooming-in model to a co-creation model: Towards a more dynamic account of classification and kinds. In C. E. Kendig (Ed.), *Natural kinds and classification in scientific practice* (pp. 59–73). Routledge.

Samuels, R. (2009). Delusion as a natural kind. In M. R. Broome & L. Bortolotti (Eds.), *Psychiatry as cognitive neuroscience: Philosophical perspectives* (pp. 49–82). Oxford University Press.

Slater, M. H. (2013). *Are species real? An essay on the metaphysics of species*. Palgrave Macmillan.

Slater, M. H. (2015). Natural kindness. *British Journal for the Philosophy of Science, 66*, 375–411.

Thomas-Peter, B. A. (1992). The classification of psychopathy: A review of the hare vs. Blackburn debate. *Personality and Individual Differences, 13*, 337–342.

Tsou, J. Y. (2013). Depression and suicide are natural kinds: Implications for physician-assisted suicide. *International Journal of Law and Psychiatry, 36*, 461–470.

Tsou, J. Y. (2016). Natural kinds, psychiatric classification and the history of the DSM. *History of Psychiatry, 27*, 406–424.

Tsou, J. Y. (2019). Philosophy of science, psychiatric classification, and the DSM. In Ş. Tekin & R. Bluhm (Eds.), *The Bloomsbury companion to philosophy of psychiatry* (pp. 177–196). Bloomsbury.

Varga, S. (2018). "Relaxed" natural kinds and psychiatric classification. *Studies in History and Philosophy of the Biological and Biomedical Sciences, 72*, 49–54.

Waters, C. K. (2017). No general structure. In M. H. Slater & Z. Yudell (Eds.), *Metaphysics and the philosophy of science: New essays* (pp. 81–108). Oxford University Press.

Waters, C. K. (2019). An epistemology of scientific practice. *Philosophy of Science, 86*, 585–611.

Zachar, P. (2000). Psychiatric disorders are not natural kinds. *Philosophy, Psychiatry & Psychology, 7*, 167–182.

Zachar, P. (2015). Psychiatric disorders: Natural kinds made by the world or practical kinds made by us? *World Psychiatry, 14*, 288–290.

Zachar, P., & Potter, N. N. (2010). Personality disorders: Moral or medical kinds—Or both? *Philosophy, Psychiatry, & Psychology, 17*, 101–117.

Chapter 11
Psychopathy, Maladaptive Learning and Risk Taking

Johanna C. Glimmerveen, Joseph H. R. Maes, and Inti A. Brazil

Abstract Individuals with psychopathy present with maladaptive tendencies that have been linked to disturbed processing of outcomes during decision making, in particular with respect to aversive outcomes. In general, individuals with psychopathy show risk-seeking behaviour, as well as excessive reward-oriented behaviour. This chapter provides an overview of theories and empirical work on maladaptive behaviour in psychopathy in the context of reinforcement learning and risky decision making. In addition, we capitalise on recent neuroscientific advances to discuss the empirical results and propose how the use of personalised rewards in experimental designs may help to create a more ecologically valid reflection of real-world maladaptive behaviour in psychopathy.

Keywords Psychopathy · Reinforcement learning · Decision making · Risk taking

11.1 Introduction

Psychopathy is a personality disorder characterised by interpersonal and emotional dysfunctions, as well as impulsive and maladaptive behaviour, with an increased risk for the development of an antisocial lifestyle. The disorder is prevalent among offender populations and is a strong predictor of various forms of recidivism (Hawes et al., 2013; Leistico et al., 2008). This tendency for psychopathic individuals to reoffend at a much higher rate than non-psychopathic offenders are indicative of a reduced ability to adapt behaviour appropriately in response to negative outcomes (e.g. incarceration). Historically, the maladaptive tendencies have been attributed to a relative insensitivity to punishment, or at least an incapacity to learn from

J. C. Glimmerveen (✉) · J. H. R. Maes · I. A. Brazil
Department of Neuropsychology and Rehabilitation Psychology, Donders Institute for Brain, Cognition and Behaviour, Radboud University, Nijmegen, Netherlands
e-mail: j.glimmerveen@donders.ru.nl

© Springer Nature Switzerland AG 2022
L. Malatesti et al. (eds.), *Psychopathy*, History, Philosophy and Theory of the Life Sciences 27, https://doi.org/10.1007/978-3-030-82454-9_11

experiences leading to aversive outcomes. The relationship between maladaptive behaviour and psychopathy has been described since the early 1800's (see Hoppenbrouwers et al., 2016, for an overview).

Through time, many experimental paradigms have been employed in an attempt to unravel and explain the core impairments associated with the maladaptive behavioural tendencies observed in psychopathy. Among the most extensively studied impairments observed in individuals with psychopathy are those related to the disturbed processing of reward and punishment in relation to different forms of learning and decision making, which have led to the development of different theories. The main goal of this chapter is to provide an overview of the empirical work on maladaptive behaviour in psychopathy, with a particular focus on reinforcement learning. First, some of the most influential models explaining disturbed learning from reward and punishment information in psychopathy will be presented: the Two-Factor Learning theory, the Low Fear model, the Response Modulation hypothesis, and the Integrated Emotion Systems account. Subsequently, an overview will be provided of the experimental paradigms that were used in the research that formed the basis for these theories. Finally, the generalisability of these study results to real-life impairments in decision making and behavioural adaptation are discussed in light of neuroscientific findings regarding the importance of considering individual variations in the appraisal of rewards and their respective values.

11.2   Theoretical Frameworks

11.2.1   Two-Factor Learning Theory

One of the influential models explaining the reinforcement learning deficits in psychopathy is, in essence, a more general theory of reward and punishment processing. The two-factor learning theory (Gray, 1987) conceptualises reward and punishment processing in terms of the behavioural activation system (BAS) and the behavioural inhibition system (BIS). The BAS responds to appetitive stimuli and serves to initiate goal-directed action in response to reward. The BIS is focused on detecting threat cues and serves to inhibit behaviours leading to aversive outcomes, such as those associated with punishment. According to the two-factor learning theory, decreased inhibition of behaviour leading to punishment (i.e., lower BIS reactivity) is associated with reduced negative arousal in response to punishment and with increased ongoing engagement in reward-seeking behaviour in the face of potential punishment. As such, modifications of this theory have been used to explain psychophysiological as well as behavioural data from studies showing disturbed reward and punishment processing in psychopathy (e.g., Fowles, 1980), such as poor aversive conditioning (Lykken, 1957; Schmauk, 1970) and passive avoidance learning (Lykken, 1957; Newman & Kosson, 1986; Newman & Schmitt, 1998; Schmauk, 1970). In these studies, psychopathic individuals showed a smaller increase in skin conductance level in anticipation of aversive stimuli, and were less able to use punishment cues to avoid aversive outcomes. Despite the fact that different aspects of the two-factor learning theory have been incorporated in psychopathy-specific models of contingency learning (e.g., *Low Fear hypothesis*, Lykken, 1995; *Response Modulation hypothesis*, Patterson & Newman, 1993), the theory itself has been losing traction over time. Part of the reason is that the theory lacks specificity and does not account for inter-individual differences stemming from variations in the aetiology of psychopathy. Perhaps more important is the fact that anxiety proneness is directly related to BIS sensitivity, but anxiety is also believed to be a key factor contributing to the heterogeneity observed across psychopathic individuals (e.g., Newman et al., 2005; Skeem et al., 2007). As such, although attractive and able to provide useful theoretical elements, the two-factor learning model is not able to explain the complex and heterogeneous nature of psychopathy. However, the notion that impaired experience of negative affective states plays a key role in understanding learning impairments in psychopathy was a cornerstone for other theories of psychopathy as well, such as the low-fear hypothesis (Lykken, 1957, 1995).

11.2.2   The Low-Fear Hypothesis

According to the low-fear hypothesis (Lykken, 1957, 1995), the most important mechanism underlying the impairments seen in psychopathy is a deficient emotional response to aversive events. The main assumption is that a deficit in experiencing fear in response to aversive outcomes impedes appropriate learning about the events leading to these bad outcomes. The belief is that psychopathic individuals already show difficulties in adapting their behaviour at an early stage of socialisation. During typical (moral) socialisation, children learn to behave appropriately by their tendency to avoid (parental) punishment, but children that are less sensitive to the negative valence of punishment are more likely to engage in the same (punished) behaviour again. Moreover, this theory assumes that the capacity to consider and act upon the negative consequences of behaviour largely relies on emotional processing. According to the low-fear theory, this explains why psychopathic individuals are more prone to take risks that non-psychopathic individuals would avoid. Although the avoidance learning deficits (Lykken, 1957; Newman & Kosson, 1986; Newman & Schmitt, 1998; Schmauk, 1970) and reduced electrodermal reactivity to threat cues (Hare, 1965; Lykken, 1957; Schmauk, 1970) observed in individuals with psychopathy can be regarded as evidence for the low-fear hypothesis, more recent advances in cognitive and neuroscientific research has indicated that it is likely that the involved mechanisms are far more complex than the low-fear theory assumes (see Hoppenbrouwers et al., 2016). In addition, Lykken (1995) has utilised constructs from the work of Fowles (1980) and Gray (1987) to adapt the original low-fear theory (Lykken, 1957). Indeed, psychopathy has been related to low fear, low anxiety, and reduced BIS reactivity (e.g., Baskin-Sommers et al., 2010; Lykken,

1995; Newman et al., 2005; Skeem et al., 2007). However, most of these studies measured BIS sensitivity using the BIS scales (Carver & White, 1994), which primarily assesses anxiety instead of fear (see Poythress et al., 2008). As anxiety and fear are two distinct constructs (Grillon, 2008), it seems unlikely that a reduced experience of fear as an emotion, rather than more basal threat processing impairments, is the core mechanism underlying psychopathic behaviour (Hoppenbrouwers et al., 2016). In addition, given that reinforcement learning requires, among others, memorising and updating affective information, theories explicitly incorporating the involvement of higher-order cognitive processes have gained more ground. One of the first cognitive theories that opposed the aetiological explanations provided by the low-fear account was the Response Modulation hypothesis developed by Newman and colleagues (Gorenstein & Newman, 1980; Newman & Kosson, 1986; Patterson & Newman, 1993).

11.2.3   Response Modulation Hypothesis

The Response Modulation (RM) hypothesis (Gorenstein & Newman, 1980; Newman & Kosson, 1986; Patterson & Newman, 1993) is an attention model postulating that psychopathic individuals have a strong preference for reward in combination with an early attentional bottleneck. When presented with both reward and punishment information, the attentional filter prevents peripheral (i.e., punishment) information from being processed as long as target (i.e., reward) information is available. When certain behaviour (e.g., robbing a bank; pressing a button in an experimental task) is rewarded under some conditions (e.g., getting away with money; earning points) and punished under other conditions (e.g., getting caught by the police; losing points), the experience of being rewarded will result in a dominant response set for exhibiting this behaviour, whereas punishment information will not have enough impact to produce avoidance behaviour. This decreased cognitive flexibility of psychopathic individuals, once a dominant response set has been established, results in perseveration of behaviour even when it is inappropriate (e.g., Lykken, 1957; Newman & Kosson, 1986; Schmauk, 1970).

However, different aspects of the model have been critiqued (see Blair & Mitchell, 2009), specifically for a lack of integration with more contemporary theories of attention. For instance, general models of top down attention (e.g., Posner & Rothbart, 2007) do not predict the automatic allocation of top down resources to peripheral information in healthy individuals, raising the question why this is considered to be an impairment in psychopathic individuals. In addition, psychopathic individuals perform comparably to healthy individuals in attentional set-shifting tasks, in which allocating attention to peripheral information is also required (Lapierre et al., 1995; Mitchell et al., 2002). Moreover, the model lacked integration with more recent neuroscientific findings about dysfunctional emotion circuitry in the brain, suggesting that abnormalities in the limbic system are fundamental to the emotional and behavioural dysregulation associated with psychopathy (e.g., Kiehl,

2006). A theory more directly incorporating such neuroscientific findings is the Integrated Emotion Systems model developed by Blair (2005).

11.2.4   Integrated Emotion Systems Model

The Integrated Emotion Systems (IES) model (Blair, 2005) is a neurocognitive model that assigns a central role to dysfunctional interactions between the amygdala and the ventromedial prefrontal cortex (vmPFC). According to the IES model, the dysfunction results in impaired reinforcement-based decision making in individuals with psychopathy under specific circumstances that are reliant on amygdala-vmPFC integration. In healthy individuals, the learning of stimulus-outcome associations, both aversive and appetitive, is dependent on the amygdala (Everitt et al., 2003; LeDoux, 2007). When an individual learns certain behaviours to gain reward or to avoid punishment (i.e., during instrumental learning), the amygdala sends the corresponding associations and expectancy information to the vmPFC. In turn, the vmPFC signals whether the expected reinforcement is present, continuously updating reinforcement expectancy representations. However, individuals with psychopathy have been found to be impaired in stimulus-outcome learning, particularly in reversal learning (e.g., Baskin-Sommers et al., 2015; Brazil et al., 2013; Budhani et al., 2006; Mitchell et al., 2002). One explanation for this impairment might be that during reversal learning, psychopathic offenders show increased vmPFC signalling when a previously rewarded response is punished, whereas healthy individuals and non-psychopathic offenders show decreased activation (Gregory et al., 2015). This suggests that the updating of expectancy representations may therefore be compromised. In addition, psychopathic individuals have also been found to be impaired in aversive conditioning (Lykken, 1957; Rothemund et al., 2012; Schmauk, 1970), and psychopathy has been associated with reduced amygdala activity during aversive conditioning (Birbaumer et al., 2005). Moreover, the transfer of reinforcement information from the amygdala to the vmPFC, which is essential for instrumental learning, is disrupted in individuals with psychopathy, as indexed by both reduced integrity of the white matter tracts and reduced functional connectivity between amygdala and vmPFC in psychopathy (Craig et al., 2009; Motzkin et al., 2011; Sundram et al., 2012; Vermeij et al., 2018).

To summarise, thus far we have described four explanatory models of the reinforcement learning deficits of psychopaths, and how they place different weights on the affective and attentional aspects of emotional information processing. The Two-Factor model and the Low-Fear model rely on the idea that impaired processing of aversive events explains the learning difficulties observed in psychopathic individuals. According to the Response Modulation hypothesis, an attentional bottleneck prevents (meaningful) peripheral information from being processed during ongoing goal-directed behaviour. The Integrated Emotion Systems (IES) model postulates that dysfunctional interactions between the amygdala and the vmPFC result in impaired reinforcement-based decision making. The outline above also illustrates how theories of reinforcement learning in psychopathy have been influenced by general technical and neuroscientific developments. Importantly, these developments also affected the way in which experimental testing is conducted. In order to be able to evaluate the findings from these experiments it is necessary to understand how these results were obtained and in which way the experimental paradigms evolved through time. As such, the next section describes the experimental paradigms most commonly used in this field, each followed by an overview of findings in psychopathic offenders.

11.3   Experimental Paradigms and Empirical Results

11.3.1   Aversive Conditioning

Aversive conditioning studies focus on autonomic reactivity to aversive stimuli, most often by measuring electrodermal responses. They are based on the principles of classic conditioning, in which an emotionally salient event (UCS) that evokes a biological (reflexive) response (UCR) is repeatedly paired with a neutral stimulus (NS). The NS becomes associated with the UCS, and is eventually transformed into a conditioned stimulus (CS) that will evoke the reflexive response (conditioned response: CR) in the absence of the UCS. During appetitive conditioning, repeated pairing of a pleasant stimulus with a neutral stimulus will eventually result in approach-related reflexes in response to the neutral stimulus, whereas pairing with unpleasant stimuli during aversive conditioning typically results in withdrawal-related reflexes. As such, abnormal conditioned reflexes may indicate deficient or excessive processing of appetitive or aversive events, or normal processing but deficient association of these unconditioned events with the NS. Physiological reactivity to conditioned threat cues can therefore be used to investigate the fear deficits that may underlie the failure to refrain from previously punished behaviour observed in psychopathic individuals.

David Lykken (1957) was the first to obtain experimental evidence for what he believed were fear-based learning deficits in individuals with psychopathy (for a contemporary view see Hoppenbrouwers et al., 2016). Lykken (1957) measured electrodermal reactivity to conditioned auditory cues for electric shock in an offender sample and found that individuals with psychopathy developed weaker anticipatory responses to threats than healthy controls. Almost a decade later, Hare (1965) reported that besides the slower acquisition of conditioned fear responses to shock cues in psychopathic offenders, these responses also generalised less to unconditioned cues than in controls. Schmauk (1970) found that anticipatory electrodermal reactivity to both shock cues and social disapproval cues was lower in psychopathic offenders than in controls, but no differences were found in anticipation of loss of money. Another two decades later, in a study by Ogloff and Wong (1990), psychopathic offenders displayed increased heart rate but no significant increase in electrodermal activity during anticipation of an aversive auditory stimulus.

Since the early days, several studies on brain activation during aversive conditioning in criminal psychopaths have been performed. Birbaumer et al. (2005) measured electrodermal responding and brain activation using fMRI during anticipation of painful pressure. Compared to healthy controls, criminal psychopaths (although non-incarcerated at the time of testing) showed reduced electrodermal reactivity and reduced brain activation in areas associated with the acquisition of conditioned threat (e.g., amygdala, orbitofrontal cortex, anterior cingulate, anterior insula; see Büchel & Dolan, 2000). However, when comparing probability and contingency ratings between the two groups, no differences emerged. This suggests that the psychopathic participants were able to predict the occurrence of harmful events from threat cues on a cognitive level, with deficits emerging during the emotional processing of this information. Similar results were obtained in a study by Rothemund et al. (2012), in which psychopathic individuals displayed deficient conditioned startle and skin conductance responses, whereas cognitive processing of the stimuli appeared intact.

In line with this central role for deficient emotional information processing, Veit et al. (2013) showed that deficient threat conditioning in psychopathic offenders, as reflected by reduced electrodermal reactivity in anticipation of electric shock, was most strongly related to the affective facet of psychopathy. On the other hand, event-related brain potentials (ERPs) showed scores on the interpersonal facet to be related to increased information processing, whereas the antisocial facet was related to decreased attention to the conditioned threat cues. Interestingly, Larson et al. (2013) showed in another fMRI study that manipulating the focus of attention could regulate the reduction in amygdala activation observed in psychopathic individuals during threat anticipation. When attention was explicitly directed to the threat cues signalling an electric shock, non-psychopathic and psychopathic offenders did not differ in amygdala activation during threat conditioning. Conversely, when attention was directed to goal-relevant non-threatening stimuli prior to the presentation of the threat cues, psychopaths displayed decreased amygdala activation and increased activation in the lateral prefrontal cortex. Schultz et al. (2016), however, observed enhanced amygdala responding to conditioned threat cues in psychopathic individuals relative to controls. Moreover, disrupted processing of conditioned threat cues in psychopathy was related to level of anxiety. BOLD activity patterns and electrodermal responses in low anxious psychopathic individuals were consistent with normal threat conditioning, whereas electrodermal responses and brain activity patterns consistent with fear inhibition were observed in high anxious psychopathic individuals.

11.3.1.1 Summary: Aversive Conditioning in Psychopathic Offenders

Offenders with high levels of psychopathy consistently show reduced autonomic responding to conditioned threat cues as indexed by electrodermal reactivity, at least when threat cues indicate physical harm (i.e., electric shock; painful pressure; loud noises) or social disapproval. However, cues indicating loss of money do not elicit abnormal autonomic responding. It might be argued that cues regarding social disapproval and loss of money require more higher-order cognitive processing, whereas the primary reflexes associated with the avoidance of physical harm may reflect a more direct measure of threat conditioning. More recently, imaging studies have provided evidence for reduced activation in brain areas associated with threat conditioning, such as the amygdala and the orbitofrontal cortex. However, the focus of attention during aversive conditioning may modulate amygdala reactivity to threat cues, with explicit direction of attention to aversive stimuli resulting in normal amygdala responses. Interestingly, one study to date showed enhanced instead of reduced amygdala activation in psychopathy and suggests that the threat-conditioning deficit pertains exclusively to high anxious as opposed to low anxious psychopathic individuals. All in all, it is evident that more research is needed to investigate how the different underlying aetiological mechanisms of psychopathy (e.g., on a factor or facet level) contribute to the observed deficiencies in threat conditioning. Other forms of associative learning have also been studied, with a focus on instrumental learning.

11.3.2 Instrumental Learning Paradigms

11.3.2.1 Passive Avoidance Learning

One of the most extensively studied instrumental learning deficits in psychopathy is passive avoidance learning. During passive avoidance learning, participants are instructed to learn by trial and error, which stimuli to respond to and which stimuli to withhold responding to. Immediately after responding to a stimulus, positive (i.e., rewarding) or negative (i.e., punishing) feedback is presented. The participant should use this feedback information to guide future behaviour during encounters with the stimuli.

Lykken (1957) found that psychopathic offenders were less successful than non-psychopathic offenders and controls in learning to avoid shock punishment. Participants were instructed to learn a sequence of 20 'choice points' in a mental maze, each consisting of four alternatives. In each choice point, one alternative was correct and one of the three incorrect alternatives gave an electric shock punishment. The manifest task was to learn to choose the (rewarded) correct alternatives, whereas the latent task was to learn to avoid the punished incorrect alternatives. Lykken (1957) found that psychopathic offenders performed significantly worse than controls on the latent task. This was the first study providing evidence for avoidance learning deficits in psychopathy. In a similar paradigm, Schmauk (1970) partly replicated this finding, as psychopathic offenders performed worse than healthy controls when electric shocks or social disapproval were used as punishment, but performed equally well when the punishment was loss of money. Also using a similar task, Schachter and Latané (1964) found that an injection with norepinephrine, which increases (emotional) arousal, improved avoidance learning in offenders with psychopathy, but not in non-psychopathic offenders. These results were regarded as evidence for deficient punishment processing, or, more general, a fear deficit.

Newman and Kosson (1986) developed a go/no-go discrimination task, in which participants were presented with eight different two-digit numbers of which half were go-stimuli (S+) and the other half were no-go stimuli (S-). Participants were instructed to learn to respond to S+ and to withhold a response to S-. The task was performed under two conditions: a punishment-only condition, in which participants only had the opportunity to learn from punishment for incorrect responses, and a reward+punishment condition, in which both correct responses were rewarded and incorrect responses were punished. Interestingly, there were no differences between psychopathic and non-psychopathic offenders in the punishment-only condition, but in the reward+punishment condition, psychopathic offenders made significantly more commission errors than the non-psychopathic offenders. There were no group differences in the number of omission errors. According to Newman and Kosson (1986), these results could not be explained by a 'simple' fear deficit, as in that case there also would have been differences in the punishment-only condition. Instead, they attributed the differences to disturbances in attentional processing and developed the RM- hypothesis (see previous section).

A number of variations of this task have been developed over the last decades. For example, to establish a dominant response set, by providing a high probability of reward for responding at the start of the task, Newman et al. (1990) gave participants a four-trial reward pre-treatment for the four S+. Again, psychopathic and non-psychopathic offenders did not differ in the number of omission errors, and psychopathic offenders did make significantly more commission errors than the non-psychopathic offender group. Newman and Schmitt (1998) replicated this finding using the reward pre-treatment, but when the groups were split in low-anxious and high-anxious subgroups, the difference was only observed in the low-anxious subgroups. When applying this variation of the paradigm in incarcerated female offenders, no group differences in either commission or omission errors were observed between psychopathic and non-psychopathic offenders, or between low-anxious and high-anxious subgroups (Vitale et al., 2011).

In addition, Newman et al. (1990) measured reflection after negative feedback, as indexed by the response time to terminate visual feedback on the screen in order to move on to the next trial. The extent to which participants slow down after punishment is generally the most predictive of passive avoidance learning (Patterson et al., 1987). Psychopathic offenders displayed less reflection after negative feedback, and when reflection required the interruption of a dominant response set, the differences between the two groups in passive avoidance errors were most profound.

In another adaptation of the go/no go-task of Newman and Kosson (1986), Arnett et al. (1993) measured autonomic responsivity to reward and punishment feedback. Low-anxious but not high-anxious psychopathic offenders displayed lower heart rate responding following punishment than following reward. Moreover, following punishment, psychopathic offenders showed weaker heart rate and skin conductance responding than non-psychopathic offenders. These results were interpreted as the first evidence of psychopaths being less reactive to punishment, as until then, there was only evidence of reduced autonomic responding in anticipation of punishment. Importantly, no behavioural differences in either commission or omission errors were observed between the psychopathic and non-psychopathic offenders. However, this might be explained by the long and variable inter-stimulus intervals (8–14 s) that were incorporated to measure autonomic responding to feedback, forcing longer time to reflect on the outcomes of previous responses.

In order to evaluate effects of differential reward and punishment value on passive avoidance learning, Blair et al. (2004) attached different values to the four different S+ and S- stimuli in the design of Newman and Kosson (1986). As expected, psychopathic offenders made more commission errors than non-psychopathic offenders. In addition, psychopaths displayed a weaker learning effect across blocks than non-psychopathic offenders. It should be noted, however, that intelligence was a significant covariate and only a modest correlation between PCL-R score and commission errors remained after controlling for IQ. Interestingly, punishment value was not related to the performance of psychopathic offenders, but non-psychopathic offenders made more commission errors as punishment level increased. In addition, both groups were more likely to respond under high reward conditions, as indexed by a decrease in omission errors. DeBrito et al. (2013) used the same task as Blair et al. (2004) in offenders with and without psychopathy and included a healthy control group. Both offender groups tended to make more commission errors than healthy controls, although this difference did not reach statistical significance. However, contrary to previous findings, there were no differences in the number of commission errors between the two offender groups. In addition, unlike the two comparison groups, psychopathic offenders were more likely to respond under the lowest reward value condition, which is a different finding than reported in Blair et al. (2004).

A portion of these studies suggests an effect of anxiety level on passive avoidance learning in psychopathic individuals. Both Lykken (1957) and Schmauk (1970) divided their psychopathic participant groups in 'primary (low-anxious) sociopaths' and 'neurotic (high-anxious) sociopaths'. In Lykken's study, the difference in performance of low anxious and high-anxious psychopaths was not statistically significant. In the study of Schmauk (1970), low-anxious psychopaths showed stronger passive avoidance learning deficits than high-anxious psychopaths in the psychical punishment condition. However, in the social punishment and the tangible punishment condition, the two psychopathic subgroups did not differ in passive avoidance learning. Other studies analysing subgroups of low-anxious and high-anxious offenders have found that the observed effects were either stronger in the subgroup of low-anxious psychopathic offenders (Arnett et al., 1993), or the difference between psychopathic and non-psychopathic offenders was only present between the low-anxious subgroups (Newman & Schmitt, 1998). On the other hand, psychopathic and non-psychopathic offenders in the study of Newman and Kosson (1986) did not differ in anxiety levels. Moreover, Kosson et al. (1990) as well as Thornquist and Zuckerman (1995), both using the go/no go-task of Newman and Kosson (1986), found PCL-R score and anxiety to be unrelated, although group differences based on anxiety level were not directly assessed. Vitale et al. (2011) did find a positive relation between PCL-R score and anxiety in female offenders, but there were no differences in passive avoidance learning between the low- and high-anxious subgroups. In addition, Newman et al. (1990), finding that low-anxious psychopathic offenders were less likely to interrupt a dominant response set to process negative feedback than low-anxious non-psychopathic offenders, reported that their results were even stronger when high-anxious participants were included. Others, such as DeBrito et al. (2013) and Blair et al. (2004), did not include measures of anxiety level. Overall, the exact role of anxiety in passive avoidance learning remains unclear.

Differences in psychopathy-related passive avoidance learning deficits have also been reported in relation to ethnic differences, but it is still unclear what underlies these observations. Moreover, ethnic differences are not consistently observed in all passive avoidance studies. Newman & Schmitt, (1998) and Thornquist and Zuckerman (1995) only observed the expected psychopathy-specific passive avoidance learning deficit in Caucasian offenders, and not in African-American or Hispanic offenders. However, Kosson et al. (1990) did observe the expected pattern of psychopathy-specific passive avoidance learning deficits in African-American offenders, but the effect was not as profound as observed in previously obtained data from Caucasian offenders, since group differences did not reach statistical significance. Importantly, when combining the data from this sample with the previously obtained data from their Caucasian offender sample, there were no effects of ethnic background on passive avoidance learning. Although not explicitly testing for ethnic differences due to a small number of non-Caucasian participants, Blair et al. (2004) reported that there were no indications that psychopathic and non-psychopathic African-American offenders performed differently than the Caucasian offenders in these respective groups. Other forms of instrumental learning, in particular reversal learning, have also been comprehensively studied in relation to psychopathy.

11.3.2.2 Reversal Learning

As discussed in the section on passive avoidance learning, psychopathy is associated with deficiencies in learning stimulus-outcome contingencies. However, psychopathic individuals also have difficulties to adapt their behaviour to changing contingencies (e.g., Baskin-Sommers et al., 2015; Brazil et al., 2013; Budhani et al., 2006; Mitchell et al., 2002). The updating of information on reward and punishment contingencies appears disturbed, which is typically studied in reversal learning paradigms. During reversal learning, participants first acquire stimulus-reinforcement associations guiding them to discriminate between rewarding and punishing stimuli. At a certain point, however, the learned reinforcement contingencies will reverse and participants have to re-learn the discrimination in order to gain reward and to avoid punishment.

In the first study on reversal learning in psychopathic offenders, LaPierre et al. (1995) used a go/no-go paradigm. In the first 50 trials a strong response set was created, by having participants learn to respond to one stimulus (a square) and to avoid responding to another stimulus (a cross). In the next 150 trials participants had to withhold their response to the square and to respond to the cross. Psychopathic offenders made more commission errors than non-psychopathic offenders in the reversal phase, but there were no differences in omission errors or reaction times. However, LaPierre et al. (1995) did not report on performance during acquisition, whether there were any rewards or punishments contingent on responding, or how feedback was provided. More recently, Brazil et al. (2013) also used a go/no-go reversal task with two stimuli, but this time including two distinct cues indicating whether a go or a no-go stimulus was more likely to follow. Halfway the task, the predictive (probabilistic) relationship between the cues and the stimuli was reversed. Participants performed the task twice: once without instructions on the predictive relationship between the cue and the stimulus (i.e., automatic learning) and once with explicit instructions on this relationship (i.e., controlled learning). Psychopathic offenders, unlike healthy controls, displayed abnormal response reversal during controlled learning (as indexed by prolonged reaction times), but this impairment was absent in the automatic learning condition. These results suggest that psychopathic individuals do not have a general response reversal deficit, but that they experience problems in behavioural adaptation when information on predictive relationships between stimuli is actively processed.

Mitchell et al. (2002) used the intradimensional/extradimensional (ID/ED) shift task to assess reversal learning, which is a multicomponent instrumental learning task. The ID/ED shift task has nine phases, in which participants have to respond to different features of presented stimuli. After a fixed number of correct trials, the task shifts to the next phase. The task starts with two different shapes and the participant has to learn to respond to shape 1 and to withhold a response to shape 2. Once the participant has learned this discrimination, the contingencies are reversed. In the next phases, new shapes and features are added, cueing participants to adapt their responding (i.e., attentional set shifting), or the newly learned contingencies are simply reversed again (i.e., response reversal). The measure of interest in this task is the number of errors within a phase, before shifting to the next phase (i.e., errors to criterion).

Mitchell et al. (2002) found that psychopathic offenders made more errors than non-psychopathic offenders in two of the four reversal phases, whereas there were no differences in attentional set shifting performance. Dargis et al. (2017) obtained similar results with the ID/ED shift task, but also found an interaction between psychopathy and childhood maltreatment history. Offenders with higher levels of psychopathic traits who had suffered a greater degree of childhood maltreatment performed worse on reversal learning. Interestingly, when controlling for childhood maltreatment history, psychopathic offenders did not differ from offenders with low and intermediate levels of psychopathy on reversal learning performance. Dolan (2012), however, found that offender groups with differing levels of psychopathic traits all performed worse than healthy controls in the reversal phases as well as the attentional shift phases, but there were no differences between the offender groups. In addition, psychopathy scores were not related to any outcome measures of the ID/ED task.

Another reversal learning paradigm was developed by Budhani et al. (2006). On each trial, participants were presented with two images. Using probabilistic feedback that was provided after choosing one of the two images, participants learned by trial and error which image was the correct choice most often and were instructed to stay with this choice until the contingencies were reversed. During reversal, psychopathic offenders made more errors and were less likely to stay with a rewarded response than controls, whereas no impairments were observed during acquisition. DeBrito et al. (2013) also applied this paradigm, but only observed differences between a combined offender group and controls in the number of errors during reversal. Similar to the previously discussed results of their passive avoidance study, which were obtained in the same sample, the subgroups of offenders that were high and low in psychopathy did not differ in response reversal performance.

Mitchell et al. (2006) designed an instrumental learning task with two reversal phases, which was presented to offender groups with differing levels of psychopathy. In the acquisition phase, participants were instructed to choose one of two stimuli presented on each trial, learning the stimulus-outcome associations by trial and error. In the first reversal phase, the contingencies of two of the four of the stimuli were reversed and in the second reversal phase the contingencies of the other two stimuli were reversed. Highly psychopathic offenders performed worse during acquisition and the second reversal phase than the group with low levels of psychopathy. The intermediate group also performed worse than the low-psychopathy group in the second reversal phase. Although there were no performance differences between the high and intermediate psychopathy group, psychopathy scores were negatively related to performance during acquisition and the second reversal phase.

In order to further disentangle the relation between psychopathy level and response reversal deficits, Gregory et al. (2015) investigated the neural basis of reversal learning in antisocial offenders with and without psychopathy. Although behavioural differences between the groups were not observed, there were remarkable group differences regarding brain activation in response to rewarded and punished responses. In psychopathic offenders, activity in the posterior cingulate cortex and anterior insula was increased in response to punished reversal errors. Additionally, offenders in this group were hyporesponsive to reward information in the superior temporal gyrus. These patterns were not seen in offenders without psychopathy and suggest that prediction error signalling and consolidation of reward information is dysfunctional in psychopathy.

11.3.2.3 Summary: Instrumental Learning Abnormalities in Psychopathy

Psychopathy has consistently been found to be related to deficits in instrumental learning based on stimulus-outcome contingencies. During passive avoidance learning, psychopathic offenders show deficits in withholding responses to avoid punishment, particularly when a dominant response set for reward has been established. However, the influence of other variables on passive avoidance learning, such as anxiety levels or ethnic background, is still unclear. Offenders with high levels of psychopathy are also impaired in adapting their behaviour to changing contingencies as indexed by reversal learning. However, performance differences between offender groups with different levels of psychopathy are not as robust as the differences found when comparing psychopathic offenders with healthy controls. In addition, other cognitive and clinical variables, such as the level of processing of predictive information or childhood maltreatment history, appear to play an important role in the severity of the reversal learning impairment in psychopathy. Also, several maladaptive behavioural outcomes share underlying mechanisms of cognitive impairments associated with psychopathy. For instance, impaired processing of predictive information, such as the probability of upcoming aversive outcomes, does not only affect reversal learning and passive avoidance, but also promotes risky decision making, which we will further discuss in the next section.

11.3.3 Risk-Taking Studies

A key characteristic of psychopathy is a strong need for stimulation; psychopathic individuals often show excessive risk behaviour, such as sexual risk taking and substance abuse. Moreover, offences committed by psychopathic individuals often have a violent and/or sexual nature and can therefore be characterised as behaviour intended to gain immediate rewards despite potential punishment. As such, risk-taking studies performed in psychopathic offenders focus on the ability to forego potential large immediate rewards for small longer-term rewards to avoid larger losses.

Risk-taking behaviour has often been explained using the framework of the somatic marker hypothesis (Bechara et al., 1994), which states that autonomic physiological reactions to learned appetitive or aversive cues rather than cognitive processes guide choices under ambiguous circumstances. The Iowa Gambling Task (IGT; Bechara et al., 1994) involves probabilistic learning using (monetary) reward and punishment information and was developed as a test of the somatic marker hypothesis. Participants are given four decks of cards, of which two are 'risky decks' involving high reward and even higher punishment magnitudes, and the other two are 'non-risky decks' involving lower reward and punishment magnitudes. Over time, selection of the non-risky decks results in the greatest accumulated reward magnitude. In healthy individuals, increased anticipatory electrodermal responses are present before choosing cards from the risky decks and these implicit, unconscious markers guide them to choose advantageously throughout the task. However, individuals with lesions in the vmPFC do not develop these anticipatory warning signals (i.e., somatic markers), resulting in impaired decision making. Individuals with psychopathy show behavioural and affective similarities with orbitofrontal patients, such as impulsivity, low empathy and impaired learning from experience, and recent research suggests that psychopathic individuals are indeed impaired in recognising their bodily sensations during stressful events (Gao et al., 2012; Nentjes et al., 2013).

The IGT has also been used to study risk behaviour in psychopathy, but studies focusing on IGT performance in psychopathic offenders primarily use behavioural measures rather than including additional indices of autonomic physiological responses. Only one (unpublished) study (Broom, 2011) included autonomic measures and found stronger electrodermal responses in psychopathic offenders compared to non-psychopathic offenders after selecting cards from the non-risky decks. Contrary to predictions based on the somatic marker hypothesis, no relation was found between psychopathy and anticipatory autonomic responses. Although this study used a modified version of the IGT including contingency reversals, overall psychopathy score was related to impaired performance throughout the task. In the previously discussed Mitchell et al. (2002) study, psychopathy was also related to impaired performance on the IGT. In line with the somatic marker hypothesis, psychopathic offenders tended to choose cards from the risky decks more often and failed to become risk averse over the course of the task. Similar results were obtained in a sample of ex-offenders (Beszterczey et al., 2013). Both PCL-R total score (reflecting overall psychopathy) and PCL-R Factor 2-score (reflecting an unstable and antisocial lifestyle) were positively related to IGT performance. Interestingly, IGT-scores strongly predicted recidivism at follow-up. Conversely, Lösel and Schmucker (2004) found no relation between psychopathy and IGT performance in offenders. However, analysing high-attentive and low-attentive subgroups revealed that low-attentive psychopathic offenders performed worse than individuals in the high-attentive psychopathic subgroup, whereas no differences were found between the non-psychopathic offender subgroups. Using both psychopathy and level of anxiety as grouping variables, Schmitt et al. (1999) found low-anxious offenders to perform worse than high-anxious offenders, but no predictive relation between psychopathy and learning in the IGT. Along the same line, Kuin and Masthoff (2016) found no relation between IGT performance and general psychopathy or specific psychopathic traits. Contradictory to other findings, Hughes et al. (2015) found psychopathy to be positively related to IGT performance. However, this study was somewhat atypical, since all three groups of participants (healthy controls, non-psychopathic and psychopathic offenders) failed to show learning over the course of the task.

Another well-known risk-taking task is the Balloon Analogue Risk Task (BART; Lejuez et al., 2002), in which participants are instructed to accumulate money or points by inflating balloons. Every button press inflates a balloon presented on a computer screen and increases the amount of money in a temporary bank. The money that has been accumulated in the temporary bank can be transferred into a permanent bank at each point in the experiment, after which a new balloon is presented. However, inflating the balloon too much will make it 'pop', resulting in a loss of all money accumulated in the temporary bank. Unlike the IGT, the BART does not involve a learning component. Although risk behaviour on the BART correlates with real-world risk behaviour also seen in psychopathic individuals such as substance abuse, gambling, stealing and unsafe sex (Lejuez et al., 2002, 2003a, b), Swogger et al. (2010) found no relation between psychopathy scores and BART performance in offenders, although there was a relation between psychopathy and self-reported real-world risk-taking behaviour. Moreover, similar to the results of Schmitt et al. (1999) obtained in the IGT, there was a negative relation between anxiety and BART performance. Snowden et al. (2017) did find a relation between psychopathy and risk taking on the BART in a mixed offender and community sample, but this effect could be largely attributed to Boldness dimension of the triarchic psychopathy model (Patrick et al., 2009) rather than to the Meanness and Disinhibition dimensions. As such, BART performance seems stronger related to psychopathy-associated fearlessness than to affective impairments and antisocial tendencies.

DeBrito et al. (2013) used the Cambridge Gambling Task (CGT; Rogers et al., 1999) to assess risk taking in psychopathic offenders. Like the BART (Lejuez et al., 2002) the CGT does not include a learning component. On each trial, participants are presented with a row of ten boxes that are either red or blue. Participants have to guess under which of the two colours a token has been hidden by betting a proportion of their earned points or money. DeBrito et al. (2013) did find controls to outperform offenders with and without psychopathy on decision-making quality (i.e., the proportion of trials the most likely colour is chosen), but there were no group differences in risk taking (i.e., the percentage of earned points that is betted in each trial) and pre-betting deliberation time. These findings suggest that both offender groups were cognitively aware of the risks associated with certain choices, but failed to adjust their behaviour accordingly. Since there were no differences between psychopathic and non-psychopathic offenders, antisociality rather than psychopathy seems to account for these performance deficits.

11.3.3.1   Summary: Risk-Taking Behaviour in Psychopathy

According to the somatic marker hypothesis, psychopathic individuals should fail to develop anticipatory warning signals towards risky events or choices. Although research findings are quite mixed, most studies suggest that psychopathy is negatively related to task performance in risk-taking tasks. However, attention may moderate the relation between psychopathy and impaired learning in the IGT. Moreover, primarily psychopathic characteristics related to lifestyle instability and antisociality appear to be related to impaired decision making under risk in the IGT, rather than interpersonal and affective psychopathic features. This makes sense as the similarities between patients with orbitofrontal lesions and those with psychopathy mostly pertain to the behavioural domain. However, research using the BART in psychopathic offenders indicates that performance on the BART is stronger related to low anxiety levels associated with psychopathy, rather than emotional deficits or antisocial and impulsive behavioural tendencies. Moreover, one study using the IGT and one study using the BART did not find performance to be related to psychopathy, but to anxiety level. Taken together, findings from studies using behavioural measures of risk taking are far from conclusive, and more research is needed to disentangle the mechanisms explaining risk behaviour in psychopathic offenders. Specifically, the role of anxiety, attention and the underlying aetiology of the psychopathy construct in offenders needs to be further clarified.

Recently, different researchers have pointed to the need for differentiation between risk and ambiguity in decision-making studies (Buckholtz et al., 2017; Maes et al., 2018). In decisions under risk, the outcomes of one's choices have known (or knowable) probabilities, whereas ambiguous decision making entails unknown (and unknowable) outcome probabilities. Where antisocial behaviour is often referred to as the outcome of 'risky' decision making, Buckholtz et al. (2017) argue that these decisions can be better classified as 'ambiguous', since the exact probabilities of aversive outcomes of criminal behaviour are, in fact, unknowable. Incorporating operationalisations of ambiguity in experimental decision-making paradigms would therefore contribute to the ecological validity of these tasks when used to study psychopathic offender populations. We will further elaborate on this topic in the next section of this chapter.

11.4 Considering Ecological Validity: Daily Life and Subjectivity of Reward Value

The studies outlined above illustrate the impairments in reward and punishment processing in psychopathy. Most importantly, these studies show deficient responding to (predictors of) punishment, especially when facing a competing reward, as well as impaired learning and decision making following reward. However, a possibly crucial limitation of these studies is that the rewards that were used may not have been ecologically valid, and consequently not relevant or motivationally significant for participants with psychopathy, particularly for those that were incarcerated. The performance of psychopathic individuals in studies that focused on feedback-based learning and risky decision making in the lab setting may, therefore, not be fully generalisable to the problems they encounter in daily life. Incorporating a variety of ecologically valid rewarding stimuli in task design could be a way to overcome this problem, such as food, small goods, or pleasant activities.

However, in this light it should be considered that psychopathic individuals might be impaired in neural coding for subjective reward value, which should normalise the values of rewards of different natures. Normalising reward values is necessary when, for instance, comparing or choosing between a chocolate bar and a movie ticket. More specifically, there is convincing evidence that in healthy individuals a subarea of the vmPFC/orbitofrontal cortex (OFC) represents the subjective value of different reward types in a common neural currency that is used to direct decision making in daily life (Levy & Glimcher, 2012). This representational system is subject-specific and has been found to be active across various tasks. The suggestion that this system is dysfunctional in individuals with psychopathy is based on findings in neuropsychological studies and in structural and functional imaging studies.

First, as already noted in previous paragraphs, psychopathy has been associated with impaired performance on neuropsychological tasks relying on the vmPFC/OFC. Imaging studies have also linked psychopathy to abnormalities in vmPFC/OFC structure and functioning. For instance, on the structural level, reductions in orbitofrontal grey matter have been observed in psychopathic individuals compared to non-psychopathic individuals (Boccardi et al., 2011; de Oliveira-Souza et al., 2008; Tiihonen et al., 2008). Moreover, in a sample of individuals with high levels of psychopathic traits, grey matter volume and cortical thickness in the OFC was reduced in those with self-reported criminal convictions compared to those without a criminal record (Yang et al., 2010). This is compatible with the finding that cortical thickness in the OFC region is inversely related to response perseveration (Yang et al., 2011), an impairment in executive functioning typically linked to antisocial behaviour as observed in criminal psychopathic individuals (Morgan & Lilienfeld, 2000; Newman et al., 1987). In addition, although there is some evidence for increased activity in the vmPFC/OFC during specific tasks (e.g., instructed lying; see Glenn et al., 2017), the majority of functional imaging studies directed at the vmPFC/OFC in psychopathy show reduced activity in these areas, using a variety of tasks tapping different underlying mechanisms. These findings include reduced medial vmPFC/OFC activity in psychopathic individuals during aversive conditioning (Birbaumer et al., 2005), during cooperation choices in the prisoner's dilemma paradigm (Rilling et al., 2007), as well as in adolescents with psychopathic tendencies during reinforcement in a passive avoidance task (Finger et al., 2011).

If coding for subjective reward value in the vmPFC/OFC is also compromised in psychopathic individuals, this would implicate that their failure to learn from negative consequences and their tendency to make suboptimal -sometimes catastrophic-choices in daily life might partly stem from an inability to weigh their behavioural options on a common scale. However, we do not know whether, or to what extent, this function of the vmPFC/OFC region is also affected, or that psychopathic individuals still have more or less intact coding for subjective value. Regarding the lab setting, there is often no evidence on whether psychopathic individuals find the rewards equally attractive as, or at least comparable to, the control groups. As such, it seems important to find a way to test this, for instance, by including reward attractiveness as an experimental variable. One way to achieve this could be to make subjective reward values more explicit, or to use tailor made rewards to ensure their motivational relevance. The question would be, then, whether psychopathic individuals' deficient responding to (potentially) punishing stimuli, as well as their disturbed processing of rewards, would remain. Considering the evidence from the studies cited above, it could be expected that the 'right' rewards and punishers would motivate psychopathic offenders to make more appropriate behavioural adaptations, both in the lab setting and in real life situations. Taking these issues into consideration in future research would be beneficial to the field of reward-based learning and decision making in psychopathy; an exciting prospect of an area yet to be further developed.

References

Arnett, P. A., Howland, E. W., Smith, S. S., & Newman, J. P. (1993). Autonomic responsivity during passive avoidance in incarcerated psychopaths. *Personality and Individual Differences, 14*(1), 173–184. https://doi.org/10.1016/0191-8869(93)90187-8

Baskin-Sommers, A. R., Curtin, J. J., & Newman, J. P. (2015). Altering the cognitive-affective dysfunctions of psychopathic and externalizing offender subtypes with cognitive remediation. *Clinical Psychological Science, 3*(1), 45–57. https://doi.org/10.1177/2167702614560744

Baskin-Sommers, A. R., Wallace, J. F., MacCoon, D. G., Curtin, J. J., & Newman, J. P. (2010). Clarifying the factors that undermine behavioral inhibition system functioning in psychopathy. *Personality Disorders: Theory, Research, and Treatment, 1*(4), 203–217. https://doi.org/10.1037/a0018950

Bechara, A., Damasio, A. R., Damasio, H., & Anderson, S. W. (1994). Insensitivity to future consequences following damage to human prefrontal cortex. *Cognition, 50*(1–3), 7–15. https://doi.org/10.1016/0010-0277(94)90018-3

Beszterczey, S., Nestor, P. G., Shirai, A., & Harding, S. (2013). Neuropsychology of decision making and psychopathy in high-risk ex-offenders. *Neuropsychology, 27*(4), 491–497. https://doi.org/10.1037/a0033162

Birbaumer, N., Veit, R., Lotze, M., Erb, M., Hermann, C., Grodd, W., & Flor, H. (2005). Deficient fear conditioning in psychopathy: A functional magnetic resonance imaging study. *Archives of General Psychiatry, 62*(7), 799–805. https://doi.org/10.1001/archpsyc.62.7.799

Blair, J. R. (2005). Applying a cognitive neuroscience perspective to the disorder of psychopathy. *Development and Psychopathology, 17*(3), 865–891. https://doi.org/10.1017/S0954579405050418

Blair, R. J. R., & Mitchell, D. G. V. (2009). Psychopathy, attention and emotion. *Psychological Medicine, 39*(4), 543–555. https://doi.org/10.1017/s0033291708003991

Blair, R. J. R., Mitchell, D. G. V., Leonard, A., Budhani, S., Peschardt, K. S., & Newman, C. (2004). Passive avoidance learning in individuals with psychopathy: Modulation by reward but not by punishment. *Personality and Individual Differences, 37*(6), 1179–1192. https://doi.org/10.1016/j.paid.2003.12.001

Boccardi, M., Frisoni, G. B., Hare, R. D., Cavedo, E., Najt, P., Pievani, M., … Tiihonen, J. (2011). Cortex and amygdala morphology in psychopathy. *Psychiatry Research: Neuroimaging, 193*(2), 85–92. https://doi.org/10.1016/j.pscychresns.2010.12.013

Brazil, I. A., Maes, J. H. R., Scheper, I., Bulten, B. H., Kessels, R. P. C., Verkes, R., & De Bruijn, E. R. A. (2013). Reversal deficits in psychopathy in explicit but not implicit learning conditions. *Journal of Psychiatry and Neuroscience, 38*, e13–e20. https://doi.org/10.1503/jpn.120152

Broom, I. (2011). *The relationship between psychopathy and performance on a modified version of the Iowa Gambling Task in offender and undergraduate student samples* (Doctoral dissertation, Carleton University Ottawa).

Büchel, C., & Dolan, R. J. (2000). Classical fear conditioning in functional neuroimaging. *Current Opinion in Neurobiology, 10*(2), 219–223. https://doi.org/10.1016/S0959-4388(00)00078-7

Buckholtz, J. W., Karmarkar, U., Ye, S., Brennan, G. M., & Baskin-Sommers, A. (2017). Blunted ambiguity aversion during cost-benefit decisions in antisocial individuals. *Scientific Reports, 7*(1), 2030. https://doi.org/10.1038/s41598-017-02149-6

Budhani, S., Richell, R. A., & Blair, R. J. R. (2006). Impaired reversal but intact acquisition: Probabilistic response reversal deficits in adult individuals with psychopathy. *Journal of Abnormal Psychology, 115*(3), 552–558. https://doi.org/10.1037/0021-843X.115.3.552

Carver, C. S., & White, T. L. (1994). Behavioral inhibition, behavioral activation, and affective responses to impending reward and punishment: The BIS/BAS scales. *Journal of Personality and Social Psychology, 67*(2), 319–333. https://doi.org/10.1037/0022-3514.67.2.319

Craig, M. C., Catani, M., Deeley, Q., Latham, R., Daly, E., Kanaan, R., … Murphy, D. G. (2009). Altered connections on the road to psychopathy. *Molecular Psychiatry, 14*(10), 946–953. https://doi.org/10.1038/mp.2009.40

Dargis, M., Wolf, R. C., & Koenigs, M. (2017). Reversal learning deficits in criminal offenders: Effects of psychopathy, substance use, and childhood maltreatment history. *Journal of Psychopathology and Behavioral Assessment, 39*(2), 189–197. https://doi.org/10.1007/s10862-016-9574-6

De Brito, S. A., Viding, E., Kumari, V., Blackwood, N., & Hodgins, S. (2013). Cool and hot executive function impairments in violent offenders with antisocial personality disorder with and without psychopathy. *PLoS One, 8*(6), e65566. https://doi.org/10.1371/journal.pone.0065566

de Oliveira-Souza, R., Hare, R. D., Bramati, I. E., Garrido, G. J., Azevedo Ignacio, F., Tovar-Moll, F., & Moll, J. (2008). Psychopathy as a disorder of the moral brain: Fronto temporo-limbic grey matter reductions demonstrated by voxel-based morphometry. *Neuro Image, 40*(3), 1202–1213. https://doi.org/10.1016/j.neuroimage.2007.12.054

Dolan, M. (2012). The neuropsychology of prefrontal function in antisocial personality disordered offenders with varying degrees of psychopathy. *Psychological Medicine, 42*(8), 1715–1725. https://doi.org/10.1017/S0033291711002686

Everitt, B. J., Cardinal, R. N., Parkinson, J. A., & Robbins, T. W. (2003). Appetitive behavior: Impact of amygdala-dependent mechanisms of emotional learning. *Annals of the New York Academy of Sciences, 985*(1), 233–250. https://doi.org/10.1111/j.1749-6632.2003.tb07085.x

Finger, E. C., Marsh, A. A., Blair, K. S., Reid, M. E., Sims, C., Ng, P., … Blair, R. J. R. (2011). Disrupted reinforcement signaling in the orbitofrontal cortex and caudate in youths with conduct disorder or oppositional defiant disorder and a high level of psychopathic traits. *American Journal of Psychiatry, 168*(2), 152–162. https://doi.org/10.1176/appi.ajp.2010.10010129

Fowles, D. C. (1980). The three arousal model: Implications of Gray's two-factor learning theory for heart rate, electrodermal activity, and psychopathy. *Psychophysiology, 17*, 87–104. https://doi.org/10.1111/j.1469-8986.1980.tb00117.x

Gao, Y., Raine, A., & Schug, R. A. (2012). Somatic aphasia: Mismatch of body sensations with autonomic stress reactivity in psychopathy. *Biological Psychology, 90*, 228–233. https://doi.org/10.1016/j.biopsycho.2012.03.015

Glenn, A. L., Han, H., Yang, Y., Raine, A., & Schug, R. A. (2017). Associations between psychopathic traits and brain activity during instructed false responding. *Psychiatry Research: Neuroimaging, 266*, 123–137. https://doi.org/10.1016/j.pscychresns.2017.06.008

Gorenstein, E. E., & Newman, J. P. (1980). Disinhibitory psychopathology: A new perspective and a model for research. *Psychological Review, 87*, 301–315. https://doi.org/10.1037/0033-295X.87.3.301

Gray, J. A. (1987). *The psychology of fear and stress*. Cambridge University Press.

Gregory, S., Blair, R. J., Simmons, A., Kumari, V., Hodgins, S., & Blackwood, N. (2015). Punishment and psychopathy: A case-control functional MRI investigation of reinforcement learning in violent antisocial personality disordered men. *The Lancet Psychiatry, 2*(2), 153–160. https://doi.org/10.1016/S2215-0366(14)00071-6

Grillon, C. (2008). Models and mechanisms of anxiety: Evidence from startle studies. *Psychopharmacology, 199*, 421–437. https://doi.org/10.1007/s00213-007-1019-1

Hare, R. D. (1965). Acquisition and generalization of a conditioned-fear response in psychopathic and nonpsychopathic criminals. *The Journal of Psychology, 59*(2), 367–370. https://doi.org/10.1080/00223980.1965.10544625

Hawes, S. W., Boccacini, M. T., & Murrie, D. C. (2013). Psychopathy and the combination of psychopathy and sexual deviance as predictors of sexual recidivism: Meta-analytic findings using the Psychopathy Checklist-Revised. *Psychological Assessment, 25*(1), 233–243. https://doi.org/10.1037/a0030391

Hoppenbrouwers, S. S., Bulten, B. H., & Brazil, I. A. (2016). Parsing fear: A reassessment of the evidence for fear deficits in psychopathy. *Psychological Bulletin, 142*(6), 573–600. https://doi.org/10.1037/bul0000040

Hughes, M. A., Dolan, M. C., Trueblood, J. S., & Stout, J. C. (2015). Psychopathic personality traits and Iowa gambling task performance in incarcerated offenders. *Psychiatry, Psychology and Law, 22*(1), 134–144. https://doi.org/10.1080/13218719.2014.919689

Kiehl, K. A. (2006). A cognitive neuroscience perspective on psychopathy: Evidence for paralimbic system dysfunction. *Psychiatry Research, 142*, 107–128. https://doi.org/10.1016/j.psychres.2005.09.013

Kosson, D. S., Smith, S. S., & Newman, J. P. (1990). Evaluating the construct validity of psychopathy in Black and White male inmates: Three preliminary studies. *Journal of Abnormal Psychology, 99*(3), 250–259. https://doi.org/10.1037/0021-843X.99.3.250

Kuin, N. C., & Masthoff, E. D. M. (2016). Investigating the relationship between psychopathic personality traits and decision making deficits in a prison population. *Journal of Forensic Psychology, 1*, 1–7. https://doi.org/10.4172/jfpy.1000104

Lapierre, D., Braun, C. M., & Hodgins, S. (1995). Ventral frontal deficits in psychopathy: Neuropsychological test findings. *Neuropsychologia, 33*(2), 139–151. https://doi.org/10.1016/0028-3932(94)00110-B

Larson, C. L., Baskin-Sommers, A. R., Stout, D. M., Balderston, N. L., Curtin, J. J., Schultz, D. H., ... Newman, J. P. (2013). The interplay of attention and emotion: Top-down attention modulates amygdala activation in psychopathy. *Cognitive, Affective, & Behavioral Neuroscience, 13*(4), 757–770. https://doi.org/10.3758/s13415-013-0172-8

LeDoux, J. (2007). The amygdala. *Current Biology, 17*(20), R868–R874. https://doi.org/10.1016/j.cub.2007.08.005

Leistico, A. M. R., Salekin, R. T., DeCoster, J., & Rogers, R. (2008). A large-scale meta-analysis relating the Hare measures of psychopathy to antisocial conduct. *Law and Human Behavior, 32*(1), 28–45. https://doi.org/10.1007/s10979-007-9096-6

Lejuez, C. W., Aklin, W. M., Jones, H. A., Richards, J. B., Strong, D. R., Kahler, C. W., & Read, J. P. (2003a). The balloon analogue risk task (BART) differentiates smokers and nonsmokers. *Experimental and Clinical Psychopharmacology, 11*(1), 26–33. https://doi.org/10.1037/1064-1297.11.1.26

Lejuez, C. W., Aklin, W. M., Zvolensky, M. J., & Pedulla, C. M. (2003b). Evaluation of the Balloon Analogue Risk Task (BART) as a predictor of adolescent real-world risk-taking behaviours. *Journal of Adolescence, 26*(4), 475–479. https://doi.org/10.1016/S0140-1971(03)00036-8

Lejuez, C. W., Read, J. P., Kahler, C. W., Richards, J. B., Ramsey, S. E., Stuart, G. L., ... Brown, R. A. (2002). Evaluation of a behavioral measure of risk taking: The Balloon Analogue Risk Task (BART). *Journal of Experimental Psychology: Applied, 8*(2), 75–84. https://doi.org/10.1037/1076-898X.8.2.75

Levy, D. J., & Glimcher, P. W. (2012). The root of all value: A neural common currency for choice. *Current Opinion in Neurobiology, 22*(6), 1027–1038. https://doi.org/10.1016/j.conb.2012.06.001

Lösel, F., & Schmucker, M. (2004). Psychopathy, risk taking, and attention: A differentiated test of the somatic marker hypothesis. *Journal of Abnormal Psychology, 113*(4), 522–529. https://doi.org/10.1037/0021-843x.113.4.522

Lykken, D. T. (1957). A study of anxiety in the sociopathic personality. *Journal of Abnormal and Social Psychology, 55*, 6–10. https://doi.org/10.1037/h0047232

Lykken, D. T. (1995). *The antisocial personalities.* Lawrence Erlbaum Associates.

Maes, J. H., Woyke, I. C., & Brazil, I. A. (2018). Psychopathy-related traits and decision-making under risk and ambiguity: An exploratory study. *Personality and Individual Differences, 122*, 190–194. https://doi.org/10.1016/j.paid.2017.10.017

Mitchell, D. G., Colledge, E., Leonard, A., & Blair, R. J. R. (2002). Risky decisions and response reversal: Is there evidence of orbitofrontal cortex dysfunction in psychopathic individuals? *Neuropsychologia, 40*(12), 2013–2022. https://doi.org/10.1016/S0028-3932(02)00056-8

Mitchell, D. G. V., Fine, C., Richell, R. A., Newman, C., Lumsden, J., Blair, K. S., & Blair, R. J. R. (2006). Instrumental learning and relearning in individuals with psychopathy and in patients with lesions involving the amygdala or orbitofrontal cortex. *Neuropsychology, 20*(3), 280–289. https://doi.org/10.1037/0894-4105.20.3.280

Morgan, A. B., & Lilienfeld, S. O. (2000). A meta-analytic review of the relation between antisocial behavior and neuropsychological measures of executive function. *Clinical Psychology Review, 20*(1), 113–136. https://doi.org/10.1016/S0272-7358(98)00096-8

Motzkin, J. C., Newman, J. P., Kiehl, K. A., & Koenigs, M. (2011). Reduced prefrontal connectivity in psychopathy. *Journal of Neuroscience, 31*(48), 17348–17357. https://doi.org/10.1523/JNEUROSCI.4215-11.2011

Nentjes, L., Meijer, E., Bernstein, D., Arntz, A., & Medendorp, W. (2013). Brief communication: Investigating the relationship between psychopathy and interoceptive awareness. *Journal of Personality Disorders, 27*, 617–624. https://doi.org/10.1521/pedi_2013_27_105

Newman, J. P., & Kosson, D. S. (1986). Passive avoidance learning in psychopathic and nonpsychopathic offenders. *Journal of Abnormal Psychology, 95*(3), 252–256. https://doi.org/10.1037/0021-843X.95.3.252

Newman, J. P., MacCoon, D. G., Vaughn, L. J., & Sadeh, N. (2005). Validating a distinction between primary and secondary psychopathy with measures of Gray's BIS and BAS constructs. *Journal of Abnormal Psychology, 114*, 319–323. https://doi.org/10.1037/0021-843X.114.2.319

Newman, J. P., Patterson, C. M., & Kosson, D. S. (1987). Response perseveration in psychopaths. *Journal of Abnormal Psychology, 96*(2), 145–148. https://doi.org/10.1037/0021-843x.96.2.145

Newman, J. P., Patterson, C. M., Howland, E. W., & Nichols, S. L. (1990). Passive avoidance in psychopaths: The effects of reward. *Personality and Individual Differences, 11*(11), 1101–1114. https://doi.org/10.1016/0191-8869(90)90021-I

Newman, J. P., & Schmitt, W. A. (1998). Passive avoidance in psychopathic offenders: A replication and extension. *Journal of Abnormal Psychology, 107*, 527–532. https://doi.org/10.1037/0021-843X.107.3.527

Ogloff, J. R., & Wong, S. (1990). Electrodermal and cardiovascular evidence of a coping response in psychopaths. *Criminal Justice and Behavior, 17*(2), 231–245. https://doi.org/10.1177/0093854890017002006

Patrick, C. J., Fowles, D. C., & Krueger, R. F. (2009). Triarchic conceptualization of psychopathy: Developmental origins of disinhibition, boldness, and meanness. *Development and Psychopathology, 21*, 913–938. https://doi.org/10.1017/S0954579409000492

Patterson, C. M., Kosson, D. S., & Newman, J. P. (1987). Reaction to punishment, reflectivity, and passive avoidance learning in extraverts. *Journal of Personality and Social Psychology, 52*(3), 565–575. https://doi.org/10.1037/0022-3514.52.3.565

Patterson, C. M., & Newman, J. P. (1993). Reflectivity and learning from aversive events: Toward a psychological mechanism for the syndromes of disinhibition. *Psychological Review, 100*(4), 716–736. https://doi.org/10.1037/0033-295x.100.4.716

Posner, M. I., & Rothbart, M. K. (2007). Research on attention networks as a model for the integration of psychological science. *Annual Review of Psychology, 58*(1), 1–23. https://doi.org/10.1146/annurev.psych.58.110405.085516

Poythress, N. G., Edens, J. F., Landfield, K., Lilienfeld, S. O., Skeem, J. L., & Douglas, K. S. (2008). A critique of behavioral inhibition scale (BIS) for investigating theory of primary psychopathy. *Personality and Individual Differences, 45*(4), 269–275. https://doi.org/10.1016/j.paid.2008.04.014

Rilling, J. K., Glenn, A. L., Jairam, M. R., Pagnoni, G., Goldsmith, D. R., Elfenbein, H. A., & Lilienfeld, S. O. (2007). Neural correlates of social cooperation and non-cooperation as a function of psychopathy. *Biological Psychiatry, 61*(11), 1260–1271. https://doi.org/10.1016/j.biopsych.2006.07.021

Rogers, R. D., Everitt, B. J., Baldacchino, A., Blackshaw, A. J., Swainson, R., Wynne, K., ... London, M. (1999). Dissociable deficits in the decision-making cognition of chronic amphetamine abusers, opiate abusers, patients with focal damage to prefrontal cortex, and tryptophan-depleted normal volunteers: Evidence for monoaminergic mechanisms. *Neuropsychopharmacology, 20*(4), 322–339. https://doi.org/10.1016/S0893-133X(98)00091-8

Rothemund, Y., Ziegler, S., Hermann, C., Gruesser, S. M., Foell, J., Patrick, C. J., & Flor, H. (2012). Fear conditioning in psychopaths: Event-related potentials and peripheral measures. *Biological Psychology, 90*(1), 50–59. https://doi.org/10.1016/j.biopsycho.2012.02.011

Schachter, S., & Latané, B. (1964). Crime, cognition, and the autonomic nervous system. *Nebraska Symposium on Motivation, 12*, 221–273.

Schmauk, F. J. (1970). Punishment, arousal and avoidance learning in sociopaths. *Journal of Abnormal Psychology, 76*, 325–335. https://doi.org/10.1037/h0030398

Schmitt, W. A., Brinkley, C. A., & Newman, J. P. (1999). Testing Damasio's somatic marker hypothesis with psychopathic individuals: Risk takers or risk averse? *Journal of Abnormal Psychology, 108*(3), 538–543. https://doi.org/10.1037//0021-843x.108.3.538

Schultz, D. H., Balderston, N. L., Baskin-Sommers, A. R., Larson, C. L., & Helmstetter, F. J. (2016). Psychopaths show enhanced amygdala activation during fear conditioning. *Frontiers in Psychology, 7*, 348. https://doi.org/10.3389/fpsyg.2016.00348

Skeem, J., Johansson, P., Andershed, H., Kerr, M., & Louden, J. E. (2007). Two subtypes of psychopathic violent offenders that parallel primary and secondary variants. *Journal of Abnormal Psychology, 116*, 395–409. https://doi.org/10.1037/0021-843X.116.2.395

Snowden, R. J., Smith, C., & Gray, N. S. (2017). Risk taking and the triarchic model of psychopathy. *Journal of Clinical and Experimental Neuropsychology, 39*(10), 1–14. https://doi.org/10.1080/13803395.2017.1300236

Sundram, F., Deeley, Q., Sarkar, S., Daly, E., Latham, R., Craig, M., ... Murphy, D. G. (2012). White matter microstructural abnormalities in the frontal lobe of adults with antisocial personality disorder. *Cortex, 48*(2), 216–229. https://doi.org/10.1016/j.cortex.2011.06.005

Swogger, M. T., Walsh, Z., Lejuez, C. W., & Kosson, D. S. (2010). Psychopathy and risk taking among jailed inmates. *Criminal Justice and Behavior, 37*(4), 439–452. https://doi.org/10.1177/0093854810361617

Thornquist, M. H., & Zuckerman, M. (1995). Psychopathy, passive-avoidance learning and basic dimensions of personality. *Personality and Individual Differences, 19*(4), 525–534. https://doi.org/10.1016/0191-8869(95)00051-7

Tiihonen, J., Rossi, R., Laakso, M. P., Hodgins, S., Testa, C., Perez, J., ... Frizoni, G. B. (2008). Brain anatomy of persistent violent offenders: More rather than less. *Psychiatry Research: Neuroimaging, 163*(3), 201–212. https://doi.org/10.1016/j.pscychresns.2007.08.012

Veit, R., Konicar, L., Klinzing, J. G., Barth, B., Yilmaz, Ö., & Birbaumer, N. (2013). Deficient fear conditioning in psychopathy as a function of interpersonal and affective disturbances. *Frontiers in Human Neuroscience, 7*, 706. https://doi.org/10.3389/fnhum.2013.00706

Vermeij, A., Kempes, M. M., Cima, M. J., Mars, R. B., & Brazil, I. A. (2018). Affective traits of psychopathy are linked to white-matter abnormalities in impulsive male offenders. *Neuropsychology, 32*(6), 735–745. https://doi.org/10.1037/neu0000448

Vitale, J. E., Maccoon, D. G., & Newman, J. P. (2011). Emotion facilitation and passive avoidance learning in psychopathic female offenders. *Criminal Justice and Behavior, 38*(7), 641–658. https://doi.org/10.1177/0093854811403590

Yang, Y., Raine, A., Colletti, P., Toga, A. W., & Narr, K. L. (2010). Morphological alterations in the prefrontal cortex and the amygdala in unsuccessful psychopaths. *Journal of Abnormal Psychology, 119*(3), 546–554. https://doi.org/10.1037/a0019611

Yang, Y., Raine, A., Colletti, P., Toga, A. W., & Narr, K. L. (2011). Abnormal structural correlates of response perseveration in individuals with psychopathy. *The Journal of Neuropsychiatry and Clinical Neurosciences, 23*, 107–110. https://doi.org/10.1176/appi.neuropsych.23.1.107

Part III
Psychopathy and Values

Chapter 12
The Value-Ladenness of Psychopathy

Marko Jurjako and Luca Malatesti

Abstract The recurring claim that the construct of psychopathy is value-laden often is not qualified in enough detail. The chapters in this part of the volume, instead, investigate in depth the role and significance of values in different aspects of the construct of psychopathy. Following these chapters, but also by offering a background to them, we show how certain values are involved in the characterisation of psychopathy, inform societal needs satisfied by this construct, and have a central role in determining whether psychopathy is a mental disorder. Moreover, we relate this description to our criticism of the view that the entrenchment of the notion of psychopathy with values renders it in principle irreconcilable with sound psychiatric theory and practice. However, we also recognize that the value-ladenness of psychopathy leaves open other important challenges. Meeting them needs addressing interdisciplinary interrelated issues that have empirical, normative, and theoretical dimensions.

Keywords Psychopathy · Value-ladeness · Mental disorder · Social constructivism · Moral capacities

12.1 Introduction

The construct of psychopathy is a particularly apt example of the general issue of the interrelation between psychiatric classification, explanation, disorder status and values (Bolton, 2008; Fulford, 1989; Graham, 2013; Sadler, 2008). Several authors have maintained, in fact, that the construct of psychopathy involves values (Malatesti & McMillan, 2014; Reimer, 2008). Some argue that such a value-ladenness implies

M. Jurjako · L. Malatesti (✉)
Department of Philosophy, Faculty of Humanities and Social Sciences, University of Rijeka, Rijeka, Croatia
e-mail: lmalatesti@ffri.uniri.hr

© Springer Nature Switzerland AG 2022
L. Malatesti et al. (eds.), *Psychopathy*, History, Philosophy and Theory of the Life Sciences 27, https://doi.org/10.1007/978-3-030-82454-9_12

that this construct cannot be adopted in sound psychiatric theory and practice. Let us call this view the *irreconcilability thesis*. Some have, for example, argued that value-ladenness undermines the scientific status of psychopathy, because it just groups individuals who do not align to certain social norms (Cavadino, 1998; Jalava et al., 2015; Mullen, 2007). Others maintain that, leaving aside the issue of the validity of the notion of psychopathy, its value-ladenness might put pressure on the idea that it is a legitimate mental disorder (Holmes, 1991).

This chapter introduces critically the role and significance of values for the construct of psychopathy. These themes are then considered in more detail in the other chapters in this part of the volume. Instead of just summarising these chapters, we relate their content to our views on how values influence and colour the ontological, epistemological, and medical status of psychopathy. While we endorse the value-ladenness of this construct, following the lead of the chapters in this part of the volume, we highlight the importance of qualifying this endorsement. In fact, often in the context of polemic debates, not enough attention has been dedicated to clarifying the relevant types of values and how they relate to the notion of psychopathy.

We maintain that this construct, even if we consider its more theoretical dimension as opposed to its practical uses, involves several types of values whose presence opens important questions concerning its nature and use. Regarding the irreconcilability thesis, we argue that value-ladenness of the construct of psychopathy, *per se*, should not be taken as a sufficient ground for succumbing to it. Instead, the nature, use and status of psychopathy should be investigated by recognising that it lies at the intersection of theoretical and practical issues that can only be satisfactorily addressed by firstly making explicit the underlying complex network of conceptual, normative, and empirical issues (Jurjako et al., 2018). These issues could represent a source of difficulties for the construct of psychopathy (Jalava et al., 2015) or a ground for a deeper appreciation of its nature (Rosenberg Larsen, 2018, 2020), and its implications for public policy, possibility of treatment options, and even philosophical theorising (Maibom, 2018).

We will proceed as follow. First, for the purposes of this chapter, we clarify the distinction between facts and values as it has been understood by philosophers and that we think is operative in several of the debates that are considered by the chapters in this part of the volume. Then we consider values that are explicit in the diagnosis of psychopathy, insofar they inform the standards of behaviour, mental life, and traits from which people with psychopathy depart. We then describe more implicit cultural preferences and needs that might be satisfied by a notion such as psychopathy. In both cases, we will investigate how these evaluative dimensions might affect the scientific credentials of the construct of psychopathy.

12.2 Facts, Values, and Value-Ladenness

At least since David Hume (1748), many draw the distinction between facts and values, although there is dissent on its significance (Putnam, 2004). Facts denote objective, and for the most part, empirically discoverable states of affairs. Values, instead, are related to our evaluations of states of affairs. This distinction is sometimes introduced by distinguishing a normative and a descriptive domain. The normative domain pertains to what *should* be the case. This is expressed in judgments to the effect that something ought to be the case or that something is forbidden to be the case, that something is valuable or non-valuable, and so forth. The paradigmatic normative judgments are those that express our moral views. Such as that we should not harm other people without a good reason, relieve of pain those who are suffering, or more generally that we should help those in need in accordance with our capabilities.

The descriptive domain pertains to supposedly value-free facts. Standardly, it is thought that science, at least ideally, aspires to discover objective facts that are expressed by descriptive judgments, such as "The measurement is showing 2 Degrees Celsius", and "Water is composed of two hydrogen and one oxygen atoms". On this view, descriptive judgments *ascribe* properties to things (or we might say they *describe* things by ascribing properties to them), while normative judgments *evaluate* things and prescribe how things ought to be.

Some judgments seem to crosscut the fact/value distinction. Aesthetic judgments ascribe values to things in a vocabulary that is commonly expressed by descriptive judgments, such as "This work of art is sophisticated". Similarly, even moral judgments are often expressed by judgments that ascribe properties to things, as when we say "Hurting people is bad". On the face of it, we ascribe the property of badness to an act of hurting someone. Nonetheless, the proponents of the strict fact/value distinction will be quick to say that these judgments, at least implicitly, contain normative prescriptions, which can be revealed by further analysis. Saying that a work of art is sophisticated normally means that the speaker has a positive attitude towards it and invites other people to share this attitude with her. Similarly, when we say that an act is bad, we are saying that in normal circumstances this act should not be performed (for discussion, see, Railton, 2003).

Saying that a term, concept, or description is *value-laden* means that values play a role or in some way influence how we describe or conceive of things. Values may enter differently in our descriptions or theories about different things. Values might have a constitutive role in how we describe things. For instance, when we describe somebody as brave, this usually means that this person tends to behave in ways that we find worthy of admiration. Accordingly, the concept of bravery essentially includes a descriptive and a value component. Alternatively, values might be only accidentally or contingently involved in our descriptive judgments. For instance, we might think that one employee is more efficient than another because she is our friend, and we think she is a good person. Here our value judgment colours our assessment of the employee in a contingent way. Insofar we have an objective way of measuring efficiency at work, we can determine it without passing on a value judgment on the employee's character.

The distinction between values and facts is at the core of what we can call the thesis of the *irreconcilability* of science and values. A line of argument for this thesis tends to portray science as a value-free enterprise and the scientist as a dispassionate searcher for objective factual knowledge. Values, thus, tend to be construed as contingently, that is, non-constitutively related to scientific judgments. Moreover, on this idealized view of science, values are typically thought of as having a negatively biasing influence. For instance, a researcher might be especially invested in receiving a positive result in her study, which might influence her to erroneously interpret the data.

Sometimes, the claim that science should steer away from values is based on a principled grounding involving the "is/ought gap". According to this line of thought, there is no straightforward logical link between facts and values. You cannot derive a statement that something *ought* to be the case from statements that something *is* the case (Hume, 1748). The significance of this gap is then used to vindicate some form of subjectivism about value. The idea would be that the objective reality is exhausted by facts, which leaves no room for objective values. The only place where values can reside is in subjective evaluative attitudes of specific persons or culturally relative attitudes (Ayer, 1970; Stevenson, 1944; cf. Railton, 2003). Accordingly, if values only reflect our subjective attitudes or those shared in a certain culture, science, conceived as an enterprise for discovering objective facts, should strive to remain value-free (Weber, 2011; for discussion, see Longino, 2004).

Even though some philosophers of science urge that the ideal of science as a value-free enterprise should be abandoned in contemporary philosophy of science (Kitcher, 2001; Longino, 2004), still it seems that the distinction between facts and values looms large in general discussions of the scientific status of psychiatry and in applied scientific study of psychopathy. In the next section, we will examine the role that values play in psychopathy research (broadly construed) and whether they support instances of the irreconcilability argument.

12.3   The Role of Values in the Construct of Psychopathy

Some authors, who have claimed that psychopathy is value laden, have not provided too detailed analyses of how this is so. This is mostly because their claim was made to prove the irreconcilability thesis. For instance, Michael Cavadino simply maintains that psychopathy offers "a prime example of moralism masquerading as medical science" (Cavadino, 1998, p. 6). Paul Mullen sees psychopathy as "a passing manifestation of the technologies of social control" (Mullen, 2007, p. 147). Understandably, for these authors, given their critical aims of reducing harmful and stigmatizing effects of labelling someone as a psychopath, it is not important to delve deeply into the values that enter in the notion of psychopathy and how they do so. However, evaluating their lines of reasoning and other arguments about the nature of psychopathy might benefit from such a clarification.

While values clearly guide and regulate the practical applications of psychiatric knowledge and inform the social reactions to psychiatric patients, values are also involved in the theoretical aspects of the discipline. Values shape the preferences of different stakeholders that, implicitly or explicitly, inform relevantly shared assumptions that guide the categorization and explanation of psychiatric constructs. These values can be of different types. Some of them concern theoretical or pragmatic preferences related to the scientific practices. The associated values, thus, dictate, for instance, criteria of simplicity that must be satisfied by the construct (see, for instance, Kitcher, 2001). Other values determine the aims pursued with the construct of psychopathy in the first place. For instance, psychopathy might be countenanced as a tool for grounding the prediction of risk and protection of the society or to extend therapeutic intervention (McMillan, this volume; Tamatea, this volume). Values can also inform the background conceptions of the normal and acceptable types of behaviour, mental lives, and personality traits, against which psychiatric constructs are salient and mental disorders defined (Bolton, 2008). Several authors have maintained convincingly that moral values, that prescribe morally permissible or impermissible ways of living, have a role of this type in the construct of psychopathy (Jalava et al., 2015; Maibom, 2014; Rosenberg Larsen, 2018, 2020; see also Schramme, 2014). A brief historical overview of the notion of psychopathy appears to confirm this.

From its very beginnings, the scientific study of psychopathy focussed on investigating something akin to a moral disorder (Sass & Felthous, 2014; Ward, 2010). One of the early predecessors of the contemporary construct of psychopathy referred to a subclass of criminals who are impulsive, uninhibited, and unremorseful about their antisocial behaviour. At the turn of the nineteenth century, in France, Phillipe Pinel described such individuals as being insane without suffering from delusions (*manie sans délire*) (see Jalava et al., 2015, ch. 1). At the same time, the American psychiatrist Benjamin Rush (2009) described such individuals as being morally deranged. Later, in the mid-nineteenth century, the British physician James C. Prichard (1837) used the term "moral insanity" to refer to all disorders that seemed to correlate with preserved intellectual capabilities and lack of delusion, but included severe impairments in normal expressions of emotions, moral dispositions, and behavioural controls.

The common idea behind these labels was to capture a disorder that selectively affects a person's moral faculties given that their cognitive functioning seemed otherwise preserved. These conceptualizations of moral insanity are echoed in the contemporary conceptualization of the construct of psychopathy (Rosenberg Larsen, 2018, 2020). Some argue that exactly this is the problem with the modern scientific study of psychopathy and that because of this value-ladenness we should not expect significant progress in psychopathy research and clinical practice. In this regard, in their book *The Myth of the Born Criminal*, Jarkko Jalava, Stephanie Griffiths, and Michael Maraun (2015), maintain that psychopathy research has not produced conclusive empirical findings about the neurobiological underpinnings of psychopathy that might later be useful for clinical purposes.

The general underlying thought exposed by Jalava, Griffiths, and Maraun (see, e.g. Jalava et al., 2015, p. 31) seems to be an instance of the irreconcilability thesis. They find the core of the problem in the early conceptualizations of moral insanity. They were based on a Judeo-Christian conception of evil or moral degeneracy that has been inherited by the modern construct of psychopathy. According to this line of thought, psychopathy is essentially defined as a deviation from Judeo-Christian conception of moral norms. These norms are not grounded in scientific research or facts. Thus, it should not be expected that research will reveal biological underpinnings of such socially or religiously defined deviations.

This criticism can be developed in at least two ways, depending on the level at which the supposedly problematic values "infiltrate" the construct of psychopathy. At the descriptive level, the defining features of psychopathy are characterized in value-laden terms. At the explanatory level, influential approaches construe the disorder status of psychopathy as grounded in the neuropsychological impairments of capacities underpinning what might generally be called our moral dispositions (Blair et al., 2005; Raine, 2019). While these criticisms capture something important about our conceptualizations of psychopathy, it is not entirely clear that their being value-laden should diminish their scientific credibility. Let us consider these objections in turn.

Analysing the dominant conceptualizations and measurements of psychopathy reveals that the construct of psychopathy is permeated with value-laden descriptions (Blackburn, 1988). The problem seems to be that if psychopathy is defined by deviations from moral or religious norms, then it is likely that as a construct it will reveal only our culturally bounded conceptions of good and evil and not a biologically grounded objective medical disorder (cf. Jefferson, 2020). For instance, the dominant measure of psychopathy applied in forensic settings, the Psychopathy Checklist-Revised (PCL-R Hare, 2003), uses 20 diagnostic items that are usually divided into four factors characterizing interpersonal styles, affectivity, impulsivity, and antisocial behaviour. These items could be seen as largely characterizing psychopathy in value-laden terms. Interpersonally psychopathic individuals are characterized as manipulative and pathological liars. They lack or do not show important moral emotions, such as empathy, remorse, and guilt when they do something wrong. Other items that refer to behaviours are explicitly defined as behavioural deviations from moral and legal norms. Most notably, they refer to traditionally understood immoral behaviours with religious overtones, such as leading a parasitic lifestyle, promiscuous sexual behaviours, and the tendency to engage in extramarital relations (for a more detailed analysis, see Jalava et al., 2015, Appendix A).

Indeed, ever since Rush introduced the concept of moral derangement, researchers have been enticed by the prospect of finding a neurobiological cause of extreme and pervasive forms of immoral or antisocial behaviour (see, e.g. Blair et al., 2005; Glenn & Raine, 2014). Rush (2009) proposed the idea that the brain has a moral faculty whose function is to discern good from bad or right from wrong. He hypothesized that morally deranged people have dysfunctional brain areas underlying this moral faculty. However, in Rush's account this idea is overlaid by religious tones and culturally dominant views of his time of what this faculty could be. He defined the moral faculty as an innate capacity to discern right from wrong, to act in accordance with these conceptions, but it also involves a Christian sense of conscience and a sense of deity (Jalava et al., 2015, p. 25). Thus, the very dysfunction of the brain areas underlying this faculty cannot be defined without a recourse to a culturally and religiously defined moral faculty. These areas and their dysfunctionality are only assumed based on a culturally bound conception of the moral faculty. If this faculty is a product of social construction, then its supposed biological underpinnings and the idea of their being dysfunctional cannot be real outside of this socially grounded construct.

However, just because a construct has been defined in terms of deviations from moral or conventional norms, it does not follow *a priori* that the construct is not sufficiently unified or interesting for research or practical purposes (Jurjako et al., 2019, 2020; Malatesti & McMillan, 2014). In fact, a construct can be value-laden and still denote a syndrome underpinned by a biological cause that is unified enough to provide fruitful explanations. An example might be the phenomenon of incest. Incestual relations in human groups are usually condemned on moral grounds (Haidt, 2001). Nonetheless, it is a plausible hypothesis that our endorsement of such moral judgments has a biological underpinning to prevent people from having progeny with reduced chances of survival (Lieberman & Smith, 2012). Thus, whether the construct of psychopathy is unified enough to be interesting for research purposes, and whether we will find a biological cause of it, remain questions that should be decided on empirical grounds.

Similarly, it seems to be an open empirical question whether there are areas of the brain that ground our psychological, including moral, capacities (see, also Jefferson, 2014). Indeed, these capacities are couched in the vocabulary whose origins are shaped by our specific cultural, religious, and other types of experiences. But this does not mean that we cannot study these concepts as referring to our psychological capacities that we employ in our daily life and transactions with other people. Some conceptions of these capacities will be discarded as obsolete, but other will survive due to their usefulness in shaping our lives. In particular, the conception of the moral faculty as a set of capacities that enable us to choose and act in accordance with our conceptions of what is good and bad seems to be a commonly employed psychological ability. As such it can be and has been studied by empirical means (see, e.g. Decety & Wheatley, 2017; Liao, 2016). Moreover, some researchers think that these capacities are grounded in discernible brain areas and their functional connections (Moll et al., 2003). Thus, it remains an open empirical question whether disturbances of these brain areas and their networks might lead to antisocial behaviour and personality types that characterize the construct of psychopathy (Raine, 2019). Of course, it might turn out that our conception of the moral faculty designates a very heterogeneous set of areas of the brain, that are not worth to be investigated as a unified scientific kind. In this case, psychopathy as a specific failure in the moral faculty would turn out not to be a good scientific or clinical category (Brzović et al., 2017). But this should be a result of scientific study, and not an immediate consequence of the fact that sometimes the biological underpinnings of purported disorders are individuated in value-laden terms (for discussion of these empirical issues, see Jalava et al., 2015; Jalava & Griffiths, 2017).

Nonetheless, even if the complete psychobiological explanation of psychopathy as is currently conceived and measured is not forthcoming (for discussion, see Jurjako et al., 2020), still this does not diminish its relevance as a clinically and socially relevant construct. This is a message of contemporary philosophy of psychology and more specific proposals in psychiatric categorization and explanation, where levels of categorisation and explanation other than the biological and neurological ones are deemed acceptable (Bermúdez, 2005; Borsboom, 2017).

Besides being manifest in the diagnostic items of measures of psychopathy, there are also other subtler ways in which values enter and might taint our conceptions of mental disorders and psychopathy. We will consider them in the next section and consider again whether they support the irreconcilability thesis.

12.4   Cultural Discourse and the Regulative Role of Psychopathy

Given that the construct of psychopathy is pervaded with values and that this does not *ipso facto* support the irreconcilability thesis, it might be fruitful to embrace and investigate its normative overtones. This can be done by focusing on general prescriptive practices in our culture. Historians have given important contributions in this direction in relation to past, and even recent, conceptualisation of psychopathy (D'Alessio et al., 2017; Parhi, 2018; Parhi & Pietikainen, 2017). Gwen Adshead (this volume) undertakes a similar task from a sociological perspective. She reflects on the underlying discourse and needs that offer, in a certain society, a niche for a category such as psychopathy.

She argues that "psychopathy" in our cultural discourses plays a regulatory role in determining what is considered normal and abnormal with respect to violent behaviour. This type of regulatory discursive function is compared with gender. In different cultural contexts, gender determines the stereotypical roles and expectations regarding male and female behaviour, which often propel studies and sometimes bias them towards searching for the biological features underpinning the gender differences. Similarly, according to Adshead, the role of the construct of psychopathy is to designate individuals and behaviours that we find socially extremely undesirable ("a paradigm of badness") and that might be seen as an extreme point on a spectrum of human cruelty. Moreover, such views then ground investigations that purport to discover differences at the neurobiological level between psychopathic and non-psychopathic brains. These attempts indicate that the role of psychopathy in cultural discourses is to condemn and provide justifications for exclusion of such individuals from social goods and to provide means of reducing undesirable behaviour.

The import of Adshead's analysis is most significant with respect to the gender issues in the study of psychopathy. She notices that there is an ambiguity in how psychopathy is conceptualized in the literature depending on gender descriptions. For instance, typical conceptualizations of female psychopathy depict them as characteristically engaging in deception and promiscuous behaviour, and as using sex for manipulation, which is in contrast to male psychopathy, which is typically characterized as more emotionally stable, dominant, grandiose, and aggressive (Cleckley, 1976; Forouzan & Cooke, 2005; for review, see Wynn et al., 2012). Accordingly, it seems that abnormality underlying violent offending among women might be based on stereotypes portraying women as passive and submissive. Adshead indicates that these stereotypes likely lead to double standards. Women are often seen as deserving more compassion than men, sometimes their sentencing is reduced or alternatively sometimes their punishment is harsher, even when they share similar developmental trajectories, experiences, and social backgrounds with male offenders. Nonetheless, Adshead indicates that such gender-based explanations of violence and cruel acts among people might miss the real cause of such behaviours. Seeing acts of violent crimes as responses to different types of social interactions, rather than as simple reflections of abnormality, and listening to the actors' narratives underlying their acts of cruelty, might help us to improve understanding of the general causes of such extreme forms of antisocial behaviour (McKeown & Thomson, 2019; see, also Mullen, 2007).

Moreover, introducing the social perspective on the causes of psychopathic violent behaviour might make more visible the cultural variations in how antisocial behaviour is exhibited across different cultures. This is the topic of the chapter by Rachel Cooper (this volume). She claims that psychopathy and its close diagnostic relative antisocial personality disorder (ASPD as defined by the Diagnostic Manual for Mental Disorders, DSM-5, (American Psychiatric Association, 2013)) are culturally relative categorizations, insofar reasonable expectations how people labelled with psychopathy and ASPD will act varies across culture. This claim is significant because it might be regarded as supporting a weaker version of the irreconcilability thesis: namely, how psychopathy or ASPD will be exhibited and experienced cannot be generalized across different cultures. This does not jeopardize or delegitimize the scientific study of psychopathy. However, it implies that typical measures of psychopathy and ASPD such as PCL-R and DSM-5 should not be employed without taking into consideration the meanings of their diagnostic items and their specific culturally bounded and value-laden interpretations.

Cooper offers two arguments to support her claims. Her first argument is based on the cultural variability of the meaning of intentional actions. DSM and PCL-R crucially use diagnostic items that refer to intentional actions. For instance, DSM uses diagnostic items such as disregard for social norms, rules, and obligations, while PCL-R uses items such as criminal versatility, parasitic lifestyle, and sexual promiscuity. However, the meaning of intentional actions varies across cultures. For instance, the meaning of sexual promiscuity is not expected to be the same in cultures where monogamy dominates and cultures where polygamy is the norm. Accordingly, it can be expected that the display of antisocial and psychopathic behaviours will vary across cultures. In line with the weaker irreconcilability thesis, the significance of this conclusion is that the maladaptive behaviours associated with psychopathy and antisocial personality disorder cannot be explained as exclusively caused by the personality types that are underlined by objective neurobiological factors.

The second argument is based on the idea that psychiatric diagnosis typically exhibits looping effects, as explained by Ian Hacking (1996). One of the main differences between classifications of humans and other natural or artificial entities is that people tend to distinctively react to how they are classified. For instance, being labelled as an alcoholic might provide a motivation to a person to stop drinking, or, alternatively, to label a child as having conduct disorder might reinvigorate the problematic behaviour. Similarly, when people are diagnosed with ASPD or psychopathy, they can be aware of what it means to receive such a diagnosis and thus this knowledge can influence their attitudes and behaviour. Moreover, this influence is expected to vary across diverse cultures. Cooper concludes that folk conceptions of what ASPD and psychopathy are, how they are measured and diagnosed will differentially influence how those labels will affect individuals' behaviour and attitudes across cultures. For instance, some people would want to avoid receiving such labels and will try to deceive psychologists who test them. In other cultural contexts, some people may take pride receiving a psychopathy label and identify with it, because in certain contexts personality traits, such as being remorseless, bold, and unempathic, might benefit them (Babiak et al., 2010).

12.5   The Values-Ladeness of the Notion of Mental Disorder

Another important area where values enter psychopathy research, and where the irreconcilability thesis figures prominently, relates to the question whether psychopathy is a mental disorder. As we saw above, social discourses applied to different domains determine what is considered normal and abnormal. Answering this question has important consequences for medical and psychiatric practice, regarding, among other things, the dedication of medical resources, justifying treatment policies and determining the expected treatment outcomes (Powell & Scarffe, 2019).

The chapter by Malatesti and Baccarini (this volume) examines the problem of deciding what is the proper way of addressing the mental disorder status of psychopathy. Whether psychopathy is a mental disorder is an intricate issue in more than one way. Psychopathy is standardly characterized as a personality disorder. However, as is the case with many personality disorders, what usually grounds the disorder status to psychopathic individuals is their deviation from moral and social norms, before we even start thinking about a possible internal biological cause (Malatesti, 2014; Sadler, 2008). In this regard, one interesting issue is whether there is something more to psychopathy that could ground the disorder status than the socially negative attitudes towards people who, in addition to having callous and unemotional traits, engage persistently in antisocial conduct. The answers to this question diverge, but, as we have argued in the previous section, there is nothing in principle stopping us from discovering that psychopathy is caused by neurobiological factors (Raine, 2019; Blair et al., 2005; Glenn & Raine, 2014; cf. Jurjako, 2019).

A more practical issue concerns the implications of applying the mental disorder label to a condition, behaviours, or a set of mental states. Here again the notion of value-ladenness looms large. In general, the concept of disorder plays multifarious and interconnected roles in biomedical research and social practices. Labelling a condition or a set of traits as disordered has medical, moral, social, and legal implications (Cooper, 2005). Commonly the concept of disorder is applied to conditions or sets of traits that should be medically, psychologically, and otherwise treated. Consequently, on the scientific side, as a society we dedicate resources for studying what we deem to be disorders and prioritise individuals with disorders when making decisions about who will receive treatments/therapies (Powell & Scarffe, 2019).

Moreover, individuals with disorders invite certain responses and attitudes from other people. We tend to sympathize and think of them as deserving an excuse if they do not conform to social norms. Sometimes, disorders invite stigmatizing attitudes, where those falling under the label are deemed dangerous for themselves and society (Jurjako et al., 2019). We also tend to hold them less responsible for what they do or at least we tend to blame them less for their misconducts than otherwise we would (Aspinwall et al., 2012). This way of thinking is formalized in the insanity defence employed by criminal laws across different countries. In fact, the insanity defence usually involves a clause stating that if a violation of the law was caused by a mental disorder then there are grounds for exculpating the offender (Malatesti et al., 2020). Accordingly, if individuals with psychopathic traits are mentally disordered this would have important consequences on how we should treat them (cf. Jefferson & Sifferd, 2018; Nadelhoffer & Sinnott-Armstrong, 2013).

However, deciding whether psychopathy is a medical disorder is permeated with problems. Some of the problems stem from general issues relating to the concept of disorder. In fact, in philosophy of psychiatry there does not seem to be a consensus on how to construe the notion of mental disorder that can ground medical research, health care, and social practices or even if we need this notion at all (Bortolotti, 2020).

Other problems relate to specific issues pertaining to the value-laden nature of psychopathy. The fascination with psychopathy comes from the obvious sense in which individuals with psychopathic traits stand out. People who lack remorse and do not show empathic reactions to other people's suffering, whose interpersonal style involves pervasive cheating and lying to satisfy their selfish goals immediately sets them apart from other individuals. In addition, psychopathy is characterized by pervasive antisocial behaviour underlined by extreme forms of reactive and instrumental aggression towards people who get in their way. Such personality traits and extreme forms of antisocial behaviour invite disapproving attitudes from the rest of the society. And this is the core of the problem. Just because we have negative attitudes towards some traits and behaviours, no matter how pervasively antisocial they are, does not imply that such traits should be thought of as pathological.

The history of psychiatry is filled with cases of abuse where psychiatry was used to enforce the values of those in power (see, for instance, Fulford et al., 1993). As a response to such misgivings, DSM 5 involves a clause stating that a "[s]ocially deviant behavior (...)" is a disorder only if it "results from a dysfunction in the individual" (APA, 2013, p. 20). Seen from this perspective, we would be justified to classify psychopathy as a disorder only if the problematic personality traits and associated behaviours stem from independently determinable behavioural, psychological, or biological dysfunction in psychopathic individuals.

Traditionally, the theories of medical disorder are divided between the normative and naturalist views (Bolton, 2008). Normativists hold that the notion of disorder is value-laden and so that our labelling somebody as disordered depends on certain negative evaluations of that person. There are several ways of cashing out the normativist position. However, the most common way is to say that disorders are things that harm people. Harm is a value-laden term, in the sense that something is harmful if we find it undesirable in some respect, where undesirability is determined according to a certain system of values. Therefore, disorders are things we find undesirable from the perspective of some system of values. Normativism was influential in the anti-psychiatrist movement. Showing that many ascriptions of mental disorders are based on our value judgments has played a critical role in curbing the power of psychiatry to impose culturally dominant values on social outcasts. Exposing the normative side of psychiatry has also played a crucial role in making explicit the underlying value-laden practices of psychiatry and provided resources to effectively criticize and improve them (Fulford et al., 1993). A particularly notable example is the exclusion of homosexuality from DSM-III, which was based on the recognition that homosexuality is not intrinsically associated with distress and therefore is not harmful to the individual in that sense (Spitzer & Williams, 1982).

However, emphasizing the normative aspect of psychiatric practices has led to extreme forms of anti-psychiatry. This attitude was famously expounded by Thomas Szasz (1974), who argued that mental disorders as such do not exist. More specifically, he argued that the notion of disorder refers to anatomically or physiologically disordered biological causes. Thus, if a psychiatric condition does not have a known biological cause it cannot be considered a medical disorder. At most it can be considered a problem in living which should not typically be treated or solved by medical means.

Other authors have argued that mental disorders are social constructions and that there is nothing more to them than the fact that they designate traits and behaviours that deviate from dominant social norms (Horwitz, 2003; Sedgwick, 1973). Such views might lead to counterintuitive consequences. For instance, in the nineteenth century doctor Samuel A. Cartwright (1851) claimed that Drapetomania is a real disease referring to the impulsive need of the slaves in the American South to run away from farms. If the normativist view that mental disorders are just deviations from socially dominant values is plausible then we would have to accept that Drapetomania was a disease in the 19th USA, while it is not in modern times solely in virtue of the changes in our value judgments.

Addressing foundational issues in psychiatry or evaluating specific constructs, however, might require adjudicating values, besides detecting them. Malatesti and Baccarini (this volume) investigate this option in the case of psychopathy. They start from the standpoint of normativism on the notion of mental disorder, investigating how the relevant moral values involved in the construct of psychopathy could be properly adjudicated to establish that it is a mental disorder. Specifically, they locate their proposal within the so-called public reason tradition of political philosophy (Gaus, 2011; Rawls, 2005). This tradition has investigated and elaborated forms of justification of evaluative standards that should guide several aspects of our social life. Their aim is also to address unreconciliatory worries that the presence of values within a psychiatric construct might be a ground for relativism or of the kind of oppression that has tarnished the history of psychiatry. Thus, their response to cases such as drapetomania is to point to the fact that this construct is illegitimate insofar it reflects the specific and morally unjustifiable perspective of a society that supports slavery. However, other dimensions of the notion of mental disorder need to be invoked to address such worries.

The naturalists have taken a different turn to oppose the abuses of psychiatry. Instead of exposing or attempting to normatively adjudicate the value-laden aspects of psychiatric theory and practice, they have concentrated on delineating psychiatry as a legitimate part of medicine by emphasizing the naturalistic or biological bases of mental disorders (Boorse, 1977). In this regard, some have attempted to delineate the notion of mental disorder as referring to objective biological causes. This objectivity is usually cashed out in terms of the notion of biological dysfunction. There are different approaches to explaining what it would take for a trait to be biologically dysfunctional. Regardless of the specifics of these accounts, the resulting view is that mental disorders are those traits, behaviours or conditions that are caused by a malfunction in some biological subsystem. According to this approach, Drapetomania was never a real disease, because it is fair to say that enslaved people who wanted to run away from slave farms did not have these desires because of some biological dysfunction.

Often it is argued that purely naturalist approaches to medical disorders are not viable because medicine is a practical discipline, and its core notion of disorder must be practical as well (Bolton, 2008; Wakefield, 2014). For instance, there are biological dysfunctions that do not cause any discernible harm to people. Such conditions would not be classified as disorders worthy of medical attention. Thus, hybrid views are proposed, according to which disorders are those conditions that cause harm in virtue of some underlying dysfunction or incapacity (Wakefield, 1992; Powell & Scarffe, 2019). Such views fit the desiderata indicated by DSM's proviso that a conflict between an individual and society cannot be a disorder unless it is caused by some dysfunction in the individual.

The legitimate aspirations of the naturalists for objectivity, as it is offered by contemporary science, regarding the notion of mental disorder, however, do not need to lead to the irreconcilability thesis about psychopathy. As Malatesti and Baccarini argue in their chapter, a proper framework for thinking about the disorder status of psychopathy should include considerations regarding the objective cause of harm. Even if one is a pure normativist about the concept of disorder, still we need to distinguish between harms that are clinically significant and deserve medical attention from those which do not deserve such attention. They argue that if harm is caused by a dysfunction or a disability that is beyond the control of an individual, then this might indicate that the condition is a disorder. Applied to the case of psychopathy, they argue that the actions of the psychopathic individuals cause harm to themselves and other people (see also Nadelhoffer & Sinnott-Armstrong, 2013). However, whether this harmful behaviour deserves clinical attention will depend on the ultimate causes of such behaviour (Jurjako, 2019; Krupp et al., 2012). This is a question that should be decided on empirical grounds. Even though the problem of the disorder status is a value-laden question, still the issue can be decided on objective grounds once we settle on the values that we deem important.

So far, we have discussed how values permeate the construct of psychopathy from the third person perspective, however we should also consider values from the first-person perspective of the psychopathic individuals. It should be recognised that they are persons who can be characterized as endorsing moral and other personal values (Glenn et al., 2017). The common thought is that the values that an individual endorses form a significant part of who he or she is as a person. This internal value-ladenness is important, especially when thinking whether we are justified in medically intervening to "cure" psychopathic individuals. This is a recurring issue that involves ethical problems surrounding personality disorders in general (Charland, 2004). If psychopathic individuals' personal outlook is, among other things, defined by their peculiar set of values, then medically intervening to change their personality might entail changing their moral and personal outlooks.

This might be problematic for empirical or conceptual reasons. It is unclear whether changing a person's personal outlook and the systems of values that define it is practically feasible (Maibom, 2014). Marga Reimer (this volume) argues that this cannot even be done in principle. Her argument is based on our current conceptions of legitimate forms of medicalization and what constitutes a personal identity of a psychopath. She maintains that in our cultural context the role of medicine is, amongst other things, to treat diseases to alleviate people's suffering. This role then countenances what it means that a condition can be medicalised. Only treatments that relieve suffering caused by a medical condition can be construed as instances of medicalization. Reimer maintains that psychopathy cannot be medicalized because it cannot be treated without changing the personal identity of psychopathic individuals.

Reimer's discussion is interesting on multiple levels. This argument unravels the deep issues related to personal identity and moral values permeating the construct of psychopathy. By defining psychopathic individuals as belonging to a class of essentially immoral or amoral people, she takes the value-laden conception of psychopathy to its extremes. The core of this view is not just that psychopathy cannot be considered a medical disorder. Its central claim is that psychopathy cannot be properly medicalised according to our actual conceptions of medicalization. This raises the question what psychopathy researchers attempt to do when they classify psychopathy as a mental disorder and try to find an effective treatment for it (Brazil et al., 2018). Are they trying to find a cure for a disease, a treatment for changing someone's evaluative attitudes, or a procedure for transforming someone's personal identity? It is likely that this chapter will spark further discussions about the role of values in defining psychopathy and its implications for how to conceive the personal identity of psychopathic individuals.

12.6 Conclusion

Following the lead of the chapters in the third part of this volume, we have highlighted the presence of values within the construct of psychopathy and their significance for its scientific status. We have argued that there is no swift *a priori* move for defending the irreconcilability thesis. The relevance of values, especially moral ones, in the categorization and explanation of psychopathy, and the underlying social motivations for having such a construct do not disqualify it from being a proper posit for psychiatric theory and practice. Instead, to evaluate the significance of values for the scientific status of the notion of psychopathy, complex interrelated theoretical, empirical, and normative issues need to be explicated and addressed.

Acknowledgements We are grateful to John McMillan for his useful comments on a previous version of this chapter.

LM and MJ's research is supported by the Croatian Science Foundation (project RAD, HRZZ-IP-2018-01-3518). MJ also thanks the University of Rijeka (project KUBIM, uniri-human-18-265) for financial support, and the hosts of the BIAS institute for their summer hospitality. Preliminary work on this chapter was an outcome of the Project CEASCRO (2014-2018, Croatian Science Foundation, HRZZ-IP-2013-11-8071).

References

Adshead, G. (this volume). Unsexed cruelty: Gender and psychopathy as regulatory discourses in relation to violent women. In L. Malatesti, J. McMillan, & P. Šustar (Eds.), *Psychopathy. Its uses, validity, and status*. Springer.

American Psychiatric Association. (2013). *Diagnostic and statistical manual of mental disorders: DSM-5* (5th ed.). American Psychiatric Association.

Aspinwall, L. G., Brown, T. R., & Tabery, J. (2012). The double-edged sword: Does biomechanism increase or decrease judges' sentencing of psychopaths? *Science, 337*(6096), 846–849. https://doi.org/10.1126/science.1219569

Ayer, A. J. (1970). *Language, truth and logic*. Dover Publications.

Babiak, P., Neumann, C. S., & Hare, R. D. (2010). Corporate psychopathy: Talking the walk. *Behavioral Sciences & the Law, 28*(2), 174–193. https://doi.org/10.1002/bsl.925

Bermúdez, J. L. (2005). *Philosophy of psychology: A contemporary introduction*. Routledge.

Blackburn, R. (1988). On moral judgements and personality disorders. The myth of psychopathic personality revisited. *The British Journal of Psychiatry, 153*(4), 505–512. https://doi.org/10.1192/bjp.153.4.505

Blair, R. J. R., Mitchell, D., & Blair, K. (2005). *The psychopath: Emotion and the brain*. Blackwell.

Bolton, D. (2008). *What is mental disorder? An essay in philosophy, science, and values*. Oxford University Press.

Boorse, C. (1977). Health as a theoretical concept. *Philosophy of Science, 44*(4), 542–573. https://doi.org/10.1086/288768

Borsboom, D. (2017). A network theory of mental disorders. *World Psychiatry, 16*(1), 5–13. https://doi.org/10.1002/wps.20375

Bortolotti, L. (2020). Doctors without 'disorders'. *Aristotelian Society Supplementary Volume, 94*(1), 163–184. https://doi.org/10.1093/arisup/akaa006

Brazil, I. A., van Dongen, J. D. M., Maes, J. H. R., Mars, R. B., & Baskin-Sommers, A. R. (2018). Classification and treatment of antisocial individuals: From behavior to bio-cognition. *Neuroscience & Biobehavioral Reviews, 91*, 259–277. https://doi.org/10.1016/j.neubiorev.2016.10.010

Brzović, Z., Jurjako, M., & Šustar, P. (2017). The kindness of psychopaths. *International Studies in the Philosophy of Science, 31*(2), 189–211. https://doi.org/10.1080/02698595.2018.1424761

Cartwright, S. A. (1851). Report on the disease and physical peculiarities of the Negro race. *The New Orleans Medical and Surgical Journal*, 89–92.

Cavadino, M. (1998). Death to the psychopath. *The Journal of Forensic Psychiatry, 9*(1), 5–8. https://doi.org/10.1080/09585189808402175

Charland, L. C. (2004). Character: Moral treatment and the personality disorders. In J. Radden (Ed.), *The philosophy of psychiatry: A companion* (pp. 64–77). Oxford University Press.

Cleckley, H. (1976). *The mask of sanity* (5th ed.). Mosby.

Cooper, R. V. (this volume). Reasons to expect psychopathy and Antisocial Personality Disorder (ASPD) to vary across cultures. In L. Malatesti, J. McMillan, & P. Šustar (Eds.), *Psychopathy. Its uses, validity, and status*. Springer.

Cooper, R. V. (2005). *Classifying madness: A philosophical examination of the diagnostic and statistical manual of mental disorders*. Springer.

D'Alessio, V., Čeč, F., & Karge, H. (2017). Crime and madness at the opposite shores of the Adriatic: Moral insanity in Italian and Croatian psychiatric discourses. *Acta Medico-Historica Adriatica : AMHA, 15*(2), 219–252.

Decety, J., & Wheatley, T. (Eds.). (2017). *Moral brain: A multidisciplinary perspective*. MIT Press.

Forouzan, E., & Cooke, D. J. (2005). Figuring out la femme fatale: Conceptual and assessment issues concerning psychopathy in females. *Behavioral Sciences & the Law, 23*(6), 765–778. https://doi.org/10.1002/bsl.669

Fulford, K. W. M. (1989). *Moral theory and medical practice*. Cambridge University Press.

Fulford, K. W. M., Smirnov, A. Y. U., & Snow, E. (1993). Concepts of disease and the abuse of psychiatry in the USSR. *British Journal of Psychiatry, 162*(6), 801–810. https://doi.org/10.1192/bjp.162.6.801

Gaus, G. F. (2011). *The order of public reason*. Cambridge University Press.

Glenn, A. L., Efferson, L. M., Iyer, R., & Graham, J. (2017). Values, goals, and motivations associated with psychopathy. *Journal of Social and Clinical Psychology, 36*(2), 108–125. https://doi.org/10.1521/jscp.2017.36.2.108

Glenn, A. L., & Raine, A. (2014). *Psychopathy: An introduction to biological findings and their implications*. NYU Press.

Graham, G. (2013). *The disordered mind: An introduction to philosophy of mind and mental illness* (2nd ed.). Routledge.

Hacking, I. (1996). The looping effects of human kinds. In D. Sperber, D. Premack, & A. J. Premack (Eds.), *Causal cognition* (pp. 351–383). Oxford University Press. https://doi.org/10.1093/acprof:oso/9780198524021.003.0012

Haidt, J. (2001). The emotional dog and its rational tail: A social intuitionist approach to moral judgment. *Psychological Review, 108*(4), 814–834.

Hare, R. D. (2003). *The Hare psychopathy checklist revised* (2nd ed.). Multi-Health Systems.

Holmes, C. A. (1991). Psychopathic disorder: A category mistake? *Journal of Medical Ethics, 17*(2), 77–85. https://doi.org/10.1136/jme.17.2.77

Horwitz, A. V. (2003). *Creating mental illness*. Univ. of Chicago Press.

Hume, D. (1748). *An enquiry concerning human understanding* (Reissued). Oxford University Press.

Jalava, J., & Griffiths, S. (2017). Philosophers on psychopaths: A cautionary in interdisciplinarity. *Philosophy, Psychiatry, and Psychology, 24*(1), 1–12.

Jalava, J., Griffiths, S., & Maraun, M. (2015). *The myth of the born criminal*. University of Toronto Press.

Jefferson, A. (2014). Mental disorders, brain disorders and values. *Frontiers in Psychology, 5*. https://doi.org/10.3389/fpsyg.2014.00130

Jefferson, A. (2020). What does it take to be a brain disorder? *Synthese, 197*(1), 249–262. https://doi.org/10.1007/s11229-018-1784-x

Jefferson, A., & Sifferd, K. (2018). Are psychopaths legally insane? *European Journal of Analytic Philosophy, 14*(1), 79–96. https://doi.org/10.31820/ejap.14.1.5

Jurjako, M. (2019). Is psychopathy a harmful dysfunction? *Biology and Philosophy, 34*(5). https://doi.org/10.1007/s10539-018-9668-5

Jurjako, M., Malatesti, L., & Brazil, I. A. (2019). Some ethical considerations about the use of biomarkers for the classification of adult antisocial individuals. *International Journal of Forensic Mental Health, 18*(3), 228–242. https://doi.org/10.1080/14999013.2018.1485188

Jurjako, M., Malatesti, L., & Brazil, I. A. (2020). Biocognitive classification of antisocial individuals without explanatory reductionism. *Perspectives on Psychological Science, 15*(4), 957–972. https://doi.org/10.1177/1745691620904160

Jurjako, M., Malatesti, L., & McMillan, J. (2018). Psychopathy: Philosophical and empirical challenges. *European Journal of Analytic Philosophy, 14*(1), 5–14.

Kitcher, P. (2001). *Science, truth, and democracy*. Oxford University Press.

Krupp, D. B., Sewall, L. A., Lalumière, M. L., Sheriff, C., & Harris, G. T. (2012). Nepotistic patterns of violent psychopathy: Evidence for adaptation? *Frontiers in Psychology, 3*. https://doi.org/10.3389/fpsyg.2012.00305

Liao, S. M. (Ed.). (2016). *Moral brains: The neuroscience of morality*. Oxford University Press.

Lieberman, D., & Smith, A. (2012). It's all relative: Sexual aversions and moral judgments regarding sex among siblings. *Current Directions in Psychological Science, 21*(4), 243–247. https://doi.org/10.1177/0963721412447620

Longino, H. E. (2004). How values can be good for science. In P. K. Machamer & G. Wolters (Eds.), *Science, values, and objectivity* (pp. 127–142). University of Pittsburgh Press.

Maibom, H. L. (2014). To treat a psychopath. *Theoretical Medicine and Bioethics, 35*(1), 31–42. https://doi.org/10.1007/s11017-014-9281-9

Maibom, H. L. (2018). What can philosophers learn from psychopathy? *European Journal of Analytic Philosophy, 14*(1), 63–78.

Malatesti, L. (2014). Psychopathy and failures of ordinary doing. *Etica & Politica/Ethics & Politics, 2*, 1138–1152.

Malatesti, L., & Baccarini, E. (this volume). The disorder status of psychopathy. In L. Malatesti, J. McMillan, & P. Šustar (Eds.), *Psychopathy. Its uses, validity, and status*.

Malatesti, L., Jurjako, M., & Meynen, G. (2020). The insanity defence without mental illness? Some considerations. *International Journal of Law and Psychiatry, 71*, 101571. https://doi.org/10.1016/j.ijlp.2020.101571

Malatesti, L., & McMillan, J. (2014). Defending psychopathy: An argument from values and moral responsibility. *Theoretical Medicine and Bioethics, 35*(1), 7–16. https://doi.org/10.1007/s11017-014-9277-5

McKeown, A., & Thomson, N. D. (2019). Psychopathy and intelligence in high-risk violent women. *The Journal of Forensic Psychiatry & Psychology, 30*(3), 484–495. https://doi.org/10.1080/14789949.2018.1560487

McMillan, J. (this volume). Re-appraising psychopathy. In L. Malatesti, J. McMillan, & P. Šustar (Eds.), *Psychopathy. Its uses, validity, and status*. Springer.

Moll, J., de Oliveira-Souza, R., & Eslinger, P. J. (2003). Morals and the human brain: A working model. *Neuro Report*, 299–305. https://doi.org/10.1097/00001756-200303030-00001

Mullen, P. E. (2007). On building arguments on shifting sands. *Philosophy, Psychiatry, and Psychology, 14*(2), 143–147.

Nadelhoffer, T., & Sinnott-Armstrong, W. P. (2013). Is psychopathy a mental disease? In N. A. Vincent (Ed.), *Neuroscience and legal responsibility* (pp. 229–255). Oxford University Press.

Parhi, K. (2018). *Born to be deviant: Histories of the diagnosis of psychopathy in Finland.* University of Oulu.

Parhi, K., & Pietikainen, P. (2017). Socialising the anti-social: Psychopathy, psychiatry and social engineering in Finland, 1945–1968. *Social History of Medicine, 30*(3), 637–660. https://doi.org/10.1093/shm/hkw093

Powell, R., & Scarffe, E. (2019). Rethinking "disease": A fresh diagnosis and a new philosophical treatment. *Journal of Medical Ethics, 45*(9), 579–588. https://doi.org/10.1136/medethics-2019-105465

Prichard, J. C. (1837). *A treatise on insanity and other disorders affecting the mind.* Haswell, Barrington, and Haswell.

Putnam, H. (2004). *The collapse of the fact/value dichotomy and other essays* (3. print). Harvard University Press.

Railton, P. (2003). *Facts, values, and norms: Essays toward a morality of consequence.* Cambridge University Press. https://doi.org/10.1017/CBO9780511613982

Raine, A. (2019). The neuromoral theory of antisocial, violent, and psychopathic behavior. *Psychiatry Research, 277*, 64–69. https://doi.org/10.1016/j.psychres.2018.11.025

Rawls, J. (2005). *Political liberalism* (Expanded ed.). Columbia University Press.

Reimer, M. (this volume). Psychopathy and personal identity: Implications for medicalization. In L. Malatesti, J. McMillan, & P. Šustar (Eds.), *Psychopathy. Its uses, validity, and status.* Springer.

Reimer, M. (2008). Psychopathy without (the language of) disorder. *Neuroethics, 1*(3), 185–198. https://doi.org/10.1007/s12152-008-9017-5

Rosenberg Larsen, R. (2018). False-positives in psychopathy assessment: Proposing theory-driven exclusion criteria in research sampling. *European Journal of Analytic Philosophy, 14*(1), 33–52.

Rosenberg Larsen, R. (2020). Psychopathy as moral blindness: A qualifying exploration of the blindness-analogy in psychopathy theory and research. *Philosophical Explorations*, 1–20. https://doi.org/10.1080/13869795.2020.1799662

Rush, B. (2009). *An inquiry into the influence of physical causes upon the moral faculty: Delivered before a meeting.* BiblioLife.

Sadler, J. Z. (2008). Vice and the diagnostic classification of mental disorders: A philosophical case conference. *Philosophy, Psychiatry, & Psychology, 15*(1), 1–17. https://doi.org/10.1353/ppp.0.0152

Sass, H., & Felthous, A. R. (2014). The heterogeneous construct of psychopathy. In T. Schramme (Ed.), *Being amoral: Psychopathy and moral incapacity* (pp. 41–68). The MIT Press.

Schramme, T. (Ed.). (2014). *Being amoral: Psychopathy and moral incapacity.* The MIT Press.

Sedgwick, P. (1973). Illness-mental and otherwise. *Hastings Center Studies, 1*(3), 19–40.

Spitzer, R. L., & Williams, J. B. (1982). The definition and diagnosis of mental disorder. In W. R. Gove (Ed.), *Deviance and mental illness* (pp. 15–31). Sage Publications.

Stevenson, C. L. (1944). *Ethics and language.* Yale University Press.

Szasz, T. S. (1974). *The myth of mental illness: Foundations of a theory of personal conduct* (Rev. ed.). Harper & Row.

Tamatea, A. J. (this volume). Humanising psychopathy, or what it means to be diagnosed as a psychopath: Stigma, disempowerment, and scientifically-sanctioned alienation. In L. Malatesti, J. McMillan, & P. Šustar (Eds.), *Psychopathy. Its uses, validity, and status.* Springer.

Wakefield, J. C. (1992). The concept of mental disorder. On the boundary between biological facts and social values. *The American Psychologist, 47*(3), 373–388.

Wakefield, J. C. (2014). The biostatistical theory versus the harmful dysfunction analysis, part 1: Is part-dysfunction a sufficient condition for medical disorder? *Journal of Medicine and Philosophy, 39*(6), 648–682. https://doi.org/10.1093/jmp/jhu038

Ward, T. (2010). Psychopathy and criminal responsibility in historical perspective. In L. Malatesti & J. McMillan (Eds.), *Responsibility and psychopathy: Interfacing law, psychiatry, and philosophy* (pp. 7–24). Oxford University Press.

Weber, M. (2011). *Methodology of social sciences*. Transaction Publishers.

Wynn, R., & Hoiseth, & Pettersen, G. (2012). Psychopathy in women: Theoretical and clinical perspectives. *International Journal of Women's Health, 257*. https://doi.org/10.2147/IJWH.S25518

Chapter 13
Unsexed Cruelty: Gender and Psychopathy as Regulatory Discourses in Relation to Violent Women

Gwen Adshead

Abstract Discourses are social conversations which use words and images to apply meaning to human social experience. As such, they may be highly influential in terms of defining and regulating ideas of what is 'normal' socially, and also what is 'good' in moral terms. In this chapter I will argue that the concept of psychopathy, like that of gender, acts as a regulatory discourse that shapes ideas about what is 'normal' and 'abnormal' in terms of violence to others. I will discuss this argument in relation to women who commit acts of serious violence.

Keywords Psychopathy · Power · Gender · Masculinity · Femininity · Female violence · Stereotypes

13.1 Introduction

In this chapter I explore the regulatory function of human discourse with reference to two specific areas of human function; namely sex and cruelty. Drawing on the work of social anthropologists, I describe how social discourse shapes relationships, especially relationships of power. This regulatory function of discourse is most obvious in relation to gender; where cultural rules emerge about who or what is 'normal' in terms of maleness and femaleness. I want to suggest that there is a kind of parallel between how the concept of 'gender' has developed with the development of the concept of psychopathy; in terms of the deployment of both sociological ideas and bioscientific ideas to validate the concept itself. In so doing, (I suggest) both gender and psychopathy then act as a kind of regulatory discourse which defines normal and abnormal behaviour; whether in relation to being male or female, or being cruel to others. I suggest that studies of violent criminality in women offer some evidence for my argument; and I conclude with some reflections about further research.

G. Adshead (✉)
Consultant Forensic Psychiatrist and Psychotherapist, West London Trust, London, UK
e-mail: g.adshead@nhs.net

© Springer Nature Switzerland AG 2022
L. Malatesti et al. (eds.), *Psychopathy*, History, Philosophy and Theory of the
Life Sciences 27, https://doi.org/10.1007/978-3-030-82454-9_13

13.2   Culture and Discourses

The study and identification of 'identity', 'difference' and 'otherness' has been the focus of research by anthropologists and sociologists in relation to a variety of human characteristics. In this section on cultural discourses of identity, I draw heavily on the work of the social anthropologist, Dame Henrietta Moore. Moore (2007 p. 10) describes culture as built up through engagement with others; an engagement which is made up of representations, both linguistic and non-linguistic. Groups of humans develop bodies of language and activities that become cultural discourses that structure and influence ideas and issues of importance to the groups involved. Group membership relies on being able to identify with certainty and lack of ambiguity those individuals who are members of 'our' group, and those who are not.

Discourses not only shape group membership; they also shape enquiry into individual selves, both physical and psychological; giving rise to cultural narratives about how things are and how they *should be*. Moore suggests that discourses and cultural narratives are fantasies that we project onto the external world to control anxiety about social relationships (2007 p. 1). They are also profoundly moral in nature insofar as they reflect social values and different kinds of need, demand and desire (p. 51). They are narratives which regulate social relationships and stress cultural values; they invite debate about boundaries and differences that do matter, may matter and don't matter in terms of group rules and identity. Biological studies also divide the world up into categories, contributing to wider public discourses about what 'normal' humans are like and how to recognise them. Both biological and cultural discourses have a regulatory influence over what is considered abnormal, deviant or transgressive; and may also offer solutions to control such deviance.

Discourses of difference and similarity (whether perceived or established) are generally considered to contribute significantly to group cultures and identities; and to act as a kind of gatekeeper to either social inclusion or exclusion. These discourses become the basis for rules and expectations about 'normality' and 'abnormality'. For example, the so-called 'medical model' is a discourse of human health and sickness which is based on the identification of an (allegedly) biological 'norm' from which 'difference' can be reliably identified and measured. The development and establishment of 'norms' for human groups is the work of many groups within society; and these 'norms' become the basis for cultural beliefs and stories about how people should function. Various individual factors, such as sex and personality are also part of the 'story' of how normal individuals function in societies.

As discourses, both gender and personality are ways of relating that underpin power relationships (Moore, 2007, p. 93), and address the discrepancy between those with power and those who are vulnerable. These cultural norms have considerable power insofar as they regulate social inclusion and exclusion; violations of cultural norms in relation to sex, personality and social relationships can lead to social condemnation and exclusion. Discourses especially shape definitions of those who can be considered vulnerable and 'fair game' for exploitation.

13.3 Gender as Regulatory Discourse: Normal and Abnormal Women

Moore (1999a, b, p. 151) describes how the distinction between sex and gender became established in social sciences in the 1970s. She defines 'gender' as the 'cultural elaboration of the meaning and significance of the natural facts of biological difference between men and women'. She goes on to describe (p. 152) how gender was studied in relation to the division of labour between men and women, and the emerging understanding that gender is a process that influences how bodies are sexually defined; with profound implications for reproduction, kinship structures and power relationships.

For both Moore and Judith Butler (1990), gender is a way of performing sexuality; which comprises genital sex assignment, erotic object choice, sexual practices and sexual identity. These aspects of human sexuality do not necessarily fit together according to social conventions (Moore, 1999a, b, p. 157) and gender discourse offers ways that the conventions 'could and *should* be subverted' (emphasis added).

Moore's work in anthropology describes how gender acts as a kind of control over the body and is central to (a) how men and women relate to each other and (b) how constructs of masculinity and femininity are developed culturally (Moore, 2007: pp. 34–36). She describes gender as 'embodied personality' which is influenced by how ideas about masculinity and femininity are constructed socially (p. 75). These discourses give rise to expectations about what 'normal' men and women are like, in terms of power relationships, class and voice (p. 35). Gender identities may not always be conscious or articulated but they are embodied.

An example of how power relationships between the sexes are defined by gender is evident in enduring traditions of 'masculinity' and 'femininity'; which act as normative accounts of what a man should be like, and what a woman should be like. The language emphasises essential difference; both bodily and in terms of personality. These gender role definitions and expectations have a powerful influence on how the sexes behave. Traditional accounts of masculinity suggest that to be male is to be active, decisive and in control; happy to lead and where necessary command. Conversely, femininity implies that to be female is to be passive, non-competitive, non-assertive and happy to follow instruction. It is obvious that in reality, these traits described above may be found in everyone at different times depending on circumstance; but the point is that these stereotypic accounts act as a regulatory discourse to control and limit behaviours by men and women.

The biological sciences have been influential as a further discourse that sustains and support 'norms' about bodies and minds; and feminist scholarship has questioned the apparent 'objectivity' of scientific method in relation to examination of sex difference (e.g. Fox Keller, 2003). For example, Joyce (2006 p. 44) reviewing debates about sex and gender comments on how apparently 'obvious facts' about physical difference are then assumed to be the basis for women's differences to men in the social sphere. If the male body is seen as the 'norm', then the female body may be seen as deviant or damaged in comparison; and the same embodied difference is then applied to the mind, establishing apparent psychological difference. However, women often describe being defined by their body size, shape and genitals; whereas men are defined in terms of their minds. Moore quotes Strathearn (p. 133) in noting that Western assumptions about the body see it as a kind of property that the mind owns in a '*discourse of possession*'; and these kinds of assumptions can be extended to the possession of female bodies by males who are enabled to exert power using their minds, which are privileged in terms of personal identity.

Much of the study of gender over the last 20 years has explored the social response to those who break or challenge sex role expectations. Butler has noted (1990, 1993) that there is a '*performative*' quality to gender, which she describes as "a reiterative citation of *regulatory norms*' (1990 p. 140: emphasis added). On this account, gender role behaviours communicate something important about where one stands, vis-à-vis social norms. By implication, there may be heavy social penalties for those who challenge those norms; including challenges to the concept of essential differences between men and women and challenges to the ways that people 'do' sex and sexuality. The most obvious example historically of challenge to gender norms is same-sex erotic bonding and identity which, by breaching traditional gender role expectations, caused such alarm that for centuries, it was illegal, confused with mental illness and dysfunctional sexuality and punishable by death. Homoerotic sexuality was only removed from Western professional accounts of mental illness diagnoses in 1976; and in several non-Western countries remains illegal and a capital crime.

There is some consensus that cultural beliefs about gender roles tend to present male-ness as the human default for normality; with considerable implications for medicine and related biological sciences. The World Health Organisation (Garcia Moreno, 1998) describes concerns that gender bias in health sciences and health provision can result in damage to the health of both women and men; because either (a) sexual differences are perceived where there are none or such difference is clinically insignificant or (b) real sex differences are ignored that may be highly relevant for understanding pathology, pharmacology and treatment outcomes.

Developments in neuroscientific research methodology have provided new ways of identifying biological and apparently essential differences between male and female brains. However, the evidence from neuroscience may be confounded by implicit as well as explicit gender bias (Fine, 2010); and it is also unclear how or whether differences between male and female *brains* provide any reliable information about differences between male and female *minds*. There are strongly held conflicting views in this field; for example, Baron-Cohen (2009) claims that neuroscience provides evidence that there are essential differences between male and female brains which have an effect on social behaviour: "The female brain is hardwired for empathy. The male brain is hardwired for understanding and building systems". In contrast Joel (2015) argues that neuroscientific studies indicate that brains are neither essentially 'male' nor 'female' but have a mosaic of features in common and difference.

An appeal to scientific method as a kind of empirical trump card in establishing sexual difference, especially in neuroscience, is a nice example of how debates about normality and abnormality are socially significant for gender role expectations. A biological difference may be culturally presented as 'natural' and inevitable; but it is not always clear what such a difference might mean socially or morally. Arguments about discrimination between different groups become complex; it may be just to respect differences that matter, but unjust to attribute social significance to differences that do not matter. The focus of social and philosophical debate is then what differences are of significance in a culture or social group; why; and who decides.

Debates about gender raise particular concerns about the use of perceived differences between the sexes to justify different social treatment. There are many historical examples of this; the most obvious being enfranchisement and the many philosophers and lawyers who argued against votes for women. Early studies of criminal rule breaking in society did not mention women because it was assumed that women did not offend (Daly & Chesney-Lind, 1988); and those that did offend were seen as 'masculine' and/or 'damaged', usually by mental illness (Stangle, 2008). Feminist criminologists noted that women who offended were judged to be doubly deviant because they had broken both criminal laws and gender norms (Heidensohn, 1996): I shall return to this issue in more depth later.

The literature on gender as a regulatory discourse is huge; and the scope of this chapter does not allow for further comprehensive review of examples or analysis of what are complex arguments. I end this section with a contemporary example of the regulatory function of gender discourse in human society. Over the last decade, there has been greatly increased interest in the possibility of choosing to change sex; and associated public discussions and debates (in the anglophone world at least) about the social significance of transsexuality. These debates how to define sexual identity and gender have been active, vociferous and even vituperative; and they illustrate how discussions of gender regulate social relationships in moral, legal and political domains. The 'Trans' debates focus not only on what it is to be sexed and gendered; they also show that it is not always clear who gets to decide the answer to these questions. Again, space does not permit a full review of this issue; but it is relevant to this chapter to note that most of the public debate has focussed on who can decide what it is to be a 'real' woman, with considerably less discussion of what it is to be a 'real' man; so that 'normal' maleness and masculinity slip away from the social and regulatory gaze. It is also noteworthy that the WHO have recently (2019) used their regulatory power to define as 'normal' the wish to change gender identity i.e. such a wish is now formally excluded from the social discourse of ill health and sickness.

13.4 Psychopathy as a Kind of Regulatory Discourse: Normal and Abnormal Cruelty

In this section, I argue that the concept of psychopathy (like that of gender) has developed into a social discourse that regulates discussion about human behaviour and social relationships. However, the focus of the psychopathy discourse is not human sexuality but human cruelty; and related questions of what kind of cruelty and social rule-breaking is 'normal' and what is 'abnormal'.

The term 'psychopath' literally means 'an abnormal mind'; and the earliest use of the term was in relation to those people who appeared 'normal' but did not keep social rules or fulfil social expectations. These men (and a few women) caused great social alarm not just because they broke the rules but because their difference of mind was not obvious or visible; they wore a 'mask of sanity' (Cleckley, 1941). In the late 1930s and early post war period, the need to identify threats to the social order was urgent; especially those people who did not care about apparently agreed social values and had no anxiety about breaching them (Chenier, 2012). Psychiatric and psychological studies were a bioscientific response to those perceived threats to social order; as described by Foucault (1978) psychiatry had a crucial regulatory role in making social rule breaking into a kind of 'illness' or mental abnormality.

The concept of 'psychopathy' focussed on those people who broke social conventions but seemed unaffected by negative social responses to them. Cleckley's paradigm study (1941) was not the first to use the term 'psychopath', nor the first psychiatric account; but it was one of the first to explore the concept in such depth using a range of case studies. Cleckley's argument was that psychopaths were not simply 'bad' but instead had some kind of profound emotional impairment; they looked mentally 'normal' but exhibited emotional shallowness which resulted in behaviour that was experienced by others as irrationally cruel. Cleckley surmised that these people did not intend to be cruel, but could not make real affective connections with others; a state of mind which was memorably summarised as 'know[ing] the words but not the music' of human interaction.

Cleckley's psychopaths were cruel, but not violent or predatory. He specifically did not use terms like 'evil' and only a subgroup of his subjects were criminal rule breakers (usually for drunken fights, theft or fraud, not violence). The question of *deliberate* cruelty was studied by Robert Hare who applied Cleckley's concepts to violent offenders detained in prison. Hare's early studies date from the late1970s; a different time sociologically and one where abnormal psychology as a biological science was gaining ground. At this time, there were considerable academic debates about psychopathy and whether it represented a 'real' disorder of mind or a medical term for 'wickedness' (Lewis, 1974; Wootton, 1978 p. 107); reflecting real social anxieties about whether or not to understand human cruelty as a kind of medical condition. A nice example of this anxiety comes from the English Court of Appeal; where the Lord Chief Justice noted that the distinction between the man who would not or could not resist a bad impulse was 'incapable of scientific proof' (R.v.Byrne).

I note here the appeal (or not) to scientific method to establish a socially significant difference, as described above in relation to sexual difference.

The novelty of Hare's approach was to apply Cleckley's concept of psychopathy to violence perpetrators who had been formally identified as legally deviant by the criminal courts. Using a checklist based on Cleckley's work, Hare found that only a third of men imprisoned for severe violence seemed to fit Cleckley's concept of psychopathy (i.e. emotionally disconnected and lacking awareness or concern about others). These 'Hare' psychopaths were subsequently shown to be more likely to violently re-offend than those who did not score highly on the checklist. It is note-worthy that this study also found that violence commission was *not* synonymous with psychopathy.

Professor Hare's work has been massively influential and applied in different penal and secure psychiatric settings around the world. His checklist (Hare, 1980) formalised the concept of psychopathy into a kind of measurable psychopathology, allowing for empirical study and a bioscientific approach to the identification of high-risk offenders. Hare's psychopaths were more violent than Cleckley's; partly because they were also persistently and polymorphously criminal. Hare's group were consciously arrogant, exploitative and contemptuous of human vulnerability, which was not always the case with Cleckley's group. Although both Hare and Cleckley describe their psychopaths as apparently 'charming', Hare argued that this charm is deployed consciously to con and deceive others; whereas Cleckley did not describe this.

To be identified as a 'Hare' psychopath, one must score highly in a number of different domains of individual and social function; including affective dysregula-tion, poor or absent interpersonal relating, impulsivity and generally antisocial atti-tudes. However, a crucial aspect of the debate about psychopathy, which is relevant to the theme of this chapter, is whether actual criminal rule breaking is essential to the nature of psychopathy (Skeem & Cooke, 2010). If it is the case that psychopaths always break the criminal law, then this is contradicted by evidence that the most persistent criminal rule breakers are not psychopaths (Vaughn et al., 2011); but offend because of cultural, social and racial pressure. If it is not the case that psy-chopathy entails criminal rule breaking, and physical harm to others, then this would suggest that the concept broadly includes aspects of personality function that are unpleasant and antisocial but not necessarily risky to others (Edens et al., 2006).

There are several accounts of psychopathy that do not entail violent cruelty or even criminality; but typically rely on three psychological personality traits. For example, Cooke, David et al. (2007) define psychopathy in terms of arrogance & deceitfulness; affective deficiency, and impulsivity combined with irresponsibility. Paulhus' group (Paulhus & Williams, 2002) describes a 'Dark Triad' which includes psychopathy but also includes narcissism and Machiavellianism (Mach for short). Narcissism is a complex concept in its own right (see Gabbard & Crisp, 2018 for a recent review) but can include contempt for others' weakness and arrogance/entitle-ment. 'Mach' is also complex; but seems to imply a kind of human capacity to cleverly play people off against one another or con people into a course of action whose significance they are not aware of. These three elements together describe a state of mind in which others are treated (as Kant might have put it) merely as a means to an end. Patrick's group (2009) describes psychopathy in terms of bold-ness, mean-ness to and dominance of others. Patrick's concept is interesting because of its hints at the importance of psychopathy in regulating social status and the pos-sibility that psychopathy might include some kind of positive psychological skill, given that boldness might in some contexts be valuable. On a similar note, Lilienfield's measure of psychopathy includes a factor called 'fearless dominance' which carries a suggestion of a kind of courage; and also hints at the importance of social ranking sensitivity to some kinds of cruelty. De Clercq et al. (2012) make reference to psychopathy as involving interactions that fall on 'a gradient of power and control'.

On these accounts, the prevalence of psychopathy is likely to be higher than previously thought; and it could be present in apparently 'normal' populations of people who do not break the law. However, establishing 'norms' for cruelty and antisocial states of mind depends on how these concepts are measured; and there are significant differences between the research studies based on populations sampled and the measure used. Studies using the Hare Psychopathy Checklist are usually based on populations in penal and secure services and the use of the checklist requires detailed examination of different sources of information from courts, prison and probation staff. However, many other studies of 'psychopathy' are based on self-report questionnaire studies in university undergraduates where rates of any criminality, let alone cruel violence, are likely to be low (although arrogance, cru-elty and contempt for others may well be present in a subgroup). Some of these studies have suggested that 'psychopathy' is normally distributed in the general population but only a sub-group with other risk factors for criminality will actually offend (Edens et al., 2006). There are reports that psychopathy can be associated with social and economic success (Babiak & Hare, 2006); although evidence for the existence of the 'successful' psychopath is 'elusive' (Smith et al., 2014).

Overall, these concepts of psychopathy appear to describe a state of mind which is assumed to be socially harmful and undesirable, even if it does not actually result in physical violence to others and formal criminal rule breaking. In theory, no one would want to be a psychopath and the conclusion seems to be that psychopathy should be controlled or eliminated if it is possible to do so. The above studies seem to suggest that psychopathy is a kind of paradigm of 'badness' that helps to set an endpoint to a spectrum of human cruelty; with psychopathy at one end, and unpleasant-but-normal cruelty at the other.

However, if psychopathy is not defined by violent criminal rule breaking and/or physical cruelty to others but only in terms of nasty mean attitudes to others, then the concept arguably has become a social 'shorthand' for a group of people whom society can justifiably condemn and exclude; as exemplified in this quote from DeLisi (2009): 'Psychopathy mirrors the *elemental* nature and embodies the *pejora-tive* essence of antisocial behaviour' (emphasis added). The reference to an 'ele-mental' quality implies a belief that antisocial behaviour is an essential aspect of some human beings; and the existence of this belief may explain the significant increase in studies of psychopathy which look for structural and functional differences in the brains of psychopaths compared with non-psychopaths (similar to the studies looking for sexual difference cited above).

A further example of psychopathy acting as a regulatory concept is in relation to children. It has been argued that an elemental cruelty in the form of 'callous and unemotional attitudes' can be identified in some children from an early age (Barry et al., 2000); and attitudes are a predictor of antisocial behaviour in adolescence and potential psychopathy. Although those who make this argument emphasise,that they are not identifying 'child psychopaths', the titles and tone of many of these studies tend to convey this concept; and these studies often refer to other studies that emphasise the influence of genes on behaviours and personality and the fixed ('elemental') nature of any dysfunction. Such studies usually de-emphasise the influence of social environments and relationships; and are then used to justify social policies that favour reduction in public welfare programs and emphasise individual responsibility and punishment. The cultural narrative of the 'bad seed' who is incorrigibly 'bad' regardless of environment is at odds with the evidence that most violent offenders have been exposed to adverse childhood environments compared to non-offenders or less violent offenders (Fox et al., 2015; Bowen et al., 2018); and Hare's evidence that most violence perpetrators are not and never have been psychopaths.

The majority of the studies cited here are firmly in the domain of neuroscience, and rely on the use of an epistemology that is (or claims to be) objective, impartial, and reliable. They are also located within the field of mental health which, as described above, seeks to distinguish the 'normal' from the abnormal; with the objective to return the abnormal to a normal state. For example, Kiehl and Hoffman (2011) in a review paper claim that psychopathy is 'an astonishingly common mental disorder' (p. 2) with an identifiable brain pathology; which needs to be 'treated in order to reduce the costs to society in terms of criminal process and detention' (p. 7). Similarly Centifanti and Garofalo (2019) argue that psychopathy is a mental disorder; and that if the public is educated about its neuroscientific basis, psychopathy will be seen as having the 'dignity' of a mental illness which may elicit more sympathy and less stigma.

But within this professional literature, there seems to be a profound ambiguity as to whether 'psychopathy' represents highly abnormal and unusual human cruelty (justifying strong legal and moral responses to it); or whether 'psychopathy' is normal and common, and everyone has some degree of psychopathy in their personality, which might even be to their advantage. This ambiguity may be related to an overlap between psychopathy and masculinity: some of the features of psychopathy described above resemble features of stereotypical positive masculinity, such as dominance, fearlessness and charm. Psychopathy has been described as evidence of the presence of a 'warrior gene', (Perbal, 2013; Fallon, 2014) where 'warrior' is a term often associated with a kind of bleak heroism. The moral ambiguity of a kind of 'hero' who is also dangerous and cruel is a frequent trope in many superhero movies and cinematic depictions of psychopathy. The ambiguity is perhaps best illuminated by an internet study of psychopathy which claimed to show that Jesus of Nazareth and Mother Theresa of Calcutta are more 'psychopathic' than a variety of US and UK politicians (Dutton, 2016).

13.5   Cruel and Unusual: Violent Female Offenders

I now turn to examining the implications of the discussion so far for those unusual women who commit acts of serious violence. There is reliable evidence that women are a minority of violence perpetrators across cultures and countries. For this reason, violent women are often seen as both 'abnormal' offenders and also 'abnormal' women. As mentioned above, early criminological discussions of violent offenders did not mention women at all; and later studies minimised the violence and cruelty of such women (Stangle, 2008), using what Allen (1987) describes as 'discursive manoeuvres' (p. 81) in terms of language and agency. Early accounts of female violence tended to emphasise the causative role of emotional distress or mental illness in female violence; or else suggested that women's violence was uniquely 'feminine' (Appignanesi, 2011).

Many criminological studies of female offenders have argued that discourses of gender and gender role stereotypes are fundamental to understanding how female violence is regulated and punished. Heidensohn (1996) argued that female offenders breach sex role stereotypes by taking action and being assertive and socially challenging; and they may be disproportionately punished because of this combined social and criminal abnormality. Van Voorhuis (2012) argues that lack of understanding about why women offend leads to detention in higher custody levels, despite lower levels of violence. Unlike male violence perpetrators, female violence perpetrators are often described in terms of their social vulnerability in patriarchal societies where they may suffer poverty and social disadvantage; or they may be described as victims of prior violence, which may make them mentally ill and at risk of acting violently (Covington, 2002; Bloom et al., 2003).

The depiction of female violence perpetrators as vulnerable and deserving of compassion is both laudable at one level and curious at another. It is laudable to support compassion for offenders as this is likely to be a positive influence on attempts at rehabilitation; but it is not clear why females should excite more compassion than male offenders who have suffered similar exposure to violence and social disadvantage. Arguments based on gender role would suggest that gender role stereotypes portray women as being passive and at the mercy of others; and perhaps being less culpable because less agentic (Widom, 1984). Allen (1987) suggested that such gender-based stereotypes may actually work in women's favour at sentencing where she describes a kind of 'chivalry' effect whereby women offenders receive less punitive sentences for the same offences as males.

One important difference between male and female violence perpetrators relates to their victims; female violence perpetrators tend to attack partners or family members, including their own children. In the UK, children under a year old are statistically at the highest risk of being homicide victims, and the perpetrators are nearly always mothers. These kinds of homicide attract extremes of social condemnation and demand an explanation; maternal mental illness explains a small proportion of offending, but juries find it hard to accept that women might kill their children for a non-psychiatric reason. (Stangle, 2008).

Psychopathy is often raised in cases of female cruelty, especially in relation to women who kill or attack the vulnerable (Seal, 2010). Early studies of psychopathy in women did not include women convicted of violence; and Cleckley's two case studies of psychopathy in women focussed on their tendency to deception and their promiscuity. In this context, it is of interest that Cleckley did not mention female violence perpetrators who were in the public eye at the time that he was writing; such as Bonnie Parker and Kate ('Ma') Barker, both of whom were convicted of fatal violence to others.

Cleckley's examples are also intriguing because his description of psychopathy in young women fits with common negative feminine gender stereotypes. Traditional accounts of 'bad' women have referred to their deceitfulness and their misuse of their sexual allure. A recent review of Cleckley's cases rated them according to current concepts of psychopathy; and found real professional disagreement about whether Cleckley's female 'psychopaths' met current criteria for psychopathy (Crego & Widiger, 2016). A significant concern is that 'promiscuity' is sometimes the only criterion for psychopathy that is reliably identified in women; suggesting that women's sexual behaviour may be pathologized in ways that men's are not (Grann, 2000; Forouzan & Cooke, 2005; Dolan & Vollm, 2009).

Studies using Hare's criteria for psychopathy usually find that only a sub-group of female offenders could be classified as psychopaths (like male violence perpetrators; Nicholls & Petrila, 2005; Nicholls et al., 2005). These women tend to have committed more acts of violence and to exhibit the same kind of callous and antisocial attitudes as the males. These female psychopaths also seem to resemble male psychopaths in terms of exposure to severe trauma (Weiler & Widom, 1996); and other reviews have suggested that although they are numerically fewer in number, female psychopaths seem similar to males in terms of their affective responses and interpersonal dysfunction. The main difference relates to the number and variety of past violent/criminal offences; the base rate for which is low in women.

A key question raised by researchers is whether the study of violent females (especially female psychopathy) can ever use criteria that are gender-neutral (Widom, 1984; Zahn-Waxler, 1993). Psychopathy may be expressed differently in the two sexes because of the influence of gender role constructs in society on personality expression (Salekin et al., 1998; Skeem et al., 2011; Beryl et al., 2014). The Beryl study reviewed all studies of psychopathy in incarcerated women, and noted the widespread use of Hare's checklist which they describe as 'a measure of psychopathy *in American men'*. (emphasis added).

Because base rates for offending generally are low in women, the study of female psychopathy also invites another look at signifiers of *non-offending* psychopathy (such as Patrick's mean-ness and boldness, Paulhus' narcissism or Lilienfield's fearless dominance). Narcissism as a feature of psychopathy is of particular interest as it often manifests differently in men than women; male narcissists being grandiose and over-bearing and female being fragile and controlling (Grijalva et al., 2015); although Grijalva et al. also advise caution about the influence of gender role stereotypes in looking at personality disorders in women. On a related note, assessing 'boldness' or 'fearless dominance' in women may entail looking for traits that are stereotypically masculine, thus risking an assumption that female 'psychopaths' are psychopaths when/because they are perceived to behave like men.

Cale and Lilienfeld (2002) suggest that female psychopaths differ from male by doing psychological not physical harm; which raises questions about how psychological harm is to be defined and where the boundaries of 'normal' psychological harm might be set for both sexes. There is also an interesting conflation in this paper of psychopathy, aggression and crime which would make it (theoretically) female violence common (which the data belies). Weizmann-Henelius et al. (2010) also conclude that women could be psychopathic without showing any kind of antisocial behaviour at all which is an approach that seems profoundly different to the ways that psychopathy is understood in males. Such an approach potentially implies that many women could be identified as 'psychopaths' if they are mean or cruel or dominant i.e. unlike traditional accounts of feminine 'niceness'.

Kreis and Cooke (2011) suggest that psychopathy in women is expressed as 'less grandiose, dominant, aggressive; more seductive, emotionally unstable'. They also comment that most violence by women is relational and violent women use impression management to deceive others and manage social relationships. However, they offer no evidence that male perpetrators of relational violence do not act in similar ways; and do not comment on evidence that many male relational violence perpetrators (especially domestic violence and stalking perpetrators) are also reported to be deceptive and plausible in impression management.

The issue of 'seductiveness' as an indicator of female psychopathy links with 'shallow and superficial charm'; another psychopathy 'trait' which may be understood differently by sex and gender. Traditional accounts of positive femininity make 'charm' a valuable aspect of being female; a trait that many women will seek to convey to others, since being charmless is therefore 'unfeminine' and may be 'abnormal'. Women offenders may find themselves in a paradox where it is antisocial *not to* attempt to be charming but attempts to do so may be interpreted as 'shallow charm' and 'manipulative seductiveness'. The signifiers for female psychopathy in this context again look quite unlike those for males; as demonstrated in these quotes from a study of psychopathy in women (Cunliffe et al. 2016):

P 176 Female psychopaths are coquettish, coy, seductive damsel in distress, dramatic, demure 'eliciting pity and sympathy' using sexuality to deceive others

P 178 an excellent example of how female psychopaths are consumed with seeking approval from others

P181 it was common for a female psychopath to characterise herself as a traditional or stay-at-home mom to cover up her parasitic behaviour

13.6   Conclusion

Moore (2007, p164) asks 'to what question is sexual difference the answer?' Similarly, it might be asked 'to what question is the psychopathy the answer?' In this chapter I have tried to suggest that psychopathy answers questions about the nature of human cruelty; and especially whether it is possible to distinguish between normal and abnormal human cruelty. Answering this question involves consideration of social and scientific discourses; and ultimately is also an ethical question insofar as it addresses a question about how we 'should' think about psychopathy.

In this chapter I have described a parallel between how the concept of gender has developed; first with social constructs of what a 'good' or 'normal' woman is; then using bioscientific 'facts' to support these distinctions; and then some current consensus that gender itself acts as a way to regulate power relationships between the sexes. In the same way, the concept of psychopathy starts as a description of 'bad' behaviour which is puzzling and abnormal insofar as most people don't act like this; it then becomes the subject of medical and neuroscientific study which appears to provide facts that validate psychopathy as a mental abnormality. The current study of psychopathy includes debate about its scope; and in that sense, it is a regulatory concept because it regulates relationships between people who are cruel and unusual and people who are not.

I have argued here that gender and psychopathy are both social discourses that regulate human behaviours and attempt to establish norms by which 'offenders can be identified in terms of departure from those norms. Both are discourses that regulate power relationships between different groups of people; masculinity typically confers power and status in societies, and people identified as 'psychopaths' may have power and liberty removed from them using socially approved restraints and constraints. Both are influential in normalising and pathologizing difference; both people who challenge gender role expectations and people labelled as 'psychopaths' may meet hostility and social exclusion. In both cases, psychiatric and neuroscientific evidence (themselves social discourses) has been used to support claims that there are 'essential' differences between groups of people in terms of sex, gender, antisocial behaviour and personality; and negative perceptions of such difference can influence public policy especially in relation to punishment and detention.

By discussing how psychopathy has been studied in violent women, I have tried to show psychopathy has relied on its neuroscientific biomedical facts to normalise regulatory gender stereotypes. The difficulty posed (and faced) by very violent women is that statistically they are highly unusual in terms of 'normal' behaviour for all citizens, both male and female. Such unusual behaviour can be explained using gender role discourse (emphasising how gender roles make people ill or vulnerable to attack) and the discourse of psychopathy (which claims that subgroups of unusually bad people exist in society and can be identified by neuroscience). But attempts to explain female violence using trauma or mental illness often result in falling back on unreliable gender role stereotypes about women as victims or women as mentally vulnerable. Trying to explain female violence using the concept of psychopathy sometimes results in a tautology of stating that these women acted in cruel and unusual ways, and didn't seem to care about it. But, as we have seen, several studies of psychopathy in women conclude that the evidence for their 'psychopathy' is either because they are not feminine enough or because of a kind of femininity which is natural but 'nasty'.

The concept of psychopathy has its roots in society's general need to regulate the behaviour of those people who are cruel to others and don't seem to care about it. Despite the extensive study of psychopathy over the last three decades, there remains a real ambiguity about whether those who pursue an individualistic or solipsistic world view are normal-but-nasty or abnormal and need containing (Skeem et al., 2016). Patey (2014) suggests that this ambiguity is heightened by the use of neuroscientific discourse to hide psychopathy's political nature as a response to global capitalism and the diminishment of social capital as a public good. Egocentricity, affective coldness and opportunism may appear 'ominous' in the consulting room or clinic (Scott, 1960) but less so in a deregulated and commodified society where social relationships are market based not value based (Sandel, 2012).

I am concerned that many studies of psychopathy claim to be morally neutral insofar as they originate within the realm of bioscientific discourse; but are actually working in a wider way that powerfully influence and control how society sees 'badness' and cruelty. Accounts based on either gender or diagnosis may pathologize 'bad' behaviour in some individuals but normalise it in others; even at the level of organisations and social policies. But as violence rates fall in most social democratic societies where there is rule of law and the economy is not based on the drugs trade (UNDOC, 2014), over simplistic appeals to discourses of gender or psychopathy as explanations for violence or cruelty are increasingly unconvincing.

As violence becomes an unusual way for humans to break the rules, it may make more sense to treat cruelty and violent offending as complex social transactions (Canter, 1994), even as a communication; rather than simply as a feature of abnormal individuals. To understand this communication, we will have to listen to the voices of all the perpetrators and be willing to hear their complex narratives of cruelty. We will need to resist the kind of discourses that appear to be respectful of individual experience but in fact categorise people in ways that stop conversation.

Cases

R.v Byrne [*1960*] 44 Cr App R 246, [*1960*] 2 QB 396.

Lord Parker CJ

Furthermore, in a case where the abnormality of mind is one that affects the accused's self – control the step between 'he did not resist his impulse' and 'he could not resist his impulse' is, as the evidence of this case shows, one which is incapable of scientific proof. A fortiori there is no scientific measurement of the degree of difficulty which an abnormal person finds in controlling his impulses. These problems which in the present state of medical knowledge are scientifically insoluble, the jury can only approach in a broad, common sense way'.

References

Allen, H. (1987). *Justice unbalanced: Gender, psychiatry, and judicial decisions* (Vol. 27, p. 81). Milton Keynes: Open University Press.

Appignanesi, L. (2011). *Mad, bad and sad: A history of women and the mind doctors from 1800 to the present.* Hachette UK.

Babiak, P., & Hare, R. D. (2006). *Snakes in suits: When psychopaths go to work.* New York: ExecuGo media.

Baron-Cohen, S. (2009). *The essential difference: Male and female brains and the truth about autism.* Basic Books.

Barry, C. T., Frick, P. J., DeShazo, T. M., McCoy, M., Ellis, M., & Loney, B. R. (2000). The importance of callous–unemotional traits for extending the concept of psychopathy to children. *Journal of Abnormal Psychology, 109*(2), 335.

Beryl, R., Chou, S., & Völlm, B. (2014). A systematic review of psychopathy in women within secure settings. *Personality and Individual Differences, 71*, 185–195.

Bloom, B., Owen, B. A., & Covington, S. (2003). *Gender-responsive strategies: Research, practice, and guiding principles for women offenders.* Washington, DC: US Department of Justice, National Institute of Corrections.

Bowen, K., Jarrett, M., Stahl, D., Forrester, A., & Valmaggia, L. (2018). The relationship between exposure to adverse life events in childhood and adolescent years and subsequent adult psychopathology in 49,163 adult prisoners: A systematic review. *Personality and Individual Differences, 131*, 74–92.

Butler, J. (1990). *Gender trouble* (2nd ed.). Taylor and Francis.

Butler, J. (1993). *Bodies that matter.* New York: Routledge.

Cale, E. M., & Lilienfeld, S. O. (2002). Sex differences in psychopathy and antisocial personality disorder: A review and integration. *Clinical Psychology Review, 22*(8), 1179–1207.

Canter, D. (1994). *Criminal shadows: Inside the mind of the serial killer* (p. 119). London: Harper Collins.

Centifanti, L., & Garofalo, C (2019). *The stigma of being labelled a psychopath: Is there value in a label?* European Society for the Study of Personality Disorders newsletter. December 2019. p. 6.

Chenier, E. (2012). The natural order of disorder: Pedophilia, stranger danger and the normalising family. *Sexuality & Culture, 16*(2), 172–186.

Cleckley, H. (1941). *The mask of sanity; An attempt to reinterpret the so-called psychopathic personality.* St Louis Miss: Mosby.

Cooke David, J., Mitchie, C., & Skeem, J. (2007). Understanding the structure of the psychopathy checklist-revised. *British Journal of Psychiatry, 190*(suppl 49), s39–s50.

Covington, S. (2002). *A woman's journey home: Challenges for female offenders and their children.* Urban Institute, Justice Policy Center.

Crego, C., & Widiger, T. A. (2016). Cleckley's psychopaths: Revisited. *Journal of Abnormal Psychology, 125*(1), 75.

Cunliffe, T. B., Gacono, C., Smith, J. M., Kivisto, A. J., Meloy, J. R., & Taylor, E. (2016). Chapter 9: Assessment of female psychopaths. In C. Gacono (Ed.), *The clinical and forensic assessment of psychopathy: A practitioner's guide* (2nd ed., pp. 167–185). New York: Routledge.

Daly, K., & Chesney-Lind, M. (1988). Feminism and criminology. *Justice Quarterly, 5*(4), 497–538.

Declercq, F., Willemsen, J., Audenaert, K., & Verhaeghe, P. (2012). Psychopathy and predatory violence in homicide, violent, and sexual offences: Factor and facet relations. *Legal and Criminological Psychology, 17*(1), 59–74.

DeLisi, M. (2009). Psychopathy is the unified theory of crime. *Youth Violence and Juvenile Justice, 7*(3), 256–273.

Dolan, M., & Völlm, B. (2009). Antisocial personality disorder and psychopathy in women: A literature review on the reliability and validity of assessment instruments. *International Journal of Law and Psychiatry, 32*(1), 2–9.

Dutton, K, (2016). http://www.ox.ac.uk/news/2016-08-23-presidential-candidates-may-be-psychopaths-%E2%80%93-could-be-good-thing

Edens, J. F., Marcus, D. K., Lilienfeld, S. O., & Poythress, N. G., Jr. (2006). Psychopathic, not psychopath: Taxometric evidence for the dimensional structure of psychopathy. *Journal of Abnormal Psychology, 115*(1), 131.

Fallon, J. (2014). *The psychopath inside: A neuroscientist's personal journey into the dark side of the brain.* Current (Penguin Group).

Fine, C. (2010). *Delusions of gender.* London: Icon Books.

Forouzan, E., & Cooke, D. J. (2005). Figuring out la femme fatale: Conceptual and assessment issues concerning psychopathy in females. *Behavioral Sciences & the Law, 23*(6), 765–778.

Foucault, M., (TRs) Baudot, A., & Couchman, J. (1978). About the concept of the "dangerous individual" in 19th-century legal psychiatry. *International Journal of Law and Psychiatry, 1*(1), 1–18.

Fox, B. H., Perez, N., Cass, E., Baglivio, M. T., & Epps, N. (2015). Trauma changes everything: Examining the relationship between adverse childhood experiences and serious, violent and chronic juvenile offenders. *Child Abuse & Neglect, 46*, 163–173.

Gabbard, G. O., & Crisp, H. (2018). *Narcissism and its discontents: Diagnostic dilemmas and treatment strategies with narcissistic patients.* American Psychiatric Pub.

Garcia Moreno, C. (1998). *Gender and health: Technical report.* Geneva: WHO. https://www.who.int/docstore/gender-and-health/pages/WHO%20-%20Gender%20and%20Health%20Technical%20Paper.htm.

Grann, M. (2000). The PCL–R and gender. *European Journal of Psychological Assessment, 16*(3), 147.

Grijalva, E., Newman, D. A., Tay, L., Donnellan, M. B., Harms, P. D., Robins, R. W., & Yan, T. (2015). Gender differences in narcissism: A meta-analytic review. *Psychological Bulletin, 141*(2), 261.

Hare, R. D. (1980). A research scale for the assessment of psychopathy in criminal populations. *Personality and Individual Differences, 1*(2), 111–119.

Heidensohn, F. (1996). *Women and crime.* Macmillan International Higher Education.

Joel, D., Berman, Z., Tavor, I., Wexler, N., Gaber, O., Stein, Y., Shefi, N., Pool, J., Urchs, S., Margulies, D. S., & Liem, F. (2015). Sex beyond the genitalia: The human brain mosaic. *Proceedings of the National Academy of Sciences, 112*(50), 15468–15473.

Joyce, M. (2006). Feminist theories of embodiment and anthropological imagination. In G. Pl & M. K. Stockett (Eds.), *Feminist anthropology: Past, present and future* (pp. 43–54). Philadelphia: University of Pennsylvania.

Keller, E. F. (2003). Gender and science. In *Discovering reality* (pp. 187–205). Dordrecht: Springer.

Kiehl, K. A., & Hoffman, M. B. (2011). The criminal psychopath: History, neuroscience, treatment, and economics. *Jurimetrics, 51*, 355.

Kreis, M. K., & Cooke, D. J. (2011). Capturing the psychopathic female: A prototypicality analysis of the comprehensive assessment of psychopathic personality (CAPP) across gender. *Behavioral Sciences & the Law, 29*(5), 634–648.

Lewis, A. (1974). Psychopathic personality: A most elusive category. *Psychological Medicine, 4*(2), 133–140.

Moore, H. (Ed.). (1999a). *Anthropological theory today.* London Polity Press.

Moore, H. (1999b). Whatever happened to women and men? In Moore, H (supra) (Ed.), *Gender and other crises in anthropology* (pp. 151–171).

Moore, H. (2007). *The subject of anthropology: Gender, symbolism and psychoanalysis.* London: Polity Press.

Nicholls, T. L., & Petrila, J. (2005). Gender and psychopathy: An overview of important issues and introduction to the special issue. *Behavioral Sciences & the Law, 23*(6), 729–741.

Nicholls, T. L., Ogloff, J. R., Brink, J., & Spidel, A. (2005). Psychopathy in women: A review of its clinical usefulness for assessing risk for aggression and criminality. *Behavioral Sciences & the Law, 23*(6), 779–802.

Patey, H. (2014). Late capitalism, psychopathy and the ontology of evil. In *I want to do bad things: Modern interpretations of evil* (pp. 51–63). Brill. www.interdisciplinary.net.

Patrick, C. J., Fowles, D. C., & Krueger, R. F. (2009). Triarchic conceptualization of psychopathy: Developmental origins of disinhibition, boldness, and meanness. *Development and Psychopathology, 21*(3), 913–938.

Paulhus, D. L., & Williams, K. M. (2002). The dark triad of personality: Narcissism, Machiavellianism, and psychopathy. *Journal of Research in Personality, 36*(6), 556–563.

Perbal, L. (2013). The 'warrior gene'and the Maōri people: The responsibility of the geneticists. *Bioethics, 27*(7), 382–387.

Salekin, R. T., Rogers, R., Ustad, K. L., & Sewell, K. W. (1998). Psychopathy and recidivism among female inmates. *Law and Human Behavior, 22*(1), 109–128.

Sandel, M. J. (2012). *What money can't buy: The moral limits of markets.* London: Macmillan.

Scott, P. D. (1960). The treatment of psychopaths. *British Medical Journal, 1*(5186), 1641.

Seal, L. (2010). *Women, murder and femininity: Gender representations of women who kill.* Springer.

Skeem, J. L., & Cooke, D. J. (2010). Is criminal behavior a central component of psychopathy? Conceptual directions for resolving the debate. *Psychological Assessment, 22*(2), 433.

Skeem, J. L., Polaschek, D. L., Patrick, C. J., & Lilienfeld, S. O. (2011). Psychopathic personality: Bridging the gap between scientific evidence and public policy. *Psychological Science in the Public Interest, 12*(3), 95–162.

Skeem, J., Monahan, J., & Lowenkamp, C. (2016). Gender, risk assessment, and sanctioning: The cost of treating women like men. *Law and Human Behavior, 40*(5), 580.

Smith, S., Watts, A. L., & Lilienfield, S. (2014). On the trail of the elusive successful psychopath. *The Psychologist, 27*, 505.

Stangle, H. L. (2008). Murderous Madonna: Femininity, violence and the myth of post-partum mental disorder in cases of maternal infanticide and filicide. *William & Mary Law Review, 699–734*(2), article 7.

United Nations Office of Drugs & Crime. (2014). *Global study of homicide: 2013.* Vienna: UNODC.

Van Voorhuis, P. (2012). On behalf of women offenders: Women's place in the science of evidence-based practice. *American Sociological Criminology, 11*(2), 111.

Vaughn, M. G., DeLisi, M., Gunter, T., Fu, Q., Beaver, K. M., Perron, B. E., & Howard, M. O. (2011). The severe 5%: A latent class analysis of the externalizing behavior spectrum in the United States. *Journal of Criminal Justice, 39*(1), 75–80.

Weiler, B. L., & Widom, C. S. (1996). Psychopathy and violent behaviour in abused and neglected young adults. *Criminal Behaviour and Mental Health, 6*(3), 253–271.

Weizmann-Henelius, G., Putkonen, H., Grönroos, M., Lindberg, N., Eronen, M., & Häkkänen-Nyholm, H. (2010). Examination of psychopathy in female homicide offenders—Confirmatory factor analysis of the PCL-R. *International Journal of Law and Psychiatry, 33*(3), 177–183.

WHO. (2019). http://www.euro.who.int/en/health-topics/health-determinants/gender/gender-definitions/whoeurope-brief-transgender-health-in-the-context-of-icd-11

Widom, C. S. (1984). Sex roles, criminality, and psychopathology. In *Sex roles and psychopathology* (pp. 183–217). Boston: Springer.

Wootton, B. (1978). *Crime and the criminal law: Reflections of a magistrate and social scientist* (Hamlyn Lectures) (2nd ed., p. 107). London: Steven & Sons.

Zahn-Waxler, C. (1993). Warriors and worriers: Gender and psychopathology. *Development and Psychopathology, 5*, 79–89.

Chapter 14
Reasons to Expect Psychopathy and Antisocial Personality Disorder (ASPD) to Vary Across Cultures

Rachel Cooper

Abstract I present two philosophical arguments that Antisocial Personality Disorder (ASPD) and Psychopathy can be expected to be culturally variable. I argue that the ways in which people with ASPD and psychopaths can be expected to act will vary with societal values and culture. In the second part of the chapter, I will briefly review some of the empirical literature on cross-cultural variation in ASPD and psychopathy and argue that it is consistent with my philosophical claims. My conclusion in this chapter is that methods of diagnosis will need to be culturally specific. A diagnostic instrument (such as the PCL-R or DSM) should not be uncritically employed in cultures that are very different from those in which it was initially developed.

Keywords Psychopathy · PCL-R · Cultural variation · Looping · Action

There is much interest in the question of whether ASPD and psychopathy vary cross-culturally. In the first part of this chapter, I contribute to the debate by presenting two philosophical arguments that ASPD and Psychopathy can be expected to be culturally variable. I argue that the ways in which people with ASPD and psychopathy can be expected to act will vary with societal values and culture. In the second part of the chapter, I will briefly review some of the empirical literature on cross-cultural variation in ASPD and psychopathy, and argue that it is consistent with my philosophical claims. My conclusion in this chapter is that methods of diagnosis will need to be culturally specific. A diagnostic instrument (such as the PCL-R or DSM) should not be uncritically employed in cultures that are very different from those in which it was initially developed.

R. Cooper (✉)
Politics, Philosophy and Religion, Lancaster University, Lancaster, UK
e-mail: r.v.cooper@lancaster.ac.uk

© Springer Nature Switzerland AG 2022
L. Malatesti et al. (eds.), *Psychopathy*, History, Philosophy and Theory of the
Life Sciences 27, https://doi.org/10.1007/978-3-030-82454-9_14

14.1 Classification and Prediction

It can be useful for scientists to categorise individuals into types because classification can enable us to predict how individuals of a type will behave. Before thinking about psychopathy and ASPD let us start by considering those areas of science in which classification does more or less successfully enable prediction. The Periodic Table provides perhaps the best example of the potential importance of classification for science. The Periodic Table provides a classificatory basis for chemistry that enables different types of substance to be classified, and *via* this classification, for them to be understood and controlled. Thus, once a scientist has determined that a particular chemical sample is lead, say, they know how it will behave and how to treat it if they wish to use it in various ways. What is more, samples of lead can be expected to have the same properties wherever, or whenever, they are found; lead in the US in the 1920s behaves the same as lead in Kenya in the 2020s. In this sort of case, classification grounds explanations and predictions, which are robust across space and time.

The predictions that are facilitated by classifications in biology are less impressive but still pretty good. Of course, the characteristics of plants do vary across time; plant species naturally evolve to take on different characteristics, and artificial breeding can lead to more rapid changes. The characteristics of plants also vary with environment; plants that are small houseplants in Northern Europe may grow to large trees in the tropics. However, despite this variation, biological classification does facilitate predictions that at least sometimes work in new contexts. Consider that all colonial powers invested heavily in botanical gardens and natural history collections as they acquired empires (Browne, 1996). The aim was to catalogue, study and control the natural world. The conditions required to grow crops could be studied in botanical gardens. Pests could be identified *via* matching specimens with those in natural history collections. Such efforts at predicting how biological organisms would behave in different environments were at least sometimes successful.

As these examples illustrate, in certain areas of science, classification can enable us to predict and control in ways that are robust across space and time. When philosophers talk about the power of such classificatory schemes they often start talking about "natural kinds". Chemical elements and biological species are commonly taken to be paradigm examples of natural kinds. Members of a natural kind are alike, and natural laws mean that members of a kind will behave similarly. Depending on the author, other conditions have also been added: Natural kinds have been claimed to be universal (in the sense of occurring everywhere), discrete, to have essential properties, and so on and so forth (see, for example, the conditions imposed by Zachar, 2000; Haslam, 2002). In earlier work I spent a lot of time talking about natural kinds (Cooper, 2005). Now it seems to me that the term "natural kind" has become so encrusted with metaphysical baggage as to be unhelpful. Instead of talking of natural kinds I will talk of "repeatables" (as also in Cooper, 2013). This makes clear the basic important idea; some entities in the world are alike, and will behave in similar ways.

As applied to mental disorders, the idea that there may be repeatables is this: if we consider individual cases of mental disorder some can be seen to be similar to each other with respect to symptoms. In some cases, patients who are grouped together on the basis of symptoms will also be alike in more fundamental ways (maybe they all have the same genetic abnormality, or all have similar levels of some neurotransmitter, or all have similar relationships with their childhood caregivers). If we group cases together on the basis of the right similarities, then the hope is that the groups that such a process generates will be inductively powerful. External validation on the basis of treatment response, family history, demographic correlates and so on can give additional reason to believe that patients who are being classified together are similar in genuinely important respects. If all goes well, a case that falls in a particular group can be expected to behave in ways that are similar to others of its class. The importance of such similarities is obvious if one thinks of treatments. The hope would be that a treatment that is found to work for one member of a class will work for others in that class too. Or, that at least is the hope.

The question of this chapter can now be framed. Is ASPD or psychopathy plausibly a 'repeatable' kind?[1] Are people diagnosed with ASPD or psychopathy likely to be such that they can reasonably be expected to be importantly similar, regardless of their culture? To look ahead to the argument to come, I will argue that we should not expect that people diagnosed with ASPD or psychopathy will 'look alike' or behave in similar ways irrespective of culture. 'Intentional actions' play a central role in the diagnosis of ASPD and psychopathy, and such actions can be expected to be radically culturally variable. Before I can develop my argument for this claim in detail, however, it is necessary to clarify exactly what is meant by 'ASPD' or 'psychopathy'.

14.2   Clarifying Concepts – What Is ASPD or Psychopathy?

So far, I have talked indiscriminately of 'ASPD' and 'psychopathy', but before progressing further, some further clarification is required. Antisocial Personality Disorder is included as a diagnosis within the *Diagnostic and Statistical Manual of Mental Disorders* (DSM), but even within the DSM the definition of Antisocial Personality Disorder is somewhat ambiguous. It might be supposed that the diagnostic criteria included in the DSM provides a clear operational definition of ASPD. A DSM-diagnosis of ASPD requires that an individual has engaged in various types of bad action (e.g. criminal acts, lying, fighting, consistent irresponsibility). The pattern of bad actions needs to be deep-seated, and to have begun in childhood. However, even if we restrict our attention to the DSM, it would be overly

---

[1] This issue is also addressed by Malatesti and McMillan (2014), who think psychopathy plausibly constitutes a 'robust scientific construct' and by Brzović et al. (2017) who argue that psychopathy may be too heterogeneous to constitute a natural kind but can instead be considered a 'pragmatic kind'.

simplistic to say that the concept of ASPD can be captured purely in terms of the DSM diagnostic criteria. The DSM includes ASPD amongst the *personality disorders*, and the DSM understanding of personality disorder is richer than the diagnostic criteria for ASPD taken in isolation might suggest. The DSM defines personality disorder as 'an enduring pattern of inner experience and behavior that deviates markedly from the expectations of the individual's culture, is pervasive and inflexible, has an onset in adolescence or early adulthood is stable over time, and leads to distress or impairment' (American Psychiatric Association, 2013, p. 645). Built into the DSM concept of personality disorder is the notion that someone who meets the diagnostic criteria for ASPD can be supposed to be a person with particular deep-seated and maladaptive personality traits; there is supposed to be something amiss with their inner psychological experience, and not simply with their outer behaviour. In addition, when considering differential diagnosis, the DSM-5 states that ASPD should be distinguished from criminal behaviour that is not indicative of a personality disorder 'Only when antisocial personality traits are inflexible, maladaptive, and persistent and cause significant functional impairment or subjective distress do they constitute antisocial personality disorder '(American Psychiatric Association, 2013, p.663). The DSM concept of ASPD thus seems to go beyond merely satisfying the explicit diagnostic criteria. The DSM takes someone with ASPD to be someone who not only acts badly for a prolonged period (as per the diagnostic criteria), but suggests in addition that they act badly *because* they have a deep-seated maladaptive personality.

ASPD is a DSM diagnosis, but 'psychopathy', though commonly discussed in the research literature, is not. The conceptual links between ASPD and psychopathy are debatable. Martyn Pickersgill, a sociologist of psychiatry, studied the ambiguities in the concepts of ASPD and psychopathy, as employed by British psychiatrists and clinical psychologists (Pickersgill, 2009). He interviewed clinicians and researchers, conducted ethnographical work at clinical conferences, and examined treatment guidelines. Pickersgill found that (at least in the UK).

1. Sometimes the terms 'ASPD' and 'psychopathy' are used interchangeably
2. Sometimes psychopathy is spoken of as if it were a severe form of ASPD
3. Sometimes psychopathy is spoken of as a character type (excitement seeking, callous, lacking remorse etc.) that might or might not lead to criminal behaviour
4. Sometimes psychopathy is treated as a construct defined by a particular diagnostic test, the PCL-R (and note that PCL-R includes items that score both bad actions and character traits)[2]

These various conceptions are distinct – sometimes radically so. Note in particular that character-type psychopaths (Definition 3) might not be psychopaths in any of the 'acting-badly' senses (Definitions 1, 2 and 4) (as noted and discussed by Skeem & Cooke, 2010). Depending on context, character-type psychopaths might act in

---

[2] Part of the reason why the PCL-R is often taken to define psychopathy, is that the PCL-R test is presented within a particular theoretical framework, making easy for readers to take the test to operationalise a concept of psychopathy.

different ways; for example Lykken (1995) hypothesised that common personality traits (notably fearlessness) can give rise to both antisocial and heroic behaviour (Smith et al., 2013 find some empirical support for this suggestion). Conversely, many people meet the DSM-5 diagnostic criteria for ASPD (which consider only whether someone has acted badly for a prolonged period) but are not character-type psychopaths (as found, for example, by Poythress et al., 2010; Ogloff et al., 2016). People engage in antisocial behaviour for a wide range of reasons, not all of which are indicative of an abnormal personality (for example, poor people living in crime-ridden environments may have few options but to engage in criminal activity). Pickersgill's finding, that the terms 'ASPD' and 'psychopath' are used in very different ways by different writers, indicates that it is important to be aware of the conceptual complexity that surrounds the use of these terms, and to be explicit about exactly what types of people one is interested in.

In the remainder of this chapter, I will develop two lines of argument that demonstrate that culture needs to be considered when trying to interpret and predict the behaviour of people with ASPD or psychopathy. In the first argument I am concerned with understandings of ASPD or psychopathy whereby people with ASPD or psychopathy are understood to be people:

- With a particular character type
- Where this character type disposes them to behave badly
- And where such people can be picked out fairly reliably on the basis of their bad actions (as in the DSM ASPD criteria, or using the PCL-R)

I will argue that diagnostic instruments, such as the PCL-R and the DSM ASPD diagnostic criteria, that seek to pick out people with a particular character type on the basis of their actions, should not be uncritically employed in cultures other than those in which they were developed.

In the second argument, I go onto argue that folk knowledge of psychiatric diagnoses affects how people who might be diagnosed behave. This means that the behaviour of people diagnosed with ASPD or psychopathy can be expected to vary with folk understandings of these diagnoses (which can again be expected to be culturally variable). This second argument is applicable to all instances in which mental health experts diagnose people who have some ideas about mental health diagnoses.

14.3   First Philosophical Argument for Cultural Variation: The Meaning of Actions Varies with Cultural Setting

The DSM diagnostic criteria for ASPD seeks to pick out a group of people who, through having committed various bad actions in the past, can safely be supposed to have a personality disorder which disposes them to commit bad actions in the future. The aims of the PCL-R (Hare, 1990), which scores someone both on the basis of past bad actions and character traits, similarly seeks to pick out 'psychopaths' – where psychopaths are understood to be people with a personality that will dispose them to act badly.

Here I will argue that seeking to pick out a type of person via using diagnostic criteria that include 'intentional actions' can be expected to be problematic across cultures. The types of people who will receive scores for 'bad actions' on the DSM or PCL-R will not be the same types of people across cultures.

Intentional actions are actions that people perform for reasons, for example waving in greeting, signing a form, punching someone. Intentional actions can be distinguished from mere behaviours, such as sneezing or tripping over, in that in the case of intentional actions an actor will be able to give a reason for their action if they are asked. For example, if I'm asked why I'm signing the form I might say that I want my son to go to football club and am signing to give permission. Classically, a reason will consist of a belief and desire pair; I want to achieve something and believe my action will help bring about my goal. In many cases, either the relevant beliefs or desires will be too obvious to be explicitly stated (so I might say 'I hit him because I wanted to hurt him' without bothering to state that I believe a punch to be painful, or a bank robber might say 'I smashed the safe because that's where the money was' without making explicit that he desires money). Note that the reasons for an action do not need to be good reasons for it to count as an intentional action; the belief may be false, and the desires foolish.

Very many of the diagnostic criteria included in the DSM for ASPD and on the PCL-R are concerned with intentional actions, for example, deceiving someone, starting a fight, promiscuous sex, robbery, intentionally leading a 'parasitic lifestyle'. I will argue that intentional actions cannot be understood as being 'caused by' an underlying personality type in anything like the way that the behaviours of chemical elements or biological species can be understood to be caused by their underlying natures. Lead has a particular melting point because of its chemical make-up. Polar bears hunt seals because of their particular physiology (which itself can be explained by their genetic make-up, which can be explained by their evolutionary history). But, a psychopathic personality-type (characterized by callousness, risk taking, lack of fear, and so on), cannot cause someone to start fights or be sexually promiscuous in anything like the same sort of way.

My argument that culture must be taken into consideration when thinking about the ways in which character-types will manifest in actions will depend on the work of Alistair MacIntyre. In *After Virtue* (1981). MacIntyre argues that intentional actions can only be understood within a narrative "settings", which are culturally specific. To illustrate this point, MacIntyre asks us to consider a man digging in his garden. In order to see the man's activity as consisting of actions as opposed to being mere bodily behaviour, we have to be able to make sense of the man's intentions. We have to perceive him as being "up to something". Suppose that when we ask him what he is doing the man says he is preparing for winter. MacIntyre says that we can only make sense of this within a particular cultural and historical framework. We have to understand that people (at least in the UK) commonly dig their gardens before the frosts come to render the soil workable in the spring. MacInytre concludes, "We cannot…characterize behaviour independently of intentions, and we cannot characterize intentions independently of the settings that make those intentions intelligible both to agents themselves and to others." (1985, p. 206). Making sense of an action thus requires placing it in a narrative sequence.

When MacIntyre talks of the "settings" that make actions intelligible he often seems to have relatively rich networks of social traditions in mind. It is plausible that certain actions only make sense within a fairly rich cultural frame. It only makes sense to intend to "get married", or to "give an inaugural lecture", or to "mark the boundary of gang territory", within certain cultural settings.

There are other actions that seem less culturally specific. It seems safe to assume that throughout history and across all societies people have intended to light fires, or pick berries, or to brush their teeth. Still, even in these sorts of case, our interpretation of what an actor 'is up to' tends to depend on exactly how an action is performed and varies with cultural setting. To illustrate, consider how the same thinly described action 'killing a swan' will be taken to be indicative of very different likely motivations depending on cultural context. In the UK swans have long been protected birds. Historically, swans were eaten only by royalty, today they live in parks and are fed bread by children. In the UK, swans cannot legitimately be seen as food. When people do catch and eat swans in the UK, this is considered to be a terrible thing. Tabloids newspapers fairly regularly run stories about immigrants catching and eating swans from parks. Such headlines stir up intense feelings, and it is often unclear whether the stories are true or myths designed to whip-up racist sentiment. The point for us here is that although in some cultures catching and eating a swan might be an innocent act of food gathering, in some cultures it is not. Even in the case of those apparently "straightforward" actions that have been done by all people in all cultures (eating, going to sleep, lighting fires) the interpretation of a particular action is often culturally specific.

How does this relate to ASPD and psychopathy? My worry is that the diagnostic criteria list involves various types of "bad" intentional action (arrestable acts, lying, promiscuous sex, starting fights, avoiding employment). Note, that in addition a diagnosis of ASPD requires a childhood history of conduct disorder, and the diagnosis of conduct disorders also depends on a list of intentional actions (truancy, stealing, starting fights).

However, as with killing swans, the meaning of such intentional actions will vary radically with cultural setting. To illustrate, let's consider "truancy", and how the meaning of skipping school varies with setting. Compare three cases of truancy, and the likely character of the child involved:

Truant 1 – is a child living in an academically competitive environment where exams are essential for getting a job. They skip school for no better reason than to play with friends. We can fairly safely assume that the child is disobedient and reckless.

Truant 2 – is a child living in a culture in which there is full employment and where employers do not care about school results. This child skips school to start a job with the knowledge of their parents (as many teenagers did in the UK in the 1950s or 60s). This child may be engaged in a behavior that is technically illegal, but they are not reckless or disobedient.

Truant 3 – is a child from a conservative religious family who skips school because her parents do not approve of some planned lessons on LGBT issues. Truant 3 is an obedient child.

My point is that the sorts of child who will 'truant' in these different cultural settings are very different types of people. The DSM includes 'truancy' amongst the diagnostic criteria for conduct disorder, but it is only in certain cultural settings that it makes sense to take truancy to be a likely indicator of an incipient personality disorder.

Similar points can be made about many of the other actions included in the diagnostic criteria for ASPD or psychopathy. Promiscuous sexual behaviour is included in the text description of ASPD in DSM-5 and in the PCL-R. In some cultures, monogamy is a widely upheld ideal. In other cultures, it may be entirely culturally acceptable for single people to engage in consensual casual sex. The type of person who has many sexual partners will be very different in these distinct cultures. 'Starting fights' is also included in both the DSM diagnostic criteria for ASPD and PCL-R, but again the types of person who are likely to 'start fights' varies with cultural setting. There are many cultures in which people participate in consensual fighting as a sort of (possibly illegal) sport (this goes on in some towns in the UK where young men will congregate at known locations on particular evenings to fight). Someone who starts fights in such a setting is very different from someone who loses their temper in an argument and lashes out. The DSM supposes that someone who starts fights is reckless and unusually violent, but there are cultural settings in which 'starting fights' need not be indicative of such a personality.

The difficulties discussed here are to some extent already acknowledged in the DSM. The DSM acknowledges that ASPD appears to be associated with low socioeconomic status in ways that may be problematic and discusses concerns that 'the diagnosis may at times be misapplied to individuals in settings in which seemingly antisocial behaviour may be part of a protective survival strategy' (American Psychiatric Association, 2013, p. 662). Clinicians are urged to 'consider the social and economic context in which behaviours occur' when assessing antisocial traits. However, despite the warning, the DSM currently provides no guidance as to exactly how social context can properly be considered when making diagnoses.

I have argued that diagnostic checklists that include intentional actions (such as the DSM-5 diagnostic criteria for ASPD, and the PCL-R) cannot be expected to be reliable across cultures. The PCL-R was originally developed to pick out likely psychopaths in male, North American prison populations. In such a population (with a broadly shared cultural understanding) it can be expected to be more or less reliable; in this population someone who engages in promiscuous sex, and has a poor work history, and starts fights, and has a parasitic lifestyle, and so on, can fairly be expected to have a psychopathic personality-type (assuming they score over the certain threshold). But, when descriptions of intentional actions are included in diagnostic criteria, then those diagnostic criteria can only be expected to be reliable in cultures very similar to that for which they were designed. Before such checklists can be used in other cultural settings recalibration is required. For example, for use in cultures where promiscuous sex between consenting adults is socially acceptable, such behaviours cannot be taken to be indicative of possible psychopathy. In devising a diagnostic check list for use in such a culture, items relating to promiscuous sex would need to be removed, and possibly replaced by other items that could more fairly be taken to be indicative of a psychopathic character-type within the culture.

14.4   Second Philosophical Argument for Cultural Variation: Looping and Psychiatric Diagnosis

In this chapter, I argue that the ways in which people with ASPD and psychopathy can be expected to behave will vary with cultural setting. One of the important relevant ways in which cultures can vary is in regard to folk beliefs about psychiatry and psychiatric diagnoses. Over the last few decades, Ian Hacking's work has stressed the importance of the fact that humans respond to being classified in ways that other classified entities do not (1986, 1988, 1992, 1995a, b). A child who is told they are stupid may stop trying at school and fall behind yet further; a diagnosis of "problem drinking" may come to motivate abstinence; a whole class of people may respond to a classification with new forms of resistance, as in "fat pride". Such interactions between classifications and behaviour mean that "human kinds" – the kinds classified by the human sciences – become moving targets. No sooner has a kind been picked out than behaviours shift and classifications have to be revised.

One of Hacking's best developed examples of such "looping effects" concerns Multiple Personality Disorder (1995a). When cases of Multiple Personality Disorder were first reported, someone with MPD would typically possess just two or three clearly distinct personalities. Over time, however, usual symptoms shifted. Hacking makes a convincing case that the shift in symptoms was in part caused by changing prototypes of the disorder being made available in the media. The media tended to report more florid cases, and over time people with MPD started to present with more and more personalities, and as their numbers increased, these personalities became more diverse, and also more fragmentary. Note that in this case Hacking's claim is not that patients intentionally copy the symptoms of publicised cases. Rather the mechanism is more subtle and subconscious, but still the consequence is that a distressed individual will most likely manifest distress in ways that are culturally recognised.

In the case of ASPD and psychopathy, in many cultures, those who might be so diagnosed are acutely aware the meanings of the diagnosis and can alter their behaviour in response to it. For example, often people with ASPD or psychopathy will not want to be diagnosed as having ASPD or psychopathy, and so will be motivated to deceive interviewers.

The PCL-R is commonly employed in forensic settings. Those who fear they may have to sit the PCL-R and want to pass it, can find out about the test and prepare for it. Copies of the PCL-R Manual circulate in prisons. Prison authorities regard the manual as 'contraband', and confiscate it when found (Hare, 1998).[3]

Abraham Gentry's (2011), *Pass the PCL-R: Your Guide to Passing the Hare Psychopathy Checklist-Revised Aka the Psychopath Test* can be purchased from Amazon. The book blurb states

> This guide will take you step by step through the test. It will explain the questions asked, what traits the clinicians are looking for, how they will score your answers, and exercises to prepare. The PCL-R is unlike any other test. Many seemingly innocent traits are scored against you!
>
> If you, your friend or client have been convicted of a felony or are facing a parole hearing soon, it is very likely the PCL-R will be given. Know the rules of the game, arm yourself with the right knowledge, and rebuild your life!

Gentry's book goes through the PCL-R item by item and gives examples of "good" and "bad" answers. Readers are told how much eye-contact with the interviewer will be considered appropriate and instructed on what to say if they are caught out lying. The author provides an email at the end for readers who want individual coaching. In cultures where those who are likely to be given a PCL-R test can access and read such books, the answers provided in diagnostic interviews are likely to vary from 'untutored' responses.

To test how easy it might be for interviewees to intentionally manipulate test results, Rogers et al. (2002) asked young offenders to either 'fake bad' or 'fake good' when being interviewed for the Psychopathy Checklist: Youth Version. The interviewers were trained in administering the PCL-YV and had access to institutional files on each interviewee (as is usual with this test). Rogers et al. found that interviewees were able to significantly manipulate their test results. In particular, adolescent offenders with a moderate score on the PCL-YV (of >14) at baseline were able to intentionally decrease their total scores by an average of 7.89 points (p. 44). Like the PCL-R, the PCL-YV has a maximum possible score of 40. A reduction of around 8 points is of great practical significance, as it would be enough to bring most interviewees below the 30 'cut-off' that is usually taken to signify psychopathy.

Folk conceptions of ASPD and psychopathy, and knowledge of diagnostic processes, will affect how people behave in diagnostic settings. More subtle and pervasive effects will also occur. The ways in which each of us acts depend in part on how we think of ourselves; I might avoid dating because I think myself unattractive, or

---

[3] Hare developed the PCL-R. He comments 'The extent to which knowledge of the scoring criteria for the PCL-R items would influence an offender's score is uncertain and in need of investigation. However, assuming that the rater is reasonably skilled and that proper scoring procedures are used, the problem may not be too serious. I doubt, for example, that a psychopathic offender, knowing how item 6, callous/lack of empathy, is scored, could convince experienced clinicians that he was a warm, caring individual and that he warranted a score of 1 or 0 on the item. PCL-R items are scored by carefully integrating interview and file information. Further, an item score reflects lifetime functioning and is not sensitive to short-term changes in personality or behavioural patterns (Hare, 1998, p.111)' I fear that Hare's optimism in the abilities of scorers may be misplaced. See later discussion of empirical study on how easy it is to 'fake' PCL-R scores by Rogers et al. (2002).

apply to foster children on the basis that I think I'll be a good parent, and so on. The actions of individuals diagnosed with psychopathy or ASPD will also be affected by how they think about themselves, and how this is coloured by their knowledge of their diagnosis.

In general, 'ASPD' and 'psychopathy' are bad labels for someone to receive. In essence, for someone to be diagnosed with ASPD or psychopathy is for them to be told that they are a bad person, and that there is little hope of them ever becoming better. I was interested in how people diagnosed with ASPD responded to being diagnosed and so I looked at posts on an online forum for people with antisocial personality disorder (at psychforums.com). There are, of course, methodological problems which mean it is unwise to uncritically accept the comments on such a site as the 'views of people with ASPD'. Online internet fora are peculiar eco-systems where some of those who post on a forum for 'people with ASPD' may well not actually have ASPD. Nevertheless, the posts give some indication of possible culturally-salient responses to diagnosis with ASPD. I first looked at the online forum in 2008, and then again more recently in 2016 and 2018. Folk knowledge of psychiatric diagnoses changes over time, and there were some interesting differences in the online discussions in 2008 as compared to later. Going through the 2008 posts, at that time, there seemed to be three basic responses to receiving an ASPD diagnosis:

(i) Challenge the diagnosis – some people who are given a diagnosis of ASPD respond to their diagnosis by refusing to believe the diagnosis. Either they give reasons for distrusting the individual clinician who diagnosed them, or they give reasons for thinking that all psychiatric diagnosis are unreliable. Such a response can be expected to decrease the chances that someone will engage with therapeutic services.

(ii) Some embrace the diagnosis. Some people who are given diagnoses of ASPD (or who self-diagnose) accept the diagnosis and take pride in being bad people. Online they swap stories about the bad things they have done, for example torturing small animals and homeless people. On occasion, discussion turns philosophical, and some present themselves as being moral relativists, or think of themselves as Nietzschean supermen. Such a response can be expected to increase the chances that someone will act badly (although some such discussion may be bluster, it can be expected that someone who takes pride in being evil does more bad things than someone who does not).

(iii) Some don't know what to do. Some discussants were diagnosed with ASPD and then did not know what to do. Some reported that mental health services became inaccessible to them once they received a personality disorder diagnosis and that they would like therapy but were unable to access it.

In 2008, there were no readily accessible narratives that outlined how one might respond well to a diagnosis of ASPD. In so far as the disorder was considered basically untreatable and indicative of someone being a bad person I could neither find online, nor imagine, how someone might integrate a diagnosis of ASPD into their conception of themselves in any positive way.

However, folk understandings of psychiatric diagnoses evolve over time, and in 2016 and 2018, the range of possible reactions to diagnosis on the forum had broadened. It is now fairly easy to find online discussion of the possibility that someone might be a "good psychopath" – that is someone who has the character-traits of psychopathy but who finds a social niche where being callous, ruthless and thrill-seeking is not a problem. The possibility that many business people, surgeons, and special forces soldiers have psychopathic-type characters but manage to lead socially acceptable lives is now widely discussed (for a review see Smith & Lilienfeld, 2013, for an example of a popular book taking such a line see Dutton & McNab, 2014).[4] Plausibly this new, more positive, discussion makes it easier for those who are diagnosed with ASPD, or psychopathy, to accept their diagnosis and simultaneously make pro-social plans that will enable them to live well. Now it is a viable possibility that someone with a psychopathic personality-type might set out to find a social niche where they might be able to live a personally rewarding and also socially-acceptable life.

How someone with ASPD or psychopathy will behave depends in part on how they come to understand their diagnosis. Folk understandings of ASPD and psychopathy are culturally variable and change over time. This means that those who want to research or treat ASPD or psychopathy need to take cultural context into account. Most obviously, those with ASPD or psychopathy who have read guides to passing diagnostic tests cannot be diagnosed in the same ways as naïve interviewees. More subtly, folk conceptions of ASPD and psychopathy, and of psychiatric diagnosis and therapy, will shape how someone with ASPD or psychopathy conceives of, and shapes, their possible future actions.

14.5 Philosophy Meets the Empirical Literature

In this chapter I have presented two philosophical arguments that the actions of those with ASPD and psychopathy can be expected to vary with culture. I have claimed this should cause problems with the use of diagnostic tests (such as the PCL-R, or the diagnostic criteria for ASPD in the DSM) when they are employed in different cultures. In this final section I consider whether my claims are consistent with the empirical evidence. Most relevant empirical studies to date have examined the PCL-R. The results can be hard to interpret, many are controversial, and new studies are always being conducted. Here I do not attempt a comprehensive review of empirical work in this area. Rather I briefly discuss some of the types of study being performed and discuss how the methods and results fit with the philosophical arguments of this chapter.

---

[4] Smith and Lilienfeld review discussions in the media and academic studies. Note that they suggest that popular claims that psychopaths 'do well' in business may be over-played.

It is well known that the proportion of people in prison populations who receive a PCL-R score of over 30 (the traditional cut-off for psychopathy) varies by culture. For example, Cooke (1995) found only 3% of Scottish male prisoners obtained such a score; much less than the 28% reported for the US. It might be that the proportion of people who develop into psychopaths varies by society. If childhood experiences have some effect on whether a child grows up to be a psychopath, then differences in child-rearing practices, and in the incidence of traumatic experiences, could be expected to lead to there being differences in the prevalence of psychopathy in different places. It would be highly surprising, however, if the incidence of psychopathy genuinely varied by a factor of 9 between Scotland and the US. It seems more likely that the difference is either an artefact of sampling, or produced by cultural bias in the methods used for diagnosis,

That some cross-cultural differences in the apparent prevalence of psychopathy is due to sampling effects is very likely. Psychopaths have to be caught prior to being counted, and so differences in judicial systems can affect apparent prevalence rates (Wernke & Huss, 2008). Another possibility is that the numbers of psychopaths found at any one location varies because psychopaths may be attracted to big, exciting cities. Cooke and Michie (1999) suggest that many Scottish-born psychopaths migrate to London and so never feature in Scottish figures.

A number of different types of study have sought to examine whether the test used for diagnosis, the PCL-R, is itself culturally-biased. Most simply, some studies have examined whether the PCL-R 'works in practice' in different cultural settings. In such studies, the key question is whether the PCL-R can be used to predict some outcome of practical importance (recidivism rates, or the likelihood of future violence) in different cultural settings. Hare et al. (2000) review studies and argue that the PCL-R has been shown to work (more or less) to predict recidivism and future violence in the U.S., Sweden and England. Studies of this type are important, but take a 'broad brush' approach, and are not suited to detecting possible issues with cultural bias at the level of particular items in the PCL-R.

Confirmatory factor analysis can be used to explore the various dimensions underlying PCL-R results in different cultural settings. Cooke et al. (2005) compared PCL-R results from Continental Europe and the US. They found similar factors in both sets of samples, suggesting that a common syndrome 'psychopathy' could be found in both Continental Europe and the US. Some researchers in the field, however, worry that factor analysis may not be an appropriate approach in this area. Factor analysis assumes that any covariation between items is produced by a 'common cause', and this assumption may be unjustified (Verschuere et al., 2018). There are also concerns that factor analytic studies of psychopathy have yielded conflicting results (Boduszek & Debowska, 2016).

Verschuere et al. (2018) employed a different statistical method, network analysis, to examine whether the same features of psychopathy seem central to the condition in the US and the Netherlands. They found some differences, with items related to irresponsibility and parasitic lifestyle being more central in the Netherlands than the US. Currently, however, there is some dispute whether network analyses of psychopathology are sufficiently replicable for much weight to be placed on them (Forbes et al., 2017).

Item Response Theory analyses can be used to estimate the association between scores on a particular item on the PCL-R (e.g. promiscuous sexual behaviour) and the latent trait that underlies the scores. As such studies can hone-in on possible cultural differences in responses to individual items on the PCL-R, these studies are of most direct relevance to the arguments of this chapter. If my claims are correct, then one should expect that Item Response studies should show that those items on the PCL-R that code for intentional actions sometimes behave differently in different cultural settings. There are some studies that suggest this is the case. In their study comparing PCL-R results from the US and Continental Europe, Cooke et al. (2005), found 'there was very little cross-cultural bias in ratings of affective symptoms of psychopathy, but distinct bias in ratings of the interpersonal and behavioral symptoms.' (p. 290) This is consistent with the argument of this chapter, as the interpersonal and behavioural symptoms, which are more culturally variant, more frequently involve intentional actions. Employing similar methodology, Shariat et al. (2010) compared PCL-R scores from Iran and North America. They found that, in Iran, the PCL-R Superficial, Deceitful and Grandiose items failed to discriminate between psychopaths and non-psychopaths. Shariat, Assadi and Norozzian suggest that socially-accepted forms of polite lying and manipulation are widespread in Iran. As such, and in line with the philosophical argument I present above, in the Iranian context, lying is not indicative of personality traits suggestive of psychopathy, and so fails to help pick out psychopaths. Note, however, that although some results from some Item Response Theory analyses are consistent with my arguments here, other interpretations of the results are also possible. I have argued that the type of personality who will act in particular ways will vary with culture, and this is one possible explanation of Shariat et al.'s findings. However, there are also other possible explanations of their findings, for example, the differences might be caused by issues relating to translating the PCL-R, or there might be cultural differences in the ways in which individuals *report* acting (as distinguished from cultural differences in the ways people actually act). As such, I don't take the empirical findings to prove my argument is correct, I note only that they are consistent with my claims.

14.6 Summary

In this chapter, I have discussed some of the ways in which culture needs to be considered when thinking about how people with ASPD or psychopathy can be diagnosed and how they can be expected to act. Culture makes a difference to the meaning of actions, and to what actions seem possible. As a result, an underlying maladaptive personality that is characteristic of ASPD or psychopathy cannot be expected to lead to the same sorts of actions irrespective of cultural setting. In addition, people who are diagnosed with ASPD or psychopathy gain knowledge of these diagnoses and alter their behaviour in response. While chemical elements have the same properties throughout history and in all places, the actions that can be taken to be characteristic of those with ASPD or psychopathy will vary with cultural setting.

I opened this chapter by asking whether ASPD or psychopathy can be considered a 'repeatable' scientific kind. I have argued that the ways in which someone with ASPD or psychopathy can be expected to behave will vary with cultural setting. This being said, nothing I have said here goes against the idea that there might be a particular psychopathic personality type (callous, thrill seeking, emotionally cold etc.) that can be found amongst all peoples at all times. However, though the personality-type might be 'repeatable' across space and time, the ways in which that personality-type can be expected to manifest in action will vary.

References

American Psychiatric Association. (2013). *Diagnostic and statistical manual of mental disorders* (5th ed.). American Psychiatric Publishing.

Boduszek, D., & Debowska, A. (2016). Critical evaluation of psychopathy measurement (PCL-R and SRP-III/SF) and recommendations for future research. *Journal of Criminal Justice, 44*, 1–12.

Brzović, Z., Jurjako, M., & Šustar, P. (2017). The kindness of psychopaths. *International Studies in the Philosophy of Science, 31*(2), 189–211.

Browne, J. (1996). Biogeography and empire. In N. Jardine, J. Secord, & E. Spary (Eds.), *Cultures of natural history* (pp. 305–321). Cambridge University Press.

Cooke, D. J. (1995). Psychopathic disturbance in the Scottish prison population: The cross-cultural generalisability of the Hare psychopathy checklist. *Psychology, Crime and Law, 2*(2), 101–118.

Cooke, D. J., & Michie, C. (1999). Psychopathy across cultures: North America and Scotland compared. *Journal of Abnormal Psychology, 108*(1), 58–68.

Cooke, D. J., Michie, C., Hart, S. D., & Clark, D. (2005). Searching for the pan-cultural core of psychopathic personality disorder. *Personality and Individual Differences, 39*(2), 283–295.

Cooper, R. (2005). *Classifying madness: A philosophical examination of the diagnostic and statistical manual of mental disorders*. Springer.

Cooper, R. (2013). Natural kinds. In K. Fulford, M. Davies, G. Graham, J. Sadler, G. Stanghellini, & T. Thornton (Eds.), *Oxford handbook of philosophy and psychiatry* (pp. 950–964). Oxford University Press.

Dutton, K., & McNab, A. (2014). *The good Psychopath's guide to success: How to use your inner psychopath to get the most out of life*. Bantam Press.

Forbes, M. K., Wright, A. G., Markon, K. E., & Krueger, R. F. (2017). Further evidence that psychopathology networks have limited replicability and utility: Response to Borsboom et al. (2017) and Steinley et al. (2017). *Journal of Abnormal Psychology, 126*(7), 1011–1016.

Gentry, A. (2011). *Pass the PCL-R: Your guide to passing the Hare psychopathy checklist-revised*. Self-published Abraham Gentry.

Hacking, I. (1986). Making up people. In T. Heller, M. Sosna, & D. Wellbery (Eds.), *Reconstructing individualism* (pp. 222–236). Stanford University Press.

Hacking, I. (1988). The sociology of knowledge about child abuse. *Nous., 22*, 53–63.

Hacking, I. (1992). World-making by kind-making: Child abuse for example. In M. Douglas & D. Hull (Eds.), *How classification works* (pp. 180–238). Edinburgh University Press.

Hacking, I. (1995a). *Rewriting the soul*. Princeton University Press.

Hacking, I. (1995b). The looping effects of human kinds. In D. Sperber & A. Premark (Eds.), *Causal Cognition* (pp. 351–394). Clarendon Press.

Hare, R. D. (1990). *The Hare psychopathy checklist –revised*. Multi-Health Systems, Inc.
Hare, R. D. (1998). The Hare PCL-R: Some issues concerning its use and misuse. *Legal and Criminological Psychology, 3*(1), 99–119.
Hare, R. D., Clark, D., Grann, M., & Thornton, D. (2000). Psychopathy and the predictive validity of the PCL-R: An international perspective. *Behavioral Sciences & the Law, 18*(5), 623–645.
Haslam, N. (2002). Kinds of kinds: A conceptual taxonomy of psychiatric categories. *Philosophy, Psychiatry, Psychology, 9*, 203–217.
Lykken, D. T. (1995). *The antisocial personalities*. Lawrence Erlbaum Associates.
MacIntyre, A. (1985). *After virtue: A study in moral theory* (2nd ed.). Duckworth.
Malatesti, L., & McMillan, J. (2014). Defending psychopathy: An argument from values and moral responsibility. *Theoretical Medicine and Bioethics, 35*(1), 7–16.
Ogloff, J. R., Campbell, R. E., & Shepherd, S. M. (2016). Disentangling psychopathy from antisocial personality disorder: An Australian analysis. *Journal of Forensic Psychology Practice, 16*(3), 198–215.
Pickersgill, M. D. (2009). NICE guidelines, clinical practice and antisocial personality disorder: The ethical implications of ontological uncertainty. *Journal of Medical Ethics, 35*(11), 668–671.
Poythress, N. G., Edens, J. F., Skeem, J. L., Lilienfeld, S. O., Douglas, K. S., Frick, P. J., Patrick, C. J., Epstein, M., & Wang, T. (2010). Identifying subtypes among offenders with antisocial personality disorder: A cluster-analytic study. *Journal of Abnormal Psychology, 119*(2), 389–400.
Rogers, R., Vitacco, M. J., Jackson, R. L., Martin, M., Collins, M., & Sewell, K. W. (2002). Faking psychopathy? An examination of response styles with antisocial youth. *Journal of Personality Assessment, 78*(1), 31–46.
Shariat, S.V., Assadi, S.M., Noroozian, M., Pakravannejad, M., Yahyazadeh, O., Aghayan, S., Michie, C., & Cooke, D., (2010). Psychopathy in Iran: A crosscultural study. *Journal of personality disorders, 24*(5), 676–691.
Skeem, J. L., & Cooke, D. J. (2010). Is criminal behavior a central component of psychopathy? Conceptual directions for resolving the debate. *Psychological Assessment, 22*(2), 433–445.
Smith, S. F., & Lilienfeld, S. O. (2013). Psychopathy in the workplace: The knowns and unknowns. *Aggression and Violent Behavior, 18*(2), 204–218.
Smith, S. F., Lilienfeld, S. O., Coffey, K., & Dabbs, J. M. (2013). Are psychopaths and heroes twigs off the same branch? Evidence from college, community, and presidential samples. *Journal of Research in Personality, 47*(5), 634–646.
Verschuere, B., van Ghesel Grothe, S., Waldorp, L., Watts, A.L., Lilienfeld, S.O., Edens, J.F., Skeem, J.L., & Noordhof, A., (2018). What features of psychopathy might be central? A network analysis of the Psychopathy Checklist-Revised (PCL-R) in three large samples. *Journal of Abnormal Psychology, 127*(1), 51–65.
Wernke, M. R., & Huss, M. T. (2008). An alternative explanation for cross-cultural differences in the expression of psychopathy. *Aggression and Violent Behavior, 13*(3), 229–236.
Zachar, P. (2000). Psychiatric disorders are not natural kinds. *Philosophy, Psychiatry, Psychology, 7*, 167–182.

Rachel Cooper is Professor in History and Philosophy of Science at Lancaster University, U.K. Her publications include *Diagnosing the Diagnostic and Statistical Manual of Mental Disorders* (Karnac, 2014), *Psychiatry and the Philosophy of Science* (Routledge, 2007) and *Classifying Madness* (Springer, 2005).

Chapter 15
Psychopathy and Personal Identity: Implications for Medicalization

Marga Reimer

Abstract My aim in this chapter is to reflect on psychopathy in connection with personal identity to achieve clarity with respect to that condition's potential for medicalization. Given plausible (if theoretically thin) accounts of psychopathy, personal identity, and medicalization, I consider the question whether psychopathy is amenable to medicalization. I argue that, not only is psychopathy not *in fact* amenable to medicalization, it is not amenable to medicalization even *in principle*. In the jargon of contemporary analytic philosophy, the medicalization of psychopathy is not "logically" or "metaphysically" possible. Importantly, the operative notion of medicalization reflects contemporary Western society's conception of medicine, a conception which is subject to change, given relevant changes in that society's understanding of the scope and limits of medicine proper. It is possible that, given conceptual changes of the relevant sort, psychopathy might prove amenable, at least in principle, to medicalization. For this to occur, society would have to adopt a *new and radically different* conception of medicine, one broadened so as to include, not only the *treatment* of patients suffering from disease (illness, or disorder) but also the *transformation* of patients (psychopaths) into literally different, or "numerically distinct," persons (non-psychopaths). However, given *today's* societal conception of medicine, which emphatically does *not* include the transformation, moral or otherwise, of patients into literally different (or numerically distinct) persons, psychopathy is not amenable to medicalization even in principle. It would thus be false, or at the very least misleading, to claim *today* that psychopathy might "someday" be medicalized.

Keywords Psychopathy · Personal identity · Medicalization, essential moral self hypothesis · Dr. Jekyll and Mr. Hyde · Person vs. human being

M. Reimer (✉)
Department of Philosophy, University of Arizona, Tucson, AZ, USA
e-mail: reimer@email.arizona.edu

© Springer Nature Switzerland AG 2022
L. Malatesti et al. (eds.), *Psychopathy*, History, Philosophy and Theory of the
Life Sciences 27, https://doi.org/10.1007/978-3-030-82454-9_15

15.1 Psychopathy, Personal Identity, and Medicalization

My central aim in this paper is to reflect on psychopathy in connection with personal identity to achieve some clarity with respect to that condition's potential for medicalization. Given plausible (if theoretically thin) accounts of psychopathy, personal identity, and medicalization, I will consider the question whether psychopathy is amenable to medicalization. I will argue that, not only is psychopathy not *in fact* amenable to medicalization, it is not amenable to medicalization even *in principle*. In the jargon of contemporary analytic philosophy, the medicalization of psychopathy is not "logically" or "metaphysically" possible. Importantly, the notion of medicalization invoked herein (and throughout the paper) reflects contemporary Western society's conception of medicine, a conception which is subject to change, given relevant changes in that society's understanding of the scope and limits of medicine proper.

It is possible that, given conceptual changes of the relevant sort, psychopathy might prove amenable, at least in principle, to medicalization. For this to occur, society would have to adopt (however gradually) a *new and radically different* conception of medicine, one broadened so as to include, not only the *treatment* of patients suffering from disease (illness, or disorder)[1] but also the *transformation* of patients (psychopaths) into literally different, or "numerically distinct,"[2] persons (non-psychopaths).[3] However, given *today's* societal conception of medicine, which emphatically does *not* include the transformation, moral or otherwise, of patients into literally different (or numerically distinct) persons, psychopathy is not amenable to medicalization even in principle. It would thus be false, or at the very least misleading, to claim *today* that psychopathy might "someday" be medicalized. For again, the truth of such a claim would require that the operative conception of medicine be one that is *radically different* from the current conception, the latter of which does *not* encompass the transformation of patients into literally different persons.[4] Thus, those who would argue *today* that psychopathy might "someday" be medicalized, may be trading (unwittingly) on an ambiguity in the notion of *medicine*.

---

[1] More concisely, if less precisely, the treatment of disease, illness, or disorder.

[2] When two things are "numerically distinct," they are ipso facto *not* the same thing, however much they might resemble one another. Consider, for example, two matching earrings indistinguishable even to their owner. The two earrings are nevertheless numerically distinct – otherwise, there would be only *one* earring.

[3] Between treatment and transformation lies "enhancement" of non-pathological conditions, such as normal attention span or normal memory. Although I do not discuss the question whether such enhancements result in the transformation of one person into a different (or numerically distinct) person, my view is that they do not, as they do not dramatically alter an individual's "moral sensibility" or *Weltanschauung*.

[4] A relative of mine, a former nurse who is now in her 90's, said in response to a 1990s commercial regarding a medication for social phobia, "That's ridiculous; that's not medical, it's just shyness." She was arguably right, given a mid-twentieth century conception of medicine. But times have, of course, changed and therewith society's understanding of the scope and limits of medicine proper. Thus, she was arguably wrong given a more contemporary conception of medicine.

Before arguing for these and related claims, it will be important to achieve some clarity with respect to the relevant notions: *psychopathy*, *personal identity*, and *medicalization*. After discussing the notion of personal identity, I will present the particular account of that phenomenon adopted herein, the "essential moral self hypothesis" of Nina Strohminger and Shaun Nichols (2014).

15.2 Three Key Concepts

Much of what I say in this section will be relatively uncontroversial. However, my adoption of Strohminger and Nichols' (2014) essential moral self hypothesis might strike the reader as controversial insofar as that hypothesis is viewed as competing with (inter alia) traditional Lockean accounts of personal identity, which tie such identity to memory, particularly *autobiographical* memory. Such misguided controversy might then spill over into the conclusions drawn with that hypothesis' assistance. I will accordingly clarify the metaphysically modest ambitions of the essential moral self hypothesis to obviate any potential misunderstanding of this sort. Later in the paper, I will go on to defend the hypothesis against objections that have recently been raised against other, more metaphysically robust, morality-based accounts of personal identity.

The concepts of *psychopathy* and *personal identity* are well-known and much discussed in the philosophical literature; theories of the two phenomena (especially the latter) abound. The concept of *medicalization*, in contrast, is largely a term of art (of social science in particular) and I adopt its standard characterization herein.[5] Before considering each of these three concepts individually, I will say something brief about the first of the two (psychopathy and personal identity) in connection with *folk intuitions* and *theory-ladenness*.

With regard to the concepts of both psychopathy and personal identity, I wish to respect, and so to preserve, pre-theoretical (or "folk") intuitions to the extent possible. I realize, of course, that even the intuitions of ordinary folk are invariably theory-laden insofar as they reflect folk theories of the phenomena in question. Thus, for example, the widespread intuition that psychopaths are *fundamentally bad* (or "evil") arguably reflects the folk theory that psychopaths are "born bad" – that they are, in other words, "bad seeds."[6] My point, however, is to minimize, to the extent possible, going beyond the inevitable theory-ladenness inherent in the intuitions of ordinary folk.[7] In this way, I am freed from potentially distracting theoretical commitments not essential to the views and arguments presented herein.

---

[5] See Conrad (2007).

[6] The idea of the psychopath as a "bad seed" originates with the 1954 novel *The Bad Seed* by William March.

[7] This is precisely what Strohminger and Nichols (2014) and Prinz and Nichols (2016) do with regard to the notion of personal identity.

At the same time, I wish to avoid any patent misconceptions associated with folk intuitions and the beliefs that underlie them. Thus, for example, ordinary folk tend to conflate psychopathy and *psychosis*, a phenomenon reflected in the stereotypical belief that those suffering from schizophrenia (a severe mental disorder character-ized by delusions and hallucinations) are prone to murderous violence.[8] Hence, the ill-founded (and millennia-old[9]) notion of a "psycho killer," a notion memorialized in the 1977 Talking Heads hit "Psycho Killer."

Relatedly, this unfortunate psychosis/psychopathy conflation is reflected in the ordinary language expression "psycho," used colloquially to refer indifferently to both psychopaths and victims of psychosis (delusions and hallucinations). Thus, the expression might be used (in colloquial speech) to characterize someone like Syd Barrett (of Pink Floyd), who appears to have suffered from schizophrenia. Yet that same expression might also be used (in colloquial speech) to characterize someone like Ted Bundy – a classic psychopath. Such patent psychosis/psychopathy confla-tions will, of course, be avoided in the proposed characterization of psychopathy, to which I now turn.

(a) *psychopath/psychopathy*

A psychopath is someone who *lacks empathy* and (consequently) the moral emo-tions grounded therein, including guilt, remorse, compassion and sympathy. By "empathy," I mean (roughly) emotional sensitivity to the suffering of others espe-cially, but not exclusively, the suffering of other human beings.[10] The suffering of one's fellow human beings tends to cause, to varying degrees, psychological dis-tress or discomfort in the empath (or non-psychopath[11]) but not in the "cold-hearted" psychopath. Additionally, psychopaths tend to be *manipulative* insofar as they exploit others for their own selfish ends; they are, in this respect, *predatorial*. Metaphorically speaking, psychopaths are "wolves in sheeps' clothing,"[12] ruthlessly preying upon others with absolutely no concern for their victims' well-being. Their "illness" or "sickness," such as it is, is not generally viewed as psychiatric or

---

[8] Frith and Johnstone (2003), chapter 8.

[9] Frith and Johnstone (2003), chapter 8.

[10] The notion of empathy at issue here, *affective* empathy, contrasts with *cognitive* empathy, which involves the capacity to discern what others are thinking or feeling. Cognitive empathy is also known as "mind-reading." While psychopaths lack affective empathy, their intact cognitive empa-thy only facilitates their predatorial exploits.

[11] From here on, I contrast psychopaths with non-psychopaths rather than with empaths. In this way, I avoid the implication that my concern is with the distinction between those who are "cold-hearted" and those who are "warm-hearted." That is not my concern; my concern is with the con-trast between predatory individuals who are devoid of empathy and everyone else.

[12] Although psychopaths look like the rest of us (like "sheep"), they are fundamentally different: metaphorically speaking, they are wolves, not sheep. This idea of psychopaths as wearing a "dis-guise" is beautifully captured in the title of Hervey Cleckley's 1941 book *The Mask of Sanity*, which alludes to the fact that, like the rest of us, psychopaths appear perfectly sane. Their "insan-ity," such as it is, does not involve *rationality*; it involves *morality*.

otherwise medical; indeed, it is commonly viewed as unambiguously *moral*, reflecting the nineteenth century conception of psychopathy as "moral insanity."[13]

One might accordingly say, and without hyperbole, that psychopaths are, "by definition," both *devoid of empathy* and *predatorial*. This is what philosophers call an "analytic" truth, on par with the claim that bachelors are, by definition, both unmarried and male. Importantly, such an analysis does not rule out, but rather rules in, so-called "white collar" psychopaths, increasingly recognized, among both theorists and the general population, as inhabiting the corporate world, particularly at the very top. These are the "snakes in suits" about which Paul Babiak and Robert D. Hare (2006) have famously written.

Because of its theoretical "thinness," the foregoing characterization of psychopathy is not inconsistent with the well-known and well-respected accounts of Hervey Cleckley (1941) and Robert Hare (1999). Their characterizations are simply *richer* than the one adopted herein but they do not, in any sense, compete with it. More precisely, while the proposed characterization of psychopathy purports to provide "necessary" conditions for psychopathy, the characterizations of Cleckley and Hare are arguably more ambitious, purporting to provide a "criterion" for psychopathy: a set of conditions that are (conjointly) both necessary and sufficient. The proposed characterization is also, and importantly, in perfect keeping with the conceptions of ordinary folk, conceptions reflected in colloquialisms like "bad seed," "rotten to the core," and "wolf in sheep's clothing." These colorful metaphors beautifully capture the common understanding of psychopaths as *fundamentally bad* (or "evil") and *predatorial*.[14]

(b) *personal identity*

Personal identity, as I understand and deploy that notion, is sometimes referred to as "diachronic identity." So conceived, personal identity is what makes a person the same person *over time*. It is, in other words, what "preserves" the identity of a person. Metaphorically speaking, personal identity underlies the "survival" of a person.[15] Traditional philosophical theories of personal identity, beginning with that of John Locke (1689/2008), link personal identity to memory, particularly *autobiographical* memory.

In this paper, however, I will be adopting a distinctively *non-traditional* account of personal identity, one recently proposed and defended by Nina Strohminger and Shaun Nichols (2014). This experimentally supported "essential moral self hypothesis" aims to capture *folk views* regarding personal identity and maintains that

---

[13] Physician James Cowley Prichard was the first to use the phrase, which he did in his 1835 *Treatise on insanity and other disorders affecting the mind*.

[14] "Evil" is arguably the (English) word that best captures the ordinary person's conception of a psychopath, where individuals like Ted Bundy and Hannibal Lecter are seen as paradigms of the phenomenon.

[15] Diachronic identity contrasts with synchronic identity, the latter of which concerns the standing properties that make a person the person they are. For more on this distinction, see Prinz and Nichols (2016).

"moral traits are more essential than any other feature" of the mind to such identity. As a diachronic account of personal identity, it claims that what makes someone "the same person" over time is their "moral sensibility."[16] To the extent that an individual retains their moral sensibility over time, they remain "the same person" over time.[17]

Despite the occurrence of the word "essential" in the hypothesis' name, it is a metaphysically thin hypothesis insofar as it does not reify persons or (in other words) "selves."[18] Selves are not construed as independently existing entities, even if ordinary folk tend to conceptualize them as "souls," "spirits," or other non-physical entities that might conceivably survive bodily death. Relatedly, it would be a mistake to see the essential moral self hypothesis as metaphysically of a piece with accounts of personal identity that purport to capture some sort of "underlying reality." As Prinz and Nichols (2016, 449) explain,

> In saying that moral continuity is more important than other factors [to personal identity], we don't mean to imply that it is the one true theory of personal identity and that other theories are false. We don't think the question of identity over time depends on some deep metaphysical fact. That is not to say we don't think the question is metaphysical. We think it is metaphysical, but not deep; that is, it doesn't depend on some hidden fact about the structure of reality. Rather, it depends on us. Facts about identity are a consequence of [our] classificatory attitudes and practices.

I do not argue directly for the essential moral self hypothesis, as I regard the experimental evidence presented by Strohminger and Nichols (2014) and Prinz and Nichols (2016) as sufficiently compelling for present purposes. Such purposes involve (as noted above) reflecting on personal identity in connection with psychopathy with the aim of coming to an understanding of psychopathy's potential for medicalization. I do, however, argue *indirectly* for the essential moral self hypothesis. I do this by showing how that hypothesis not only accommodates reflections on personal identity in connection with psychopathy, but also withstands objections that have been levelled against other, more recent, morality-based accounts of personal identity, including the "good true self" account of Julian De Freitas et al. (2018).

(c) *medicalization*

I conceptualize medicalization in much the standard way,[19] as involving a multi-factorial and otherwise complex process by which human conditions and problems come to be viewed as medical conditions, and thus become the subject of medical study, diagnosis, prevention, and treatment. Previously, such conditions might have

---

[16] Strohminger and Nichols' "moral sensibility" appears to be similar to Maibom's (2014) *Weltanschauung* or "moral outlook." In the psychopath, such an outlook is arguably lacking all together. See Sect. 15.8 of the present paper for details.

[17] This way of looking at personal identity suggests that it can be retained (or not) in varying degrees over time.

[18] Human beings are persons insofar as they have a "moral sensibility" (Strohminger & Nichols, 2014) or "moral outlook" (Maibom, 2014).

[19] This standard conceptualization has its origins in the work of Conrad (2007).

been regarded as *moral* (as in alcohol use disorder) or *social* (as in social anxiety disorder). So conceived, examples of medicalized phenomena abound and include (among many others): ADHD, PTSD, sleep disorders, obesity, infertility, anorexia nervosa, PMS, menopause, alcohol use disorder, and social anxiety disorder. Is medicalization a *good* thing or a *bad* thing? As characterized herein, it is neither.[20]

One cannot, in my view, informatively discuss psychopathy's potential for medicalization without first considering its potential for *effective medical treatment*. Psychopathy's medicalization would, of course, be greatly facilitated were effective, and specifically *medical*, treatments already available. Sadly, if unsurprisingly, they are not – but could this change? Or, is genuinely *medical* treatment for psychopathy simply not possible, given the ostensibly moral (and "deeply personal"[21]) nature of the condition? I explore these and related issues in the remainder of the paper.

15.3 Treating a Patient and Thereby Transforming a Person

My interest in this section, and throughout much of the remainder of the chapter, is with hypothetically *curative medical treatment* whereby the once psychopathic patient no longer fits the characterization of psychopathy proposed above. They are (post treatment) neither devoid of empathy nor predatorial. In this way, the successfully treated former psychopath fails to fit the proposed characterization "twice over," as it were. Thus, while NN is a psychopath *before* treatment, they are no longer a psychopath *after* treatment. My reason for focusing on curative medical treatment is simple. Such focus promises to yield a clearer picture of the relation between psychopathy and personal identity and therewith a better understanding of psychopathy's potential for medicalization.

An analogy involving another well-known (but far more ordinary) chronic condition can vivify the issues at hand and help set the stage for the discussion that follows. MM is an asthmatic patient who currently takes medication (Advair and albuterol) to treat her asthma, a medical condition from which she suffers. Despite regular use of effective asthma medication, MM remains asthmatic. However, were a *truly curative* treatment for this chronic lung disease ever to become available, receiving such treatment would (by hypothesis) *cure* MM of her asthma: she would no longer be asthmatic. Importantly, MM (the particular human being so-named) would commence her curative treatment as a particular *person* and, once treatment had been completed, would emerge as the *very same person* – albeit as one who no longer suffers from asthma. Clearly, curative treatment for MM's asthma would not

---

[20] In contrast, "over-medicalization" is sometimes used to describe the phenomenon whereby human conditions that are *not* genuinely medical come to be wrongly so regarded.

[21] Psychopathy is often construed as a disorder *of personality* resembling, in certain respects, the DSM cluster B personality disorders. See Reimer (2010) and Reimer (2013) for more on these disorders.

transform her into someone else: into a *literally different* (or "numerically distinct") person – however much it might improve her lung capacity.

My interest here is with the possibility of a similarly curative, if (science) futuristic, treatment for psychopaths, for patients "suffering from" psychopathy. I am particularly interested in how such hypothesized treatment might affect the *personal identity* of the psychopathic patient. Would their personal identity be *preserved*, as with the former asthmatic patient cured of her asthma, or would curative treatment somehow *disrupt or disturb* that identity? Would the psychopathic *person* (vs. human patient[22]) even *survive* such treatment?

Let us view such questions against the backdrop of the essential moral self hypothesis. Given that hypothesis (adopted and defended herein), the idea of a *person* being cured of their psychopathy is potentially problematic insofar as it suggests that the treatment process would involve a *single person* (or "self"): a person who begins treatment as a *psychopath* and, assuming the treatment "works," ends treatment as a *non-psychopath*, thereby "personally surviving" the treatment process. Thus, psychopathic patient NN undergoes treatment for their psychopathy upon completion of which *that same person* is no longer a psychopath.

My initial reaction to such a view is to consider it in light of an alternative, and potentially more promising, view. Might not patient NN, a particular human being, be one person (a psychopath) before treatment and another person (a non-psychopath) after treatment? To see the force of this question, just imagine that Hannibal Lecter is not a fictional character but a real person. Imagine further that he is successfully cured of his psychopathy. Indeed, suppose that, post treatment, Lecter is positively *empathic*. Some might be inclined to say that the envisioned transformation would be *so* dramatic that Lecter would be, quite literally, a *different person* after curative treatment. Others, however, might dispute this, insisting that Lecter is indeed the same person from start to finish, just a *dramatically changed* person.

Despite the naturalness of the latter characterization, it is open to challenge on the grounds that it conflates persons (or selves) and human beings. Thus, although the hypothetical treatment process obviously involves just a single human being, Hannibal Lecter, to the extent that the treatment is genuinely curative, it arguably involves *two persons*, one at either end of the treatment process. Although Lecter is, like asthmatic patient MM, one and the same human being before and after treatment, unlike MM, he is arguably one person before treatment and another person, a numerically distinct person, after treatment.

This would appear to be in keeping with the essential moral self hypothesis. Indeed, according to one of the studies used to test that hypothesis (Strohminger & Nichols, 2014), participants claimed (on average) that eliminating the trait of psychopathy in a human subject by means of a pill (a "magic bullet") would change the person an incredible 73%! If a successfully treated psychopathic patient were to

---

[22] As I emphasize below, it is important to distinguish between the patient qua *human being* and the patient qua *person*.

change to such a dramatic extent, perhaps the treatment process would indeed yield a numerically distinct person – as an opposed to the same, but dramatically changed, person. After all, it would be as if the successfully treated psychopathic patient underwent a "Mr. Hyde to Dr. Jekyll" transformation, a transformation naturally construed as involving *two different persons* rather than a single, but dramatically changed, person. Intuitions to the contrary could perhaps be explained (as suggested above and below) by a conflation of *persons* (or selves) and *human beings*.

Let us therefore consider whether Robert Louis Stevenson's 1886/2008 novella "The Strange Case of Dr. Jekyll and Mr. Hyde" might help provide an intuitive, and potentially insightful, understanding of the hypothetical transformation of a psychopath (like Mr. Hyde) into a non-psychopath (like Dr. Jekyll).

15.4   The Strange Case of Dr. Jekyll and Mr. Hyde

Despite its underlying philosophical and psychopathological dimensions, Stevenson's famous novella is not particularly difficult to comprehend. Indeed, even pre-teens can understand it; it is a fascinating and especially imaginative story, much like Mary Shelley's 1823 novel *Frankenstein*. In the context of the present paper, however, I see Stevenson's story as providing the basis for a philosophical thought-experiment apt for probing intuitions regarding hypothetical cases of medically cured psychopaths. Indeed, Stevenson's curious tale mirrors, in important respects, the kind of scenario, the conceivability of which lies at the heart of this paper's central argument: that curative medical treatment of the psychopathic patient would have the effect of transforming that patient, a particular *human being*, into a numerically distinct *person*. This would mean that a single human being could, conceivably, "house" (sequentially) two different persons: one a psychopath, the other a non-psychopath.

In the story, Dr. Jekyll ingests a secret serum – indeed, a serum "prescribed" by a physician: by himself. This serum transforms Dr. Jekyll from a non-psychopath into psychopath: from the good Dr. Jekyll into the evil (and murderous) Mr. Hyde. Later in the story, Mr. Hyde ingests the same serum so that he can escape capture by transforming himself into Dr. Jekyll, whose physical appearance is markedly different from his own.

Dr. Jekyll and Mr. Hyde would appear to be *different persons*: different persons who occasionally transform themselves into one another by means of a secret serum. Yet they manage to share a single human body which undergoes physical transformations reflective of the underlying psychological ones. In this respect, Stevenson's story mirrors Oscar Wilde's *The Picture of Dorian Gray* (1890), in which grotesque changes in a portrait of protagonist Dorian Gray mirror the moral decline of that eternally young and beautiful man.

How is the transformation of Mr. Hyde into Dr. Jekyll (or vice versa) any different, in terms of its *conceivability*, from the transformation of psychopathic NN into non-psychopathic NN? It is true that, in Stevenson's story, the psychological transformations are accompanied by changes in physical appearance, but these changes are presumably in the same *human being* and, importantly, are not essential to the issue at hand. Indeed, one can easily imagine a modified version of the Stevenson story in which no overtly physical changes accompany the psychological transformations. Mr. Hyde might thus be viewed (in the modified story) as a *doppelganger*: an evil person who happens to look exactly like a good person: Dr. Jekyll. Mr. Hyde is thus the "evil twin." So amended, the story would still pump the relevant sort of intuitions: intuitions to the effect that curative medical treatment of a psychopathic human being would not only effectively, but also *literally*, transform that patient, that human being, into a *different person* – not metaphorically, hyperbolically, or otherwise non-literally. There would thus be one human being "inhabited," sequentially, by two distinct persons.[23] Such intuitions comport well with the essential moral self hypothesis, which sees "moral sensibility" as crucial to diachronic identity: to what preserves the identity of persons over time.

I thus see the fictional case of Dr. Jekyll and Mr. Hyde as providing a helpful model for understanding the conceptual possibility of a curative medical treatment for psychopathy whereby a particular human patient (NN) might be transformed from one person (a psychopath) into a numerically distinct person (a non-psychopath). The human patient NN would, of course, remain the same human being, something reflected in the continued post-treatment use of their pre-treatment name.[24] In light of these considerations, the characterization of the good Dr. Jekyll and the evil Mr. Hyde as "the same but different" is an apt one. They are the same *human being*, yet arguably (very) different *persons*.

15.5   If You Become a Psychopath, Do You Die?

The theoretically important distinction (sketched and illustrated above) between *human beings* and *persons*, would appear to undermine a rhetorical argument recently put forth by Christina Starmans and Paul Bloom in their 2018 piece "If You Become Evil, Do You Die?" There, they ask the reader:

---

[23] The fact that Jekyll and Hyde appear to have some degree of shared consciousness reinforces the appropriateness of the analogy to psychopaths imaginatively transformed into non-psychopaths. For there is no obvious reason to suppose that the morally transformed psychopathic patient NN would not have memories of NN's pre-treatment experiences. Some such memories would, however, likely differ dramatically with regard to their affective components, as the non-psychopathic NN would likely recall, with horror, their pre-treatment attitudes and behaviors. For more on this issue, see the conclusion of the present paper.

[24] This was the case with Phineas Gage, who naturally kept his name after his "person-transforming" accident. In contrast, Jekyll and Hyde have different names, despite naming the same human being because they are assumed by others to be different human beings, something reinforced by their decidedly different physical features.

Have you ever encountered someone, either in real life or in fiction, who started off good but then became immoral? If so, did the person disappear? Did their body become a shell, now occupied by a different individual? (Starmans & Bloom, 2018: 566).

Although Starmans and Bloom assume that the answer to these (rhetorical) questions is "no," the clear comprehensibility of Stevenson's story suggests that such an assumption is mistaken. After all, when the "good" Dr. Jekyll is transformed into the "evil" Mr. Hyde, what exactly happens to him? *Where does he go?* It would be natural for the reader of the story to say, "The good Dr. Jekyll disappeared as his body was taken over by the evil Mr. Hyde." In contrast, there would no inclination to say, simply, that Dr. Jekyll "died."

It would thus appear that there is indeed a fictional case of the very sort the existence of which Starmans and Bloom are skeptical. Moreover, the clear comprehensibility of this "strange case" suggests that if someone "started off good" (or non-psychopathic, as with Dr. Jekyll) and then became evil (or psychopathic, as with Mr. Hyde), then that good person would in *some* sense "disappear." Of course, their *body* would not disappear, even if it changed markedly, as it does in Stevenson's story. Moreover, that body would, in *some* sense, be "occupied" by a different person. After all, there is presumably always a human brain situated within the shared cranium of Dr. Jekyll and Mr. Hyde, a brain "occupied" (at sequential times) by both a good person and an evil person. Interestingly, the operative senses of "disappear" and "occupy" are naturally interpreted as *literal*. After all, insofar as a person's identity has its existential base in the brain, one can think of the neural states that underlie (or perhaps even *constitute*) NN's pre-treatment psychopathic personal identity as having literally "disappeared." They are no longer there or anywhere else. Such is trivially true of any brain state or process that might be said to have "changed." Where do such states and processes go when they change and thus "disappear"? Nowhere; changing brain states and processes are not the kinds of things that "go" anywhere; to claim otherwise, is to be guilty of a category mistake.[25]

Similarly, one can think of the neural states that underlie (or constitute) NN's post-treatment non-psychopathic personal identity as "occupying" NN's body '– specifically, his cranium, which houses his brain and therewith, presumably, the neural states that underlie his post-treatment personal identity. Thus, although talk about "disappearing persons" and "occupied bodies" might initially sound too much like science fiction to be taken seriously, charitably interpreted, such talk would appear to be not only coherent, but accurate as well.

---

[25] Similarly, the patient who asks, "Where did my 115/75 blood pressure go?" or "Whatever happened to my 18 BMI?" speaks nonsense if they believe that their previous blood pressure or BMI is, literally, somewhere to be found – or that something has "happened to" it.

15.6   Turning the Tables on Starmans and Bloom

Ironically, the points Starmans and Bloom make rhetorically (if unsuccessfully) against the "good true self" account of De Freitas et al. (2018), can be levelled against their own account of personal identity. That account explicitly allows for cases where "it may be thought that a person ceases to exist," while they (qua particular human being) remain alive although not, alas, well. This is what might happen, according to Starmans and Bloom, in cases of severe dementia. While Starmans and Bloom title their response to De Freitas et al., "If You Become Evil, Do You Die?," an appropriate rejoinder to that veiled rhetorical argument might be an equally rhetorical:

If you become severely demented, do you die?

The answer to both questions should be equally obvious. In neither case do you (a particular human being, NN) *die* but in both cases you are arguably no longer the *person* you were before you became evil or severely demented. That person has "disappeared"; they are "no longer around." They have "ceased to exist."

In the case of severe dementia, patient NN is perhaps a fragment of the person they once were, despite being the same living, breathing, human being. In the case of moral transformation, patient NN is not the person they were before treatment despite being the same human being; the treatment has transformed them into another person. Where, then, is the "previous" person? They have "ceased to exist," although without a dead body to show for their "demise."[26]

Importantly, I am not attempting to discredit Starmans and Bloom's claim that, in cases of severe dementia, a person might "cease to exist." In fact, I am sympathetic to that claim. What I am trying to do here is discredit a *parallel* argument against the proposed view of personal identity. I do this by claiming that, if sound, that rhetorical argument, "If you become evil, do you die?" would discredit their own, quite plausible, view that there are cases where a person may "cease to exist" (or to "disappear") even when the human being that person previously "occupied" is still alive.

15.7   Taking Transformation Seriously Means Taking it *Literally*

According to Starmans and Bloom (2018), the end-product of profound moral transformation is not a numerically distinct person; it is rather "one person who has changed dramatically" (2018, 567). This is surely true in some cases of what might be conceptualized as "moral transformation," as when a high school bully sees the

---

[26]This would make perfect sense if (as per the essential moral self hypothesis), NN's personal identity is a matter of their "moral sensibility" which might conceivably come and go with Jekyll-to-Hyde and Hyde-to-Jekyll transformations.

error of their ways after effective counseling. However, I question whether it is true in cases where the moral transformation, only hypothetical at this point, involves the transformation from a *psychopath* into a *non-psychopath*. Such cases would involve going from having no *Weltanschauung* or (in other words) "moral outlook" to having one. Thus, the difference between the "before person" and the "after person" is arguably one of *kind*, rather than of degree. I would thus be inclined to say, just as in the case of Mr. Hyde and Dr. Jekyll, that there are indeed *two persons* inhabiting, at sequential times, one and the same human being (body or cranium).

Starmans and Bloom's likely response to this sort of view is not difficult to anticipate. Their discussion of the famous (and factual) case of Phineas Gage makes clear that they would interpret talk of "two persons" in such and similar cases (including hypothetical cases involving morally transformed psychopaths) as unambiguously metaphorical, hyperbolic, or otherwise non-literal. They agree with de Freitas et al. (2018) that, when those who knew Gage *before* his accident described him as "no longer Gage" *after* that accident, they were not just casually indicating that he was no longer a "nice guy." Presumably, they meant to convey something to the effect that Gage no longer had the important traits that they had always associated with him. However, as they go on to claim,

> …we see this as similar to saying, 'I'm just not myself today', which obviously cannot be meant literally, and illustrates that we typically use this type of language to talk about changes, not obliteration. (Starmans & Bloom, 2018, 567)

Presumably, no one would deny that we often resort to non-literal language when talking about changes in personality. We have all heard people say such things as, "She's a totally different person than she was when I was friends with her in high school." There is no question that some of this talk is intended metaphorically, hyperbolically, or otherwise non-literally. But let us not forget the "strange case" of Dr. Jekyll and Mr. Hyde, while imagining that, at some point in the unforeseeable future, psychopathic individuals like Ted Bundy or Hannibal Lecter are able to enter a medical facility, obtain a doctor's prescription for a curative "magic bullet" treatment to which they adhere,[27] emerging 10 days later, with the kind of ordinary everyday "moral sensibility" characteristic of a non-psychopath.

The question is: How should we *conceptualize* such transformations? We certainly might say that NN (more precisely, the human being so-named) is "no longer the person he used to be" or that he is "a different person entirely." On the proposed view, it is sometimes reasonable to take such statements *literally*.[28] Such a view is

---

[27] Since psychopaths have no problem with their lack of empathy or predatorial behavior, it is unlikely that such adherence would be voluntary.

[28] In some cases, it may be clear whether the speaker's utterance is intended literally or non-literally. In other cases, this might not be the case but not because the speaker's communicative intentions are unknown but because the speaker *herself* might not have a clear sense of whether her utterance is best interpreted literally, non-literally, or somewhere in between. Indeed, there is no reason to suppose that natural language utterances are always straightforwardly literal or non-literal. There is surely a great deal of indeterminacy with regard to such things. That is part of what makes language so flexible.

supported, individually and conjointly, by at least three things: the truly dramatic nature of the transformations in question, the striking resemblance of such patently moral transformations to Hyde-to-Jekyll transformations, and the intuitive plausibility of the experimentally supported essential moral self hypothesis, interpreted as the metaphysically innocuous hypothesis that it is.[29] Moreover, any resistance to literal readings of "different persons" talk can be explained away (as we have seen) by distinguishing between *persons* and *human beings*, where the former are seen as "inhabiting," not only metaphorically but also literally, the latter.

If it makes sense to suppose that Dr. Jekyll is indeed a *different person* than Mr. Hyde, despite sharing the same human body, then it should make sense to suppose of a hypothetically (morally) transformed "new" Hannibal Lecter that he is *not the same person* as the "old" Hannibal Lecter. He is, of course, the same *human being* before as after his moral transformation but again, he is arguably *not* the same *person*.

The fact that Hyde-to-Jekyll (and Jekyll-to-Hyde) transformations are accompanied by striking *physical transformations* is irrelevant here. Not only is spatiotemporal continuity present in these transformations, it is easy to imagine (as noted earlier) that such physical transformations do *not* occur. We might still say that Mr. Hyde was a "different person" than Dr. Jekyll, intending it literally, with Mr. Hyde conceptualized, perhaps, as Dr. Jekyll's doppelganger or "evil twin."

Starmans and Bloom (2018) suggest that De Freitas et al. (2018) are committed to the implausible-sounding claim that "individuals who become immoral literally cease to exist" (2018, 567). I do not see this. De Freitas et al. are perhaps committed to the claim that individuals who become immoral are, literally, no longer the *persons* they once were. As to *where* the persons they once were are, I see no problem with saying what De Freitas, et al. say: they have "ceased to exist," they are (in other words) "no longer around." This would make perfect sense on the essential moral self hypothesis, which sees persons as individuated by their "moral sensibility" – or "moral insensibility," as the case may be.

15.8 Fake Persons: The "No Person" Theory of Psychopathy

One interesting possibility, alluded to above but not yet discussed explicitly, is that psychopaths are not actually persons to begin with because they do not have the (neurologically-based) moral wherewithal to be, or to naturally evolve into, persons. They are "constitutionally" incapable of being (or becoming) persons. They do not have (in other words) an essential moral self, a moral sensibility, a set of moral

---

[29] As to whether Phineas Gage's moral transformation was a transformation in personal identity, I am officially neutral, as I am with respect to whether post-injury Gage was a psychopath or even had genuine psychopathic traits. My point in invoking his case is to encourage reflection on the sort of "different person" talk used to characterize his transformation. In contrast to Starmans and Bloom (2018), I view it as a mistake to automatically assimilate such talk to patently non-literal utterances of (e.g.) "I am not myself today."

beliefs and attitudes, something akin to Maibom's (2014) *Weltanschauung* or (in other words) "moral outlook." They are *fundamentally amoral.*[30] Colloquially speaking, psychopaths "have no soul," they are "bereft of a conscience," they lack a "moral compass." In this respect, psychopaths are perhaps importantly different from ordinary everyday "bad people."

Is this way of thinking about psychopaths compatible with the essential moral self hypothesis of personal identity adopted herein? I believe that it is. In fact, it is a possibility explicitly brought up by Strohminger and Nichols (2014) in their discussion of psychopaths. We should therefore consider seriously whether psychopaths might not be *persons* in some substantive sense of that notion.

At first, such a possibility might sound incredibly counter-intuitive. When we say things like, "Ted Bundy was a terrible person," we certainly take ourselves to be speaking not only truthfully but *literally*. However, it is important to appreciate that, while the notion of *person* adopted herein is grounded in the ordinary notion of that term, it is nonetheless a *theoretical* notion, one that implicitly avoids conflating persons and human beings, people, or "individuals," construed as potential bearers of ordinary proper names. We can therefore easily soften the counter-intuitiveness of the claim that psychopaths are not persons by pointing out that they can still be accurately conceptualized (in the context of the present paper) as *human beings*, *people*, or *individuals*. They are simply human beings (people, individuals) who are, as one might say, "bereft of a conscience."

What, then, happens to the idea that a curative "magic bullet" treatment for psychopathy would involve the transformation of one person into another, numerically distinct, person? The central idea here could easily be retained but would require different language for its expression. Thus, instead of saying of a successfully treated former psychopath that they are a "different person" than they were before treatment, we might instead say that they are a "real person" now.[31] There is nothing problematic about saying such things and intending them literally. In fact, by speaking in terms of someone's having become a "real" person, one would be signaling that they have been a person *all along* – in *some* sense of "person"[32] but not in another. Having acquired a soul (conscience, moral compass), they have become a "real" person, an "authentic" person. They are no longer someone who merely bears a striking physical resemblance to genuine persons; they are no longer a "fake" person.

My only reservation about talking in such terms is that doing so is not only unnecessary but is also at odds with our pre-theoretical notion of *person*, according to which Ted Bundy was a person, a real (vs. fictional) person, albeit one who was quite possibly amoral and certainly psychopathic. We can still retain this common-sense notion of *person*, while maintaining that personal identity is to be construed

---

[30] Cleckley (1941) appears to construe psychopaths in this way.

[31] Importantly, transformations of non-persons into persons are no more within the bounds of medicine proper than are transformations of persons into numerically distinct persons.

[32] Perhaps a human being or an individual – an entity apt for naming.

in terms of "moral sensibility" or *Weltanschauung*. If we construe moral sensibility as a *spectrum notion*, having no such sensibility (having "moral insensibility") can be understood in terms of being situated on one endpoint of that spectrum; that is where psychopaths lie. If we construe a *Weltanschauung* as a "moral outlook," we can view psychopaths as having an "empty" *Weltanschauung*.

15.9   Psychopathy Medicalized? Not So Fast

In the introduction to this chapter, I stated that my aim was to reflect on the nature of both psychopathy and personal identity to better understand psychopathy's potential for medicalization. I also indicated that I would argue that the medicalization of psychopathy was simply not possible, not even in principle. In the jargon of contemporary analytic philosophy, there is no "possible world" in which psychopathy has been medicalized, *given our current conception of the scope and limits of medicine*. Importantly, this conception does *not* accommodate Hyde-to-Jekyll transformations in "moral sensibility" or *Weltanschauung*.

In setting the stage for a defense of the proposed view, I have devoted the bulk of the present paper to arguing for two (related) claims. The *first* claim is that a scenario in which psychopaths undergo *curative medical treatment* for their psychopathy is conceptually possible, a view supported by appeal to the easy understandability of Stevenson's (1886) "The Strange Case of Dr. Jekyll and Mr. Hyde." The *second* claim is that such hypothetical curative medical treatment would involve the transformation of the psychopath into a *numerically distinct* person: a non-psychopath.[33] This latter claim, also made vivid by appeal to Stevenson's novella, was noted to be in comportment with the experimentally supported essential moral self hypothesis of Strohminger and Nichols (2014). The idea that transformations in personal identity would be involved in curative medical treatment of psychopathy was defended against arguments presented by Starmans and Bloom (2018) in an effort to discredit the views of De Freitas et al. (2018) regarding the essential link between personality identity and morality.

The first of these two claims might be thought, ironically, to support a view *contrary to* the view it's intended to (help) support: namely, the view that the medicalization of psychopathy is not possible even *in principle* – let alone in fact. We can, after all, imagine a future, albeit one that is arguably unforeseeable, in which psychopaths (the Mr. Hydes of the world) are cured by being transformed, via some "special serum," into non-psychopaths (the Dr. Jekylls of the world). It is, moreover, easy to imagine that these curative treatments are prescribed by *medical professionals* and that there is *widespread societal awareness* of this practice. The question is: Why is *this* not a "possible world" in which psychopathy has been

---

[33] If one understands psychopaths as non-persons, the transformation would be from a non-person to a person.

medicalized – and why doesn't the (logical) possibility of such a world *discredit* the proposed view that psychopathy is not, even in principle, amenable to medicalization?

Despite its rhetorical force, the foregoing argument is entirely without *logical* force, as it is vitiated by an equivocation on the notion of *medicine*. In asking whether the "medicalization" of psychopathy is conceptually (logically, metaphysically) possible, we need to invoke *our* society's current conception of medicine, not that of some science-futuristic and purely hypothetical society. Importantly, the operative conception of medicine must be contemporary *society's* conception, not that of the specifically *medical* community and certainly not that of any contemporary philosophical theorist of medicine or psychopathy.

How, then, do we go about discerning society's current conception of medicine against which we can then assess psychopathy's potential for medicalization? Although it might sound flip to respond with, "Just google the definition of *medicine*," there is much to recommend such a response. Standard dictionaries, such as Merriam-Webster's or the Oxford English Dictionary (both readily available online), are dictionaries of *common usage*; they purport to catalog how ordinary speakers use words – what such speakers *mean* by those words when they use them in ordinary everyday conversations. Let us therefore have a look at the entries for *medicine* in each of these two respected sources.

Merriam Webster

> medicine: the science or art of dealing with the maintenance of health and the prevention, alleviation, or cure of disease

Oxford English Dictionary

> medicine: the science or practice of the diagnosis, treatment, and prevention of disease

Before considering psychopathy's amenability to medicalization in light of these two standard definitions, a potential problem concerning the key notion of *disease* must be addressed. Insofar as psychopathy is amenable to being understood as a condition that falls within the bounds of medicine proper, it is more naturally conceptualized as an *illness* (of the mind) or as a *disorder* (of personality) than as a *disease* (of the body). The reasons for this are no doubt complex, perhaps having to do with a conflation of paradigms of disease (malaria, strep throat, influenza) with disease more generally. However, such reasons need not be explored before addressing the problem at hand. We can simply agree to construe *disease* broadly, so as to encompass conditions more naturally conceptualized as *illnesses* or *disorders*. In this way, we might appear to give the strategic advantage to those who would claim that psychopathy is a genuinely *medical* condition: a disease, *broadly construed*. However, even with such a broadened understanding of *disease*, the idea that psychopathy is amenable to medicalization runs into trouble. In fact, a construal of *disease* that accommodates conditions more naturally thought of as "illnesses" or "disorders," only magnifies that trouble.

As I have argued elsewhere against the idea that psychopathy is a *disorder* (Reimer, 2008), I will focus here on the idea that it might be a *disease* or an *illness*. The latter notion connotes a *subjective sense of suffering*. Thus, while it sounds odd to say that one is feeling "diseased," it sounds perfectly fine to say that one is feeling "ill." Illness, in the ordinary everyday sense of that notion, is paradigmatically a *felt* experience. One comes to the conclusion that one may be ill because one *feels* ill. In the jargon of contemporary medicine, illness is typically (or at least stereotypically) *ego-dystonic*. Although the term *disease* does not convey a subjective sense of suffering in the way that the term *illness* does, its etymology, which it wears on its linguistic sleeve, does. Indeed, the modern English term "disease" derives from the Middle English term "disease," meaning *lack of ease*. And yet, the psychopath, being immune to the characteristically unpleasant moral emotions, leads an inner life of comparative ease. Such individuals are not "plagued" by guilt, or remorse; they are not "overcome" by compassion or sympathy for those who suffer. Their "condition" is not ego-*dystonic*, it is rather ego-*syntonic*.

For these reasons, the idea that psychopathy is amenable to medicalization is likely to be met with considerable resistance – after all, psychopaths do not *feel ill* on account of their psychopathy, nor do they experience a *lack of ease* on account of that putative disorder. Indeed, it sounds remarkably odd to say that Ted Bundy (for example) "had" psychopathy or that he "suffered from" it. It is far more natural to say, simply, that Ted Bundy *was* a psychopath. That is the *kind of person* he was.

This way of speaking about psychopaths – referring to them by their pathology – reflects (without entailing) the idea that psychopathy is *essential* to the personal identity of the psychopath. We see psychopathy as central to what makes a psychopath the person they are. In this way, we reveal a tacit endorsement the essential moral self hypothesis of Strohminger and Nichols (2014).

These ideas comport with what I have been claiming in regard to hypothetical Hyde-to-Jekyll transformations. Such transformations would, in ridding the human subject of their unfortunate "condition," change them into different persons – *literally*. Although this might sound incredible, properly understood, it is not. One need only appreciate the distinction between the patient, a particular *human being*, and the *person*, construed in terms of "moral sensibility" or *Weltanschauung*. Surely, no one thinks of medicine as concerned with the transformation of patients into literally different persons.[34] Rather, one thinks of contemporary medicine as concerned with the treatment of patients who "suffer from" disease, illness, or disorder. Medical treatment, if successful, does not transform the patient into a different person, but rather restores the patient to their default "healthy self." Such restoration is possible, factually, and conceptually, because patients are conceptualized as *distinct from* the unhealthy conditions from which they suffer. In the case of psychopathy, however, to the extent that we conceive of that condition as a disease, the patient *is* the disease – they cannot be separated, factually or even conceptually, from that

---

[34] Nor does anyone think of medicine as involving the transformation of non-persons into persons (for those who are sympathetic to the idea of psychopaths as non-persons).

condition, the successful cure of which would transform them into someone else – *literally*!

We would thus appear to have a pair of powerful objections to the (logical) possibility of psychopathy's medicalization. *First*, psychopathy is not naturally construed as a disorder, disease, or illness: that is, as a condition from which one "suffers," an ego-dystonic condition. *Second*, psychopathy is not a condition that can be cured or otherwise eliminated without the person (vs. the human patient) "disappearing" or "ceasing to exist." In this respect, psychopathy is *not* akin to paradigmatically medical (vs. moral) disorders, illnesses, or diseases. Such paradigm medical disorders can in theory, if not in fact, be cured or otherwise eliminated while the personal identity of the human patient (construed in terms of the essential moral self hypothesis) remains intact.[35]

15.10 Shifting Conceptions of Personal Identity

The arguments of this paper depend crucially on the experimentally supported essential moral self hypothesis of Strohminger and Nichols (2014). As noted above, this hypothesis purports to capture *contemporary folk views* regarding what makes a person the "same person" over time. As its authors point out, "the question of identity over time" does not depend on "some deep metaphysical fact," on "some hidden fact about the structure of reality." Rather, "it depends on us." Facts about personal identity, according to Strohminger and Nichols, are a "consequence of [our] classificatory attitudes and practices." Because of this, even if the hypothesis does faithfully reflect *today's* (Western) "classificatory attitudes and practices,"[36] it might not capture *tomorrow's*. Those attitudes and practices might conceivably change for any number of reasons.

Suppose, then, that within a couple of centuries, these classificatory attitudes and practices do indeed change. Suppose, further, that they are better captured by some alternative account of personal identity: perhaps a Lockean autobiographical memory account or a "narrative criterion" account.[37] Given either sort of account, the medically induced transformation of psychopaths into non-psychopaths might *not* involve transformations in personal identity. Or so it would appear. After all, if the psychopathic patient's autobiographical memories or self-told life-narrative

---

[35] What about a cure for Alzheimer's? Mightn't it transform the severely demented patient into another person – literally? No; it would return them to their authentic self. They would become the full (vs. fragmented) person they once were.

[36] This is of course open to dispute. Other theorists might claim, based on a different set of experiments, that some other conception of personal identity better captures folk views about what such (diachronic) identity involves.

[37] According to a narrative criterion account of personal identity, what makes an action, experience, or psychological trait attributable to given individual, is its apt incorporation into the self-told story (or narrative) of their life.

survives their Hyde-to-Jekyll transformation, might not they themselves "personally survive" it?

Suppose the folk inhabiting either of these two futuristic societies are asked to consider a scenario (a philosopher's "possible world") in which psychopaths are cured of their psychopathy via some "magic bullet" physician-prescribed treatment. From *their* perspective, hypothetically *our* own perspective within a couple of hundred years, we have a situation in which psychopathy might well be viewed as a *disease* from which the patient might be cured without disturbing or disrupting their personal identity. Wouldn't *this* be a counter-example to the central claim of the present paper: that psychopathy is *not* amenable to medicalization, *even in principle*? Although it might initially appear to be, any such appearance would be deceiving.

Consider the autobiographical memories of a morally transformed former psychopath, or consider the "self-told life narrative" of such an individual. Let us begin with the former. A dramatic moral change, a Hyde-to-Jekyll change, would likely have a truly profound impact on the affective component of such memories and narratives. Pre-treatment events, especially those involving behaviors, attitudes, and experiences characteristic of psychopaths, would be remembered by the post-treatment former psychopath very differently – with *disgust* and *horror*, rather than with *satisfaction* or *glee*. This sort of scenario is nicely captured in Dr. Jekyll's abhorrence at the acts perpetrated by Mr. Hyde, acts the good doctor remembers "first-personally" through a consciousness (at least partially) shared with Mr. Hyde. Thus, although Dr. Jekyll remembers those acts, he remembers them with profoundly *negative* affect, reflecting his transformed (or newly formed) "moral sensibility."

Similarly, the *pre-treatment* life-defining experiences of a former psychopath would not cohere or "fit" with their *post-treatment* self-told life narrative. It would be as though there were missing "chapters" between the last chapter of the *pre-treatment* self-told life narrative and the first chapter of the *post-treatment* self-told life narrative. How could the former psychopath successfully integrate their present *non-psychopathic* attitudes, behaviors, and experiences with their past, and profoundly different, ones? It would be as though the former psychopath had two profoundly different, if overlapping, lives: a past life and a present life. This metaphorical understanding of such futuristic scenarios is nicely captured in the idea that such scenarios would involve *two different personal identities*.

Thus, our "moral sensibility" arguably contributes significantly to both how we remember our past and how we might attempt to integrate the past "chapters" of our lives with those of the present. When the psychopath is (hypothetically) endowed with a "moral sensibility," it thus becomes doubtful whether they would *personally survive* the resultant moral transformations. If they did not, then any such transformation would arguably *not* be medical, given our current understanding of the scope and limits of medicine, which emphatically does *not* accommodate literal transformations in personal identity.

In sum, whether one construes personal identity in terms of moral sensibility, autobiographical memory, or self-told life narratives, the (medical) curation of psychopathy would arguably eliminate the psychopath and replace them with another person: a non-psychopath. Any such cure would accordingly not be "medical," given our current understanding of the scope and limits of medicine. It would then seem to follow that hypothetical scenarios of the sort we have been considering do *not* suggest that psychopathy is amenable, if only in principle, to medicalization.

Acknowledgement I would like to thank John McMillan and Luca Malatesti for their enormously helpful comments on earlier drafts of this chapter.

References

Babiak, P., & Hare, R. (2006). *Snakes in suits: When psychopaths go to work.* HarperCollins Publishers.

Cleckley, H. (1941). *The mask of sanity.* C.V. Mosby Company.

Conrad, P. (2007). *The medicalization of society: On the transformation of human conditions into treatable disorders.* Johns Hopkins University Press.

De Freitas, J., Cikara, M., Grossman, I., & Schlegel, R. (2018). Moral goodness is the essence of personal identity. *Trends in Cognitive Sciences, 22*(9), 739.

Frith, C., & Johnstone, E. (2003). *Schizophrenia: A very short introduction.* Oxford University Press.

Hare, R. (1999). *Without conscience: The disturbing world of the psychopaths among us.* Guilford Press.

Locke, J. (1689/2008). An essay concerning human understanding. Oxford University Press.

March, W. (1954). *The bad seed.* Dell Publishing.

Maibom, H. (2014). To treat a psychopath. *Theoretical Medicine and Bioethics, 35,* 31–42.

Prichard, J. (1837). *A treatise on insanity and other disorders of the mind.* Haswell, Barrington, and Haswell.

Prinz, J., & Nichols, S. (2016). Diachronic identity and the moral self. In J. Kiverstein (Ed.), *The Routledge handbook of philosophy of the social mind* (pp. 449–464). Taylor and Francis.

Reimer, M. (2008). Psychopathy without (the language of) disorder. *Neuroethics, 1,* 185–198.

Reimer, M. (2010). Moral aspects of psychiatric diagnosis: The cluster b personality disorders. *Neuroethics, 3,* 173–184.

Reimer, M. (2013). Moral disorder in the DSM-IV? The cluster b personality disorders. *Philosophy, Psychiatry, and Psychology, 20*(3), 203–215.

Starmans, C., & Bloom, P. (2018). If you become evil, do you die? *Trends in Cognitive Sciences, 22,* 566–568.

Stevenson, R. L. (1886/2008). *The strange case of Dr. Jekyll and Mr. Hyde.* Oxford University Press.

Strohminger, N., & Nichols, S. (2014). The essential moral self. *Cognition, 131*(1), 159–171.

Wilde, O. (1890/2006). *The picture of Dorian gray.* Oxford University Press.

Chapter 16
The Disorder Status of Psychopathy

Luca Malatesti and Elvio Baccarini

Abstract In this chapter, we investigate whether psychopathy is a mental disorder. We argue that addressing this question requires engaging, at least, with three principal issues that have conceptual, empirical, and normative dimensions. First, it must be established whether current measures of psychopathy individuate a unitary class of individuals. By this we mean that persons classified as psychopaths should share some relevant similarities that support explanation, prediction, and treatment. Second, it must be proven that psychopathy harms the person who has it. Third, it must be established that the harm associated with psychopathy is relevant for the ascription of disorder status. Regarding this latter issue, we argue that psychopathy should be considered a disorder if its harmfulness derives from certain incapacities or limited capacities. These incapacities should affect basic competences that are justifiably required for conducting a preferable type of life. Within this framework, we tentatively advance the hypothesis that some normatively justified conclusions and empirical evidence about psychopathy, that needs nonetheless to be further investigated, might support the claim that people with psychopathy have a mental disorder.

Keywords Psychopathy · Mental disorder · Harm in psychiatry · John Rawls · Reinforcement learning · Normative justification · Normativism in psychiatry

L. Malatesti (✉) · E. Baccarini
Department of Philosophy, Faculty of Humanities and Social Sciences, University of Rijeka, Rijeka, Croatia
e-mail: lmalatesti@ffri.uniri.hr

© Springer Nature Switzerland AG 2022
L. Malatesti et al. (eds.), *Psychopathy*, History, Philosophy and Theory of the Life Sciences 27, https://doi.org/10.1007/978-3-030-82454-9_16

16.1   Introduction

Several authors have investigated whether psychopathy is a mental disorder (Graham, 2013; Jurjako, 2019; Krupp et al., 2012, 2013; Leedom & Almas, 2012; Malatesti, 2014; Nadelhoffer & Sinnott-Armstrong, 2013; Reimer, 2008a). This issue lies at the intersection of interrelated theoretical, empirical, and practical concerns. The morally justified response to the crimes, antisocial behaviours, and personality traits of individuals with psychopathy depends significantly on whether they are mentally disordered. The stances of those who interact with these individuals might be significantly affected by knowing that they are mentally disordered. If the crimes of the psychopathic offenders are explained by a mental disorder, there might grounds for legal exculpation (Jefferson & Sifferd, 2018; Malatesti et al., 2020; Nadelhoffer & Sinnott-Armstrong, 2013). Moreover, the disorder status of psychopathy could motivate and justify therapeutic concern and intervention, more tolerant attitudes towards the shortcomings of this kind or personality, and, in general, less hurtful interactions with individuals with this condition (Tamatea, this volume).

In this chapter, we highlight three main issues, that have theoretical, empirical, and normative dimensions, that need to be addressed to solve the problem of the disorder status of psychopathy. First, it must be established whether current measures of psychopathy individuate a unitary class of individuals. By this we mean that psychopathic individuals should share some relevant similarities that support explanation, prediction, and treatment. Second, it must be established that psychopathy harms the person who has it. Third, it must be shown that the harm associated with psychopathy is relevant for the ascription of disorder status. Regarding this latter issue, we argue that psychopathy should be regarded as a disorder if its harmfulness derives from certain incapacities or limited capacities. These incapacities should affect basic competences that can be justifiably regarded as necessary requirements for conducting a preferable type of life.

Besides offering a framework for investigating the problem of the disorder status of psychopathy, we advance more substantive claims towards its solution. One is that we should use a normative justification of the basic competences required for a preferable type of life and of the kind of harmful impairment of these competences that should ground the status of mental disoder. Following the Kantian tradition, we argue that this justification should offer reasons that all rational persons can accept when exercising their rationality. This normative justification is important because we argue that mental disorders are not merely the objects of sophisticated human curiosity, as might be the posits of theoretical science (cf. Boorse, 2014). Their individuation is motivated by the need to prove the legitimacy of treatment and the rights to be treated and receive social support.

Another of our more substantive claims is that empirical evidence suggests that individuals with psychopathy are affected by very contextual incapacities or limited capacities that might explain their impulsivity. Although the ecological significance of these results should be investigated further, we cautiously suggest that these incapacities might be, within the framework that we advance, a ground for regarding psychopathy as a mental disorder.

In the chapter we proceed as follows. First, we briefly introduce the notion of psychopathy as measured by the *Psychopathy Checklist Revised* PCL-R (Hare, 2003), that is one of the most widely used measures of this construct. We briefly survey some dimensions of the debate on whether this measure shows that psychopathy is a unitary construct. In the third section, we argue that being harmful to the individual is a necessary requirement for a mental disorder. In addition, we claim that individuals with psychopathy are harmed by the societal responses to their characteristic behaviours. In the next section, we argue that these harms, insofar they derive from deviations from morally justifiable social standards, might be relevant for assigning mental disorder status. However, as we show in the fifth section, these deviations can ground such a status only if they derive from what we call *basic* incapacities to lead a preferable form of life. We claim that the individuation of these incapacities is a normative task. In the sixth section, we argue that this task is to be achieved by means of rational considerations that are not sectarian, that is, that do not impose illegitimately the values of certain individuals on others. Specifically, we conclude with a normative justification of this type that the capacities required for legal or moral accountability are basic in the relevant sense. With a brief review of the relevant empirical literature, we finally address the issue whether individuals with psychopathy suffer incapacities of this type. Although psychopaths do not have incapacities that should exculpate them for their criminal or antisocial behaviour, there is evidence of incapacities, to be further investigated, that might ground the mental disorder status of their condition.

16.2   The Unity of Psychopathy

Robert Hare's Psychopathy Checklist-Revised (PCL-R) (Hare, 2003) is surely a prominent measure in the contemporary scientific study of psychopathy (Patrick, 2018), that is also employed in forensic psychiatry (DeLisi, 2016). Thus, several studies concerning the behavioural, functional, neurocognitive, neuroanatomy and genetics of psychopathy, as its significance for crime management, use it. However, it is important to remark that it is not the only measure of psychopathy (Fowler & Lilienfeld, 2013). Moreover, a measure of a construct should not be confused with the construct itself (Cooke, 2018).

The PCL-R is used to evaluate a subject on 20 items: (1) glib/superficial charm, (2) grandiose sense of self-worth, (3) need for stimulation/proneness to boredom, (4) pathological lying, (5) conning/manipulativeness, (6) lack of remorse or guilt, (7) shallow affect, (8) callous/lack of empathy, (9) parasitic lifestyle, (10) poor behavioural controls, (11) promiscuous sexual behaviour, (12) early behavioural problems, (13) lack of realistic long-terms goals, (14) impulsivity, (15) irresponsibility, (16) failure to accept responsibility for one's own actions, (17) many short-term marital relationships, (18) juvenile delinquency, (19) revocation of conditional release, and (20) criminal versatility. The PCL-R is applied *via* semi-structured interviews and extensive study of the history of the subject. For each element in the list, there is a score ranging from 0 to 2 points; the largest total score is thus 40 points. When a subject scores 30 (25 in some studies) or more points he/she is considered psychopathic.

Establishing the disorder status of psychopathy requires investigating whether the current measures of this condition individuate a unitary class. This means that individuals with psychopathy should share similarities that support some degree of explanation of their behaviours, mental states, and traits. Moreover, and more importantly from a practical perspective, these similarities should enable effective treatment and relevant predictions.

Some authors argue that the construct of antisocial personality in general (Agich, 1994; Charland, 2004, 2006) or psychopathy (Jalava et al., 2015, Appendix A) are value-laden. This because they involve some irreducible moral evaluations that concern preferable personality traits and behaviours. This is a plausible conclusion that is confirmed by the items of PCL-R given above (Hare, 2003). Individuals with psychopathy stand out from the others, for example, insofar they violate the law, exploit, manipulate, and lie to others. All these behaviours are condemned from moral or legal standpoints. As Marga Reimer states perspicuously, it is obvious that psychopathy involves a "moral pathology" (Reimer, 2008b, p. 187).

Some think that the value-ladenness of psychopathy shows that it is not unified as a scientific or medical category should be (Cavadino, 1998; Gunn, 1998; Mullen, 2007). As Michael Cavadino puts it, the construct of psychopathy exemplifies a case "of moralism masquerading as medical science". Thus, he recommends that "[p]erhaps we should strip away the mask completely, and for the term 'psychopath' substitute the word 'bastard'" (Cavadino, 1998). If these criticisms are sound, individuals with psychopathy, besides contravening certain shared moral and cultural norms, would have nothing in common that is relevant for scientific study, and a possible ground for treatment or predictions. This would also imply that there is not much point in asking whether psychopathy is a mental disorder.

In any case, it seems that the value-ladenness of psychopathy does not imply, by itself, a lack of the unity that is needed in theoretical or clinical contexts. In fact, others argue that empirical evidence validates the diagnosis of psychopathy, by showing that it correlates significantly with characteristic phenomena (Hare, 2003). Behavioural experiments in controlled conditions show that individuals with psychopathy perform characteristically with experimental paradigms aimed at measuring emotional responses, instrumental learning, and decision making (Blair, 2013; Blair et al., 2005; Brazil & Cima, 2016; Patrick, 2018). In addition, it seems that high scores on PCL-R individuate a class of individuals that are significantly more likely to reoffend than controls (Leistico et al., 2008; Wallinius et al., 2012). There are also empirical studies suggesting that individuals with psychopathy manifest anatomical and physiological peculiarities in their prefrontal cortex, especially in the orbitofrontal and ventrolateral cortex, the hippocampus, and the amygdala (for a survey, see Brazil & Cima, 2016; Poeppl et al., 2019 is a recent meta-analysis).

The significance of these empirical results for the unity of the construct of psychopathy, or the validity of its measures, depends, of course, also on the notion of unity adopted. This is a complex theoretical issue that is dealt in psychometrics, with different notions of validity (see Sellbom et al., this volume), but that has also sparked discussion in philosophy of science, in terms of natural or other kinds (Brzović & Šustar, this volume; see also Haslam, 2014). For instance, some, although they recognise that psychopathy is not unified at the level of underlying neurological, genetic, and cognitive mechanisms, think that the evidence supports the conclusion that psychopathy, as measured by PCL-R, is pragmatically unitary enough for forensic or other practical uses (see Brzović et al., 2017; Maibom, 2018; Malatesti & McMillan, 2014a; Reydon, this volume). Others, however, claim that PCL-R lacks the predictive powers that motivated its adoption in forensic settings (Larsen et al., 2020). In addition, as for other syndrome based categorisations in psychiatry that do not refer to aetiology, there are proposals to revise the current category of psychopathy and improve its usefulness for treatment with bio-cognitive data (see Brazil et al., 2018; Jurjako et al., 2020).

Even if the current classification of psychopathy characterises a class of individuals that is significant enough for theory or practice, the question whether psychopathy is a mental disorder is left unanswered. The answer depends on the notion of mental disorder that is adopted. Interestingly, experts who are adamant on their support of the validity of the category of psychopathy, have different views on its mental disorder status. For instance, Robert Hare, regards individuals with psychopathy as different from other individuals but not as mentally disordered. By assuming that a mental disorder should involve some evolutionary biological dysfunction, he argues that psychopathy is not a pathology because it results from an evolutionarily adaptive strategy (Hare, 2013, p. vii). On the other hand, the neuropsychologist James Blair, talks of "disordered" or "impaired" cognitive and neurological mechanisms that explain or correlate with psychopathy (Blair, 2013). This might be taken to suggest that he, with the other scientists that endorse this type of language, regards the condition as disordered or pathological (Reimer, 2008b). We have, thus, to turn to the task of accounting for the notion of mental disorder that will inform the remainder of our investigation.

16.3   Harm as a Necessary Condition for Disorder

There is some consensus that at least the practical notion of mental disorder, to be used in clinical and in other applied contexts, might include harm to the subject as a necessary requirement (Kingma, 2014). In fact, several authors have argued that a medical disorder is a condition that, at least as a necessary requirement, must negatively affect the individual (Bolton, 2008; Cooper, 2002; Glover, 1970; Graham, 2013; Reznek, 1987; Wakefield, 1992, 2014; Wakefield & Conrad, 2020). The plausible reason for associating harm to the notion of disorder is to capture a central aspect of the medical practice that is directed at alleviating or remedying the harmfulness of disorders (Cooper, 2013, 2020; Glover, 1970, pp. 120–124). In addition, even those who defend a completely naturalist account of disorder, understood as a theoretical notion that is used by pathologists and other scientists for research purposes, juxtapose to it practical notions, that are adequate for clinical and other applicative contexts, that involve harm as a relevant criterion (Boorse, 2014).

The notion of harm that is relevant for that of disorder clearly needs to be clarified. Minimally, a condition to be harmful to its bearer must be unpleasant or undesirable or must involve a deprivation of something that is pleasant or desirable. But the notion of harm needs to be more specific than this. Some authors have claimed that which type of harm is relevant for mental disorder should be judged based on the culture of the person (Wakefield, 1992, p. 384). However, this might lead to a problematic relativism. Different societies might have different views on what constitutes harm, and some of these views might not be morally justified. Thus, it has been objected that relating harm to the values of specific cultures implies implausibly that someone might be cured of a mental disorder simply by moving to a society with a different culture (Gert & Culver, 2004, p. 420).

Gert and Culver have offered an account to individuate the kind of harm relevant for the notion of mental disorder. The core tenet of their position is that:

> for the reactions to others to any deviance, either physical or mental, to make a deviant condition to count as a disorder, the reactions must be universal human responses, not merely the response of those in a particular society (Gert & Culver, 2004, p. 423)

They argue, in fact, that any rational individual in any society would recognise, absent special reasons to think otherwise, "death, pain, disability, and loss of freedom as harms" (Gert & Culver, 2004, p. 420). They think that such a list of universal harms, that were recognised in definition of disorder in the DSM-IV (American Psychiatric Association, 1994), can stand the challenge of relativism.

We think that Gert and Culver's account involves a valuable insight. This is the view that harm in this context might be involving distress or disability or loss of freedom. By distress we can think of pain or other unpleasant experiences. Disability or loss of freedom, instead, relate to interferences with capacities that are needed for living a satisfactory and rewarding life. Let us therefore move to consider whether and how psychopathy is a harmful condition for the subject.

Individuals with psychopathy are harmed differently by their condition, but paramount are the harms due to the response of society to their typical behaviours and personality traits (Nadelhoffer & Sinnott-Armstrong, 2013). The antisociality associated with the condition, when issuing in criminal behaviour, might bring them in contact with the law and lead to incarceration and other forms of punishment and restriction. These are surely a harmful consequences of the condition, given that restriction of freedom is undesirable. Similarly, their interpersonal style, characterised by manipulativeness, lying and exploitation of others, can harm them by alienating others. Finally, especially with reference to so-called unsuccessful individuals with psychopathy, their impulsivity might eventually cause failures in education and work, insofar they conflict with the norms that regulate the relevant social institutions. However, we must establish whether the harms caused by the societal response to the deviance of individuals with psychopathy are relevant for ascribing disorder status.

It might be objected that harmful societal responses are not legitimate components of a mental disorder. In fact, some have argued that disorder is a source of harm *internal* to the individual (Boorse, 1977; Gert & Culver, 2004; Wakefield, 1992, 2014). An ethical consideration for such an "internality" is that society, with the support of the medical or psychiatric establishment, has seriously harmed certain groups of individuals just for their deviant behaviour.[1] Let us consider, for instance, the cases of running slaves (Willoughby, 2018), homosexuals (Bayer, 1987), and political dissidents (van Voren, 2009). Clearly the proper response in these cases is that there was something wrong in the social practices and norms, not in these individuals. Associating psychiatrically relevant harm to societal harmful responses risks supporting repressive societies and psychiatric abuses.

It seems, thus, that proving whether individuals with psychopathy are harmed by their condition as in a mental disorder, requires investigating whether the problem is with them or with society. In the next section, we show that certain harms caused by social responses to individuals with psychopathy can be legitimate necessary criteria for a mental disorder. These harmful societal responses, however, must be properly qualified.

16.4   Situating the Problem

An important insight is that the decision whether the source of harms is "internal" to the person depends also on the normative decision about whether it is more reasonable to change society than the individual. Recently, this point has been forcefully advanced by Rachel Cooper:

> I suggest that whether we count a problem as an internally located disorder or as an externally located environmental problem, depends on whether we think it best to attempt to ameliorate the situation by altering the individual or the environment. This depends on what types of intervention might be possible, but also on whether we think that any possible environmental accommodations are reasonable or not. Determining which environmental adjustments would be reasonable depends on a range of considerations—practical and economic, but also ethical and political (Cooper, 2020, p. 157).

Thus, showing whether psychopathy involves harm that is relevant for disorder, requires finding more compelling reasons for changing individuals with psychopathy as opposed to changing society to reduce the harm associated with their condition. We are left with the problem of establishing whether there are such reasons.

At least some traits in the characterisation of psychopathy deviate from widely and firmly accepted standards. Let us consider the laws or social practices which

---

[1] Another requirement of internality is explanatory; it is usually formulated by requiring the presence in the subject of a disordered cause, dysfunction, impairment, or incapacity. We consider this issue in Sect. 16.5.

support punishment or forms of social restraint of individuals with psychopathy who significantly hurt others. These norms are more solidly endorsed than those that prescribe harmful social responses to running slaves, homosexuals, or political dissidents. While all societies enforce, in some form, the former norms, all constitutional democracies firmly condemn the latter ones.

For other traits associated with psychopathy, as, for instance, the lack of or diminished empathy or the grandiose sense of their self that associate to antisociality, we might have some prudential reasons for regarding them as problems with the psychopath. In fact, insofar they do not issue in criminal behaviour, there might be no explicit normative reasons, as the codified laws, for finding the problem within the psychopath. In addition, these features might be even helpful to the individuals with psychopathy. However, these features, insofar issue in behaviour that lack consideration for the wellbeing of others, might attract justified forms of harm, as social exclusion, in contexts where they cannot be legally sanctioned (as in friendship, sentimental relationship, and so on). Thus, we might have reasons for thinking that also these traits attract harm because of a problem within the psychopath (we argue for this conclusion, although in a different context, in Baccarini & Malatesti, 2017). Thus, also this kind of harm, in accordance with our proposal, might be relevant for determining that a condition is a mental disorder.

A condition that is harmful due to justifiable principles, even if it is unitary in a significant sense, is not, however, a mental disorder. People, who might even share some significant psychological or even neural characteristics, expose themselves, by engaging in chosen recreational activities or professions, to considerable risks and harms. Consider extreme sports or engaging in risky legitimate financial ventures. In these cases, there are no practical, legal, or ethical reasons that recommend changing the relevant physical or social environment to a point to render the activity completely harmless. Closer to the case of psychopathy, many criminals or antisocial individuals are taken to be legitimate targets of harmful societal responses, such as the restriction of freedom. But they are not regarded as disordered and in need of a cure. Thus, it seems that a further ingredient is needed to establish that psychopathy, if unitary and justifiably harmful, amounts to a disorder. We investigate it in the next section.

16.5 Clinically Significant Harmful Incapacities

According to an influential view, the presence of a dysfunction in the subject is a necessary requirement for ascribing him or her a disorder. Some have argued that this must be a biological dysfunction (Boorse, 1997; Wakefield, 1992). In DSM IV, amongst the explicit requirements for being a disorder, besides biological dysfunctions, also psychological or behavioural dysfunction apper as conditions for having a disorder (Stein et al., 2010). Similarly, the general definition of mental disorder in DSM 5 refers to a notion of dysfunction (Amoretti & Lalumera, 2019). Leaving aside the important differences on the notion of dysfunction in these accounts, let us focus on their underlying motivations.

The notion of dysfunction satisfies several important desiderata for that of disorder. First, it allows distinguishing mere departures from certain behavioural or psychological standards, even the justified ones, from symptoms of an underlying problem or cause. Second, dysfunction offers an objective (or less value-laden) component of the notion of disorder as opposed to that of harm. According to some, the notion of dysfunction represents, within a disorder, a proper target for scientific study (Boorse, 2014; Wakefield, 1992). Moreover, invoking the notion of dysfunction points to a problem within the individual that arises, completely or partly, beyond her control. Moreover, this problem cannot be dealt with by her ordinary resources and, thus, requires a special medical intervention (Fulford, 1989).

Recently Jerome Wakefield's account of mental disorder, as a harmful biological dysfunction, has been adopted to assess the disorder status of psychopathy. It has thus been argued that psychopathy is a disorder insofar it is a harmful and dysfunctional condition, in the terms of an adaptationist view of biological function (Nadelhoffer & Sinnott-Armstrong, 2013). However, others claim that psychopathy is not biologically dysfunctional and, thus, is not a mental disorder (Hare, 2013; Jurjako, 2019; Krupp et al., 2013; Lalumière et al., 2001). Thus, whether psychopathy is a disorder is hostage to whether it is a biologically dysfunctional condition. However, some considerations discourage setting the problem in these terms.

Some authors have maintained, convincingly, that the current speculative nature of evolutionary psychiatric explanations renders them unpractical for the pressing practical decisions that rely on the notion of mental disorder (Bolton, 2008, pp. 160–161; Graham, 2013, pp. 121–126). Some have even criticised the theoretical relevance of the notion of biological function and dysfunction in capturing that of mental disorder (Cooper, 2007, pp. 33–35). Without dismissing the theoretical importance of debates on the biological dysfunctionality of psychopathy and their relevance for deciding its disorder status, we suggest investigating another route to establish whether the condition is a disorder or a mere harmful deviance from justified societal norms.

The disorder status of psychopathy can be investigated by figuring out whether certain relevant incapacities or diminished capacities, that would set psychopathic offenders apart from other criminals, associate with this condition. The notion of incapacity here would point, as that of dysfunction, to some objective cause or impairment within the individuals with psychopathy that is completely or partly beyond their control. If an individual is harmed by the incapacity or a limited capacity to align to societally justified standards that apply to her, then we have a *prima facie* ground for thinking that she has a disorder.

It is important, though, to recognise that not all the harmful incapacities characterise mental disorders. George Graham, for instance, considers:

> being incapable of climbing a mountain in the winter or wrestling with a bear in the woods and this may cause losses for a person (and risk death) if they are in a situation in which they must do these things but can't. (Graham, 2013, p. 115)

Clearly, we need a narrower notion of incapacity that has some serious clinical significance. We agree with Graham (2013) that the incapacities relevant for disorder affect *basic* competencies, that we can justifiably regard as necessary for a preferable type of life. In the next section, we tackle the issue of proving what are the basic harmful incapacities that, at least in the case of psychopathy, are relevant for establishing that this condition is a mental disorder.

16.6   A Philosophical Justification

Some philosophers of psychiatry have correctly recognised that an explicit normative justification is needed to establish the basic capacities whose harmful impairment is involved in mental disorders (Graham, 2013; Megone, 1998, 2000). In fact, intuitions shared in a certain culture or by most people cannot be taken as a fair ground. As in the case of harm, the cultural and personal variability of intuitions might lead to relativism. Moreover, being our focus on the practical dimensions of the notion of disorder, intuitions might lead to *sectarian* forms of repression, consisting of imposing the values of a specific group on others. In fact, some people might legitimately depart or dissent from common evaluative standards. Therefore, individuals should be protected from unfair medicalisation and the related social responses that can severely affect their lives. A justification of evaluative standards that are protective of all is the firmer normative ground to avoid relativism and sectarianism.

Christopher Megone has offered a philosophical account of the incapacities relevant for psychiatry that is based on an Aristotelian metaphysical view (1998, 2000). According to this approach, each kind of organism flourishes when it functions, that is, it exists and develops, by following its characteristic life cycle. A "good organism" is thus equipped with the biological and psychological features that allow it to flourish. Megone, with Aristotle, thinks that: "The human function is ... the life of the fully rational animal" (Megone, 1998, p. 56). Human beings to flourish need, amongst other things, the capacity of reason. This capacity, for instance, is exercised in acquiring beliefs rationally and acting rationally upon them. Mental disorders are conditions that involve disabilities and impairments that interfere with the flourishing of human beings, insofar they negatively affect the development and exercise of reason.

Shane N. Glackin (2015) has argued convincingly that Megone's justification is not respectful of the pluralism in the democratic liberal societies. In our terminology, Megone's approach is sectarian. In a democratic liberal society, in fact, people might have different views on the standards relevant for judging, especially when this is relevant to their lives, what should count as disease, disability, or disfigurement. Megone's Aristotelian proposal fails because its allegiance to the:

> teleological notion of the human person[...] sets a standard for what is to be counted as 'fully' human – as fully realising the ergon, or function, appropriate to humans as such – with which some humans may have legitimate grounds for disagreement (Glackin, 2015, 9).

We think that this criticism finds an appropriate support in the later political philosophy of John Rawls (2005). He stressed that a free society respects the reasonable pluralism because people, due to the free use of reason, diverge on their evaluative judgments.

We suggest, thus, to use a normative justification of the basic capacities for the notion of disorder based on Rawls's interpretation of Immanuel Kant (Rawls, 2000, pp. 143–325).[2] In this perspective, as Rawls explains, the appropriate reasoning procedure that establishes evaluative standards should be followed freely and autonomously by all the agents who exercise their rationality. In this context, we extend this Kantian method to establish the preferable form of life and the basic capacities that it demands.[3] The central requirement here is that decisions that derive from this method must be fair to the capacities and perspective of each person. It is important, however, to recognise that in all societies there are capacities that are particularly rewarded, and, thus, provide more advantages to their possessors (and others that cause analogous disadvantages). The procedure, thus, does not aim at justifying a society where capacities do not confer certain advantages; its point is to rule out unfair advantages.

Adopting this type of justificatory procedure has important consequences for our present discussion. First, it excludes relativism, and offers an objective ground, insofar it recognises the role of consensus among all rational individuals in establishing evaluative standards (Rawls, 2000, pp. 243–247). Second, it is not sectarian, given that the evaluative standards do not depend on peculiar worldviews or perspectives that are alien to some individuals. This justification is thus a safeguard against repressive or discriminatory psychiatric classifications and practices (Baccarini & Lekić Barunčić, 2018). In fact, all individuals in their exercise of rationality should determine the basic incapacities that are relevant for mental disorder. Therefore, the model offers a justification of evaluative standards that protects the autonomy of agents, which is particularly important for the clinical practice in a free society. Let us illustrate the procedure by considering two cases.

The justification should offer good reasons to all the individuals that exercise their rationality and those whose incapacities are at issue. Let us consider, for instance, the proposal that extreme aggressivity is a prerequisite for a preferable life for all individuals. Then, those individuals who are more empathic and less aggressive than recommended are declared as suffering of a basic incapacity that needs

---

[2] Rawls has endorsed the Kantian frame in his political philosophy. In later work, however, he explicitly reshaped it and disburdened it from its metaphysical weight (Rawls, 2005).

[3] Graham (2013) justifies his selection of basic capacities by relying as well on Rawls's work. But his justification is substantially different from ours, because he uses and adaptation of Rawls's celebrated "original position" thought experiment (Rawls, 1971). Grahams argues, thus, that the basic capacities are those that would be selected by an idealised rational person who does not know her psychological capacities and other characteristics. We, instead, select the basic capacities with a justification that would be acceptable, with a free use of reason, also by the individuals who lack or might lack them. Space limitations do not allow to explain why we think that our justification is preferable to Graham's. We have done so in talks at conferences (Baccarini & Malatesti, 2019), we hope to publish soon our discussion.

cure and certain forms of social exclusion. This is not justified, because there are no valid reasons that can be offered to the individuals who oppose to be medicalised and, because of this, to be discriminated. The sectarian view that extreme aggressivity is a prerequisite for a preferable life cannot be justified to them. On the other hand, consider declaring as basic the capacities for literacy needed to have a preferable life in our society. This is not sectarian and discriminatory towards those who lack or are impaired in these capacities. In fact, we can justify to them the society that adopts these basic capacities. Such a society is rationally justified to all its members, if it fairly distributes its advantages to all. In fact, thanks to literacy, societies have achieved extraordinary political and technical improvements, that increase the quality of life for all their members. The incapacities to read and write are thus basic and needed for a preferable form of life.

In assessing the disorder status of psychopathy, we should consider whether individuals with psychopathy lack or are impaired in capacities assumed to be relevant for legal and moral responsibility (this strategy is suggested in Malatesti & McMillan, 2014b). Broadly speaking, these are capacities needed to appreciate moral or legal considerations concerning what is or is not permissible and being motivated by them to act or refrain from acting (Fischer & Ravizza, 2000; Yannoulidis, 2012). Impairments of this type, in fact, would satisfy the requirements that we have set out so far for the notion of mental disorder. First, these incapacities would cause, by explaining the antisocial and criminal behaviour of individuals with psychopathy, the justified harmful societal responses. Second, these incapacities would completely or partly exculpate them for this behaviour, to setting them apart from the ordinary offenders. Finally, let us see how we can justify these capacities as basic.

In general, to function legal systems requires *de facto* that people can understand the law and act consequently (Robinson, 2000). In societies that require some basic capacities for legal and moral responsibility, people who lack them are relatively disadvantaged. However, despite these negative effects, we can justify to them, let us call them the *addressees* of the justification, that these capacities are basic. A society with such a requirement optimizes social cooperation and its benefits, and, thus, can be advantageous to all. We can justify this to the addressees by showing that the only two alternative scenarios are not rationally justifiable. Let us consider, in fact, a society where the capabilities for moral understanding and control are not considered preferable for anyone, and thus are not encouraged, developed, and valued. In this society, everyone, including the addressees, is at a serious risk of being harmed and treated unfairly. Alternatively, let us consider a scenario where the addressees are an exception, that is the capabilities for responsibility are not basic for them. Now, all other individuals in that society have no reason to accept such asymmetry. Thus, the type of justification that we recommend would fail short. Having minimal moral capacities is, thus, a basic condition for being included in a social cooperation. Thus, the issue of the disorder status of psychopathy could be related with that of the moral and legal responsibility of psychopathic offenders, that has been widely investigated (Kiehl & Sinnott-Armstrong, 2013; Malatesti & McMillan, 2010).

Experts, however, are divided on the issue whether there is a strong correlation between psychopathy and incapacities that would, in general, exculpate individuals with psychopathy for their antisociality or crimes. Some authors argue for such a strong claim in relation to moral responsibility (Kennett, 2010; Levy, 2007; Malatesti, 2010) or for the legal one (Fine & Kennett, 2004; Gillett & Huang, 2013; Litton, 2008, 2013; Morse, 2008). Recent reviews of the empirical literature have concluded, however, that individuals with psychopathy are not so incapacitated to be legally or morally exculpated (Jalava & Griffiths, 2017; Jefferson & Sifferd, 2018; Jurjako & Malatesti, 2018; Maibom, 2008). For example, evidence for the claim that individuals with psychopathy lack moral understanding, a key capacity required for accountability, is in some studies by James Blair and collaborators (Blair, 1995, 1997). Not all recent experimental studies, however, have replicated these findings (Aharoni et al., 2012, 2014). Moreover, some studies challenge the idea that psychopathic individuals have profound moral incapacities (Borg & Sinnott-Armstrong, 2013; Cima et al., 2010; Glenn et al., 2009; Larsen et al., 2020; Marshall et al., 2018). Similarly, it has been argued that results concerning the peculiarities in psychopaths concerning reward and punishment processing, that impact on instrumental learning and decision making (Glimmerveen et al., this volume; Koenigs & Newman, 2013) do not show generalized incapacities across possible real-life scenarios that could support, in general, exculpation (Jurjako & Malatesti, 2018). The results, in fact, depend on contextual factors specific to the experiments (Baskin-Sommers et al., 2015; Brazil et al., 2013; Hamilton et al., 2015; Koenigs & Newman, 2013).

However, we can hypothesise that these peculiarities in reward and punishment processing might affect basic competences in harmful ways to individuals with psychopathy. They could have a role in explaining certain criminal or antisocial behaviours, at least in specific circumstances, and thus attract the justified harmful reactions discussed above. In addition, these peculiarities might harm the psychopath if they negatively affect their legitimate life plans, as keeping a job, being educated, having a stable sentimental and social life, and so on. A society that rewards learning capacities and the capacity to control impulsive behaviour can have wide and general advantages for all, because these are conditions for stable and well-ordered social cooperation. The standards that reward such capacities are, thus, properly justified within our framework. However, the hypothesis that these peculiarities in reinforcement learning ground the disorder status of psychopathy is hostage to empirical research. In particular, the ecological significance of these impairments in real life, as opposed to experimental circumstances, and their role in the antisocial and criminal behaviour of individuals with psychopathy, as the cognitive processes underlying them, needs to be investigated further (Glimmerveen et al., 2018, this volume).

16.7   Conclusion

Establishing whether psychopathy is a mental disorder involves addressing some complex conceptual and empirical issue. By relying on what we take to be, in this case, a plausible theoretical proposal concerning the notion of mental disorder, we have disentangled some of the most fundamental issues. We presented these problems as centring around the following concepts of: unity, harm, normative justification of harm, and normative justification of basic capacities for having a preferable type of life.

Each of these notions needs clarification and generates theorical, normative and empirical problems. Establishing in which sense a disorder is a unitary construct involves balancing theoretical aspirations and practical clinical realities and needs. Further, the technical problem is open to establish whether psychopathy, as currently measured, satisfies this account of unity. Similarly, assuming that harm is a component of mental disorder, we are left with the problem of specifying this notion. We have suggested that, at least in the case of psychopathy, a principal source of harm for people with this condition appears to be the societal response to their antisocial and criminal behaviour, but also to features of their personality.

The empirical evidence does not point to a strong correlation between psychopathy and severe incapacities that would undermine legal or moral accountability. However, individuals with psychopathy manifest peculiarities in instrumental learning and decision making that can have a role, in certain contexts, in their antisociality and impulsivity that attract justified harm. These peculiarities thus, might amount to the clinically significant incapacities of a mental disorder. However, even if our idealised procedure of justification of the relevance of these incapacities for mental disorder is accepted, to firmly support the hypothesis that psychopaths suffer them, further empirical research is needed. In any case, we hope that we have offered a plausible and useful framework for investigating whether psychopathy is a mental disorder.

Acknowledgements We are grateful to the interdisciplinary international "coalition" formed by Cristina Amoretti, Inti A. Brazil, Marko Jurjako, John McMillan, and Elisabetta Lalumera for extremely useful comments on previous versions of this chapter. This work has been supported in part by the University of Rijeka under the project number uniri-human-18-151. LM's preliminary work on this chapter was an outcome of the project *Classification and explanations of antisocial personality disorder and moral and legal responsibility in the context of the Croatian mental health and care law* (CEASCRO) (2014–2018, Croatian Science Foundation, HRZZ-IP-2013-11-8071). LM and EB's further work on the chapter is an outcome of the project *Responding to antisocial personalities in a democratic society* (RAD) (Croatian Science Foundation, HRZZ-IP-2018-01-3518).

References

Agich, G. J. (1994). Evaluative judgment and personality disorder. In J. Z. Sadler, O. P. Wiggings, & M. A. Schwartz (Eds.), *Philosophical perspectives on psychiatric diagnostic classification* (pp. 233–245). Johns Hopkins University Press.

Aharoni, E., Sinnott-Armstrong, W., & Kiehl, K. A. (2012). Can psychopathic offenders discern moral wrongs? A new look at the moral/conventional distinction. *Journal of Abnormal Psychology, 121*(2), 484–497. https://doi.org/10.1037/a0024796

Aharoni, E., Sinnott-Armstrong, W., & Kiehl, K. A. (2014). What's wrong? Moral understanding in psychopathic offenders. *Journal of Research in Personality, 53*, 175–181. https://doi.org/10.1016/j.jrp.2014.10.002

American Psychiatric Association. (1994). *Diagnostic and statistical manual of mental disorders: DSM-IV* (4th ed.). American Psychiatric Association.

Amoretti, M. C., & Lalumera, E. (2019). A potential tension in DSM-5: The general definition of mental disorder versus some specific diagnostic criteria. *The Journal of Medicine and Philosophy, 44*(1), 85–108. https://doi.org/10.1093/jmp/jhy001

Baccarini, E., & Lekić Barunčić, K. (2018). Parental selecting and autism. *Etica & politica = Ethics and politics, 20*(1), 21–34. https://doi.org/10.13137/1825-5167/22583

Baccarini, E., & Malatesti, L. (2017). The moral bioenhancement of psychopaths. *Journal of Medical Ethics, 43*(10), 697–701. https://doi.org/10.1136/medethics-2016-103537

Baccarini, E., & Malatesti, L. (2019, October 21). *A normative justification of the standards involved in the construct of psychopathy*. The many faces of personality disorder. An Interdisciplinary Conference of the Understanding Personality Disorders Network, Institute of Applied Psychology, Jagiellonian University, Kraków (Poland).

Baskin-Sommers, A. R., Brazil, I. A., Ryan, J., Kohlenberg, N. J., Neumann, C. S., & Newman, J. P. (2015). Mapping the association of global executive functioning onto diverse measures of psychopathic traits. *Personality Disorders, 6*(4), 336–346. https://doi.org/10.1037/per0000125

Bayer, R. (1987). *Homosexuality and American psychiatry: The politics of diagnosis*. Princeton University Press.

Blair, R. J. R. (1995). A cognitive developmental approach to morality: Investigating the psychopath. *Cognition, 57*(1), 1–29. https://doi.org/10.1016/0010-0277(95)00676-P

Blair, R. J. R. (1997). Moral reasoning and the child with psychopathic tendencies. *Personality and Individual Differences, 22*(5), 731–739. https://doi.org/10.1016/S0191-8869(96)00249-8

Blair, R. J. R. (2013). Psychopathy: Cognitive and neural dysfunction. *Dialogues in Clinical Neuroscience, 15*(2), 181–190.

Blair, R. J. R., Mitchell, D. R., & Blair, K. (2005). *The psychopath: Emotion and the brain*. Blackwell.

Bolton, D. (2008). *What is mental disorder? An essay in philosophy, science, and values*. Oxford University Press.

Boorse, C. (1977). Health as a theoretical concept. *Philosophy of Science, 44*(4), 542–573. https://doi.org/10.1086/288768

Boorse, C. (1997). A rebuttal on health. In J. M. Humber & R. F. Almeder (Eds.), *What is disease?* (pp. 1–134). Humana Press. https://doi.org/10.1007/978-1-59259-451-1_1

Boorse, C. (2014). A second rebuttal on health. *The Journal of Medicine and Philosophy: A Forum for Bioethics and Philosophy of Medicine, 39*(6), 683–724. https://doi.org/10.1093/jmp/jhu035

Borg, J. S., & Sinnott-Armstrong, W. (2013). Do psychopaths make moral judgments? In K. A. Kiehl & W. Sinnott-Armstrong (Eds.), *Handbook on psychopathy and law* (pp. 107–128). Oxford University Press.

Brazil, I. A., & Cima, M. (2016). Contemporary approaches to psychopathy. In M. Cima (Ed.), *The handbook of forensic psychopathology and treatment* (pp. 206–226). Routledge.

Brazil, I. A., Maes, J., Scheper, I., Bulten, B., Kessels, R., Verkes, R., & de Bruijn, E. (2013). Reversal deficits in individuals with psychopathy in explicit but not implicit learning conditions. *Journal of Psychiatry & Neuroscience, 38*(4), E13–E20. https://doi.org/10.1503/jpn.120152

Brazil, I. A., van Dongen, J. D. M., Maes, J. H. R., Mars, R. B., & Baskin-Sommers, A. R. (2018). Classification and treatment of antisocial individuals: From behavior to biocognition. *Neuroscience & Biobehavioral Reviews, 91*, 259–277. https://doi.org/10.1016/j.neubiorev.2016.10.010

Brzović, Z., Jurjako, M., & Šustar, P. (2017). The kindness of psychopaths. *International Studies in the Philosophy of Science, 31*(2), 189–211. https://doi.org/10.1080/02698595.2018.1424761

Brzović, Z., & Šustar, P. (this volume). In fieri kinds: The case of psychopathy. In L. Malatesti, J. McMillan, & P. Šustar (Eds.), *Psychopathy. Its uses, validity, and status*. Springer.

Cavadino, M. (1998). Death to the psychopath. *The Journal of Forensic Psychiatry, 9*(1), 5–8. https://doi.org/10.1080/09585189808402175

Charland, L. C. (2004). Character: Moral treatment and the personality disorders. In J. Radden (Ed.), *The philosophy of psychiatry: A companion* (pp. 64–77). Oxford University Press.

Charland, L. C. (2006). Moral nature of the DSM-IV cluster B personality disorders. *Journal of Personality Disorders, 20*(2), 116–125; discussion 181–185. https://doi.org/10.1521/pedi.2006.20.2.116

Cima, M., Tonnaer, F., & Hauser, M. D. (2010). Psychopaths know right from wrong but don't care. *Social Cognitive and Affective Neuroscience, 5*(1), 59–67. https://doi.org/10.1093/scan/nsp051

Cooke, D. J. (2018). Psychopathic personality disorder: Capturing an elusive concept. *European Journal of Analytic Philosophy, 14*(1), 15–32. https://doi.org/10.31820/ejap.14.1.1

Cooper, R. V. (2002). Disease. *Studies in History and Philosophy of Science Part C: Studies in History and Philosophy of Biological and Biomedical Sciences, 33*(2), 263–282. https://doi.org/10.1016/S0039-3681(02)00018-3

Cooper, R. V. (2007). *Psychiatry and philosophy of science*. Acumen Publishing Limited.

Cooper, R. V. (2013). Avoiding false positives: Zones of rarity, the threshold problem, and the DSM clinical significance criterion. *The Canadian Journal of Psychiatry, 58*(11), 606–611. https://doi.org/10.1177/070674371305801105

Cooper, R. V. (2020). The concept of disorder revisited: Robustly value-laden despite change. *Aristotelian Society Supplementary Volume, 94*(1), 141–161. https://doi.org/10.1093/arisup/akaa010

DeLisi, M. (2016). *Psychopathy as unified theory of crime*. Palgrave Macmillan.

Fine, C., & Kennett, J. (2004). Mental impairment, moral understanding and criminal responsibility: Psychopathy and the purposes of punishment. *International Journal of Law and Psychiatry, 27*(5), 425–443. https://doi.org/10.1016/j.ijlp.2004.06.005

Fischer, J. M., & Ravizza, M. (2000). *Responsibility and control: A theory of moral responsibility* (1. paperback ed). Cambridge University Press.

Fowler, K. A., & Lilienfeld, S. O. (2013). Alternatives to psychopathy checklist-revised. In K. A. Kiehl & W. P. Sinnott-Armstrong (Eds.), *Handbook on psychopathy and law* (pp. 34–57). Oxford University Press.

Fulford, K. W. M. (1989). *Moral theory and medical practice*. Cambridge University Press.

Gert, B., & Culver, C. (2004). Defining mental disorder. In J. Radden (Ed.), *The philosophy of psychiatry: A companion* (pp. 415–425). Oxford University Press.

Gillett, G., & Huang, J. (2013). What we owe the psychopath: A neuroethical analysis. *AJOB Neuroscience*. https://doi.org/10.1080/21507740.2013.783647

Glackin, S. N. (2015). Three Aristotelian accounts of disease and disability. *Journal of Applied Philosophy, 33*(3), 311–326. https://doi.org/10.1111/japp.12114

Glenn, A. L., Iyer, R., Graham, J., Koleva, S., & Haidt, J. (2009). Are all types of morality compromised in psychopathy? *Journal of Personality Disorders, 23*(4), 384–398. https://doi.org/10.1521/pedi.2009.23.4.384

Glimmerveen, J. C., Brazil, I. A.., Bulten, B. H. (Erik), & Maes, J. H. R. (2018). Uncovering naturalistic rewards and their subjective value in forensic psychiatric patients. International Journal of Forensic Mental Health, 17(2), 154–166. https://doi.org/10.1080/14999013.2018.1452081.

Glimmerveen, J. C., Maes, J. H. R., & Brazil, I. A. (this volume). Psychopathy, maladaptive learning and risk-taking. In L. Malatesti, J. McMillan, & P. Šustar (Eds.), *Psychopathy. Its uses, validity, and status*. Springer.

Glover, J. (1970). *Responsibility*. Routledge and Kegan Paul.

Graham, G. (2013). *The disordered mind: An introduction to philosophy of mind and mental illness (second edition)*. Routledge.

Gunn, J. (1998). Psychopathy: An elusive concept with moral overtones. In T. Millon, E. Simonsen, M. Birket-Smith, & R. D. Davis (Eds.), *Psychopathy: Antisocial, criminal, and violent behavior* (pp. 32–39). Guilford Press.

Hamilton, R. K. B., Hiatt Racer, K., & Newman, J. P. (2015). Impaired integration in psychopathy: A unified theory of psychopathic dysfunction. *Psychological Review, 122*(4), 770–791. https://doi.org/10.1037/a0039703

Hare, R. D. (2003). *The Hare psychopathy checklist revised* (2nd ed.). Multi-Health Systems.

Hare, R. D. (2013). Foreword. In K. A. Kiehl & W. P. Sinnott-Armstrong (Eds.), *Handbook on psychopathy and law* (pp. vii–ix). Oxford University Press.

Haslam, N. (2014). Natural kinds in psychiatry: Conceptually implausible, empirically questionable, and stigmatizing. In H. Kincaid & J. A. Sullivan (Eds.), *Classifying psychopathology* (pp. 11–28). The MIT Press.

Jalava, J., & Griffiths, S. (2017). Philosophers on psychopaths: A cautionary in interdisciplinarity. *Philosophy, Psychiatry, and Psychology, 24*(1), 1–12.

Jalava, J., Griffiths, S., & Maraun, M. (2015). *The myth of the born criminal*. University of Toronto Press.

Jefferson, A., & Sifferd, K. (2018). Are psychopaths legally insane? *European Journal of Analytic Philosophy, 14*(1), 79–96. https://doi.org/10.31820/ejap.14.1.5

Jurjako, M. (2019). Is psychopathy a harmful dysfunction? *Biology and Philosophy, 34*(1), 5. https://doi.org/10.1007/s10539-018-9668-5

Jurjako, M., & Malatesti, L. (2018). Neuropsychology and the criminal responsibility of psychopaths: Reconsidering the evidence. *Erkenntnis, 83*(5), 1003–1025. https://doi.org/10.1007/s10670-017-9924-0

Jurjako, M., Malatesti, L., & Brazil, I. A. (2020). Biocognitive classification of antisocial individuals without explanatory reductionism. *Perspectives on Psychological Science, 15*(4), 957–972. https://doi.org/10.1177/1745691620904160

Kennett, J. (2010). Reasons, emotion, and moral judgment in the psychopath. In L. Malatesti & J. McMillan (Eds.), *Responsibility and psychopathy: Interfacing law, psychiatry, and philosophy* (pp. 243–259). Oxford University Press.

Kiehl, K. A., & Sinnott-Armstrong, W. P. (2013). *Handbook on psychopathy and law*. Oxford University Press.

Kingma, E. (2014). Naturalism about health and disease: Adding nuance for progress. *Journal of Medicine and Philosophy, 39*(6), 590–608. https://doi.org/10.1093/jmp/jhu037

Koenigs, M., & Newman, J. P. (2013). The decision-making impairment in psychopathy: Psychological and neurobiological mechanisms. In K. A. Kiehl & W. Sinnott-Armstrong (Eds.), *Handbook on psychopathy and law* (pp. 93–106). Oxford University Press.

Krupp, D. B., Sewall, L. A., Lalumière, M. L., Sheriff, C., & Harris, G. T. (2012). Nepotistic patterns of violent psychopathy: Evidence for adaptation? *Frontiers in Psychology, 3*. https://doi.org/10.3389/fpsyg.2012.00305

Krupp, D. B., Sewall, L. A., Lalumière, M. L., Sheriff, C., & Harris, G. T. (2013). Psychopathy, adaptation, and disorder. *Frontiers in Psychology, 4*. https://doi.org/10.3389/fpsyg.2013.00139

Lalumière, M. L., Harris, G. T., & Rice, M. E. (2001). Psychopathy and developmental instability. *Evolution and Human Behavior, 22*(2), 75–92. https://doi.org/10.1016/S1090-5138(00)00064-7

Larsen, R. R., Jalava, J., & Griffiths, S. (2020). Are psychopathy checklist (PCL) psychopaths dangerous, untreatable, and without conscience? A systematic review of the empirical evidence. *Psychology, Public Policy, and Law, 26*(3), 297–311. https://doi.org/10.1037/law0000239

Leedom, L. J., & Almas, L. H. (2012). Is psychopathy a disorder or an adaptation? *Frontiers in Psychology, 3*. https://doi.org/10.3389/fpsyg.2012.00549

Leistico, A., Salekin, R. T., DeCoster, J., & Rogers, R. (2008). A large-scale meta-analysis relating the Hare measures of psychopathy to antisocial conduct. *Law and Human Behavior, 32*, 28–45.

Levy, N. (2007). The responsibility of the psychopath revisited. *Philosophy, Psychiatry, & Psychology, 14*(2), 129–138. https://doi.org/10.1353/ppp.0.0003

Litton, P. (2008). Responsibility status of the psychopath: On moral reasoning and rational self-governance. *Rutgers Law Journal, 349*, 349–392.

Litton, P. (2013). Criminal responsibility and psychopathy: Do psychopaths have a right to excuse? In K. A. Kiehl & W. Sinnott-Armstrong (Eds.), *Handbook on psychopathy and law* (pp. 275–296). Oxford University Press.

Maibom, H. L. (2008). The mad, the bad, and the psychopath. *Neuroethics, 1*(3), 167–184. https://doi.org/10.1007/s12152-008-9013-9

Maibom, H. L. (2018). What can philosophers learn from psychopathy? *European Journal of Analytic Philosophy, 14*(1), 63–78.

Malatesti, L. (2010). Moral understanding in the psychopath. *Synthesis Philosophica, 24*(2), 337–348.

Malatesti, L. (2014). Psychopathy and failures of ordinary doing. *Etica & Politica/Ethics & Politics, 2*, 1138–1152.

Malatesti, L., Jurjako, M., & Meynen, G. (2020). The insanity defence without mental illness? Some considerations. *International Journal of Law and Psychiatry, 71*, 101571. https://doi.org/10.1016/j.ijlp.2020.101571

Malatesti, L., & McMillan, J. (Eds.). (2010). *Responsibility and psychopathy: Interfacing law, psychiatry, and philosophy*. Oxford University Press.

Malatesti, L., & McMillan, J. (2014a). Defending psychopathy: An argument from values and moral responsibility. *Theoretical Medicine and Bioethics, 35*(1), 7–16. https://doi.org/10.1007/s11017-014-9277-5

Malatesti, L., & McMillan, J. (2014b). Two philosophical questions about psychopathy. In *Criminal behaviours. Impacts, Tools and Social Networks* (pp. 277–300). https://www.bib.irb.hr/755795

Marshall, J., Watts, A. L., & Lilienfeld, S. O. (2018). Do psychopathic individuals possess a misaligned moral compass? A meta-analytic examination of psychopathy's relations with moral judgment. *Personality Disorders: Theory, Research, and Treatment, 9*(1), 40–50. https://doi.org/10.1037/per0000226

Megone, C. (1998). Aristotle's function argument and the concept of mental illness. *Philosophy, Psychiatry, and Psychology, 5*(3), 187–201.

Megone, C. (2000). Mental illness, human function, and values. *Philosophy, Psychiatry, and Psychology, 7*(1), 45–65.

Morse, S. J. (2008). Psychopathy and criminal responsibility. *Neuroethics, 1*(3), 205–212. https://doi.org/10.1007/s12152-008-9021-9

Mullen, P. E. (2007). On building arguments on shifting sands. *Philosophy, Psychiatry, and Psychology, 14*(2), 143–147.

Nadelhoffer, T., & Sinnott-Armstrong, W. P. (2013). Is psychopathy a mental disease? In N. A. Vincent (Ed.), *Neuroscience and legal responsibility* (pp. 229–255). Oxford University Press.

Patrick, C. J. (Ed.). (2018). *Handbook of psychopathy* (2nd ed.). The Guilford Press.

Poeppl, T. B., Donges, M. R., Mokros, A., Rupprecht, R., Fox, P. T., Laird, A. R., Bzdok, D., Langguth, B., & Eickhoff, S. B. (2019). A view behind the mask of sanity: Meta-analysis of aberrant brain activity in psychopaths. *Molecular Psychiatry, 24*(3), 463–470. https://doi.org/10.1038/s41380-018-0122-5

Rawls, J. (1971). *A theory of justice*. Harvard University Press.

Rawls, J. (2000). *Lectures on the history of moral philosophy*. Harvard University Press.

Rawls, J. (2005). *Political liberalism (expanded edition)*. Columbia University Press.

Reimer, M. (2008a). Psychopathy without (the language of) disorder. *Neuroethics, 1*(3), 185–198. https://doi.org/10.1007/s12152-008-9017-5

Reimer, M. (2008b). Psychopathy without (the language of) disorder. *Neuroethics, 1*(3), 185–198. https://doi.org/10.1007/s12152-008-9017-5

Reydon, T. (this volume). Psychopathy as a scientific kind: On usefulness and underpinnings. In L. Malatesti, J. McMillan, & P. Šustar (Eds.), *Psychopathy. Its uses, validity, and status.* Springer.

Reznek, L. (1987). *The nature of disease.* Routledge & Kegan Paul.

Robinson, D. N. (2000). Madness, badness, and fitness: Law and psychiatry (again). *Philosophy, Psychiatry, & Psychology, 7*(3), 209–222.

Sellbom, M., Lilienfeld, S. O., Latzman, R. D., & Wygant, D. B. (this volume). Assessment of psychopathy: Addressing myths, misconceptions, and fallacies. In L. Malatesti, J. McMillan, & P. Šustar (Eds.), *Psychopathy. Its uses, validity, and status.* Springer.

Stein, D. J., Phillips, K. A., Bolton, D., Fulford, K. W. M., Sadler, J. Z., & Kendler, K. S. (2010). What is a mental/psychiatric disorder? From DSM-IV to DSM-V. *Psychological Medicine, 40*(11), 1759–1765. https://doi.org/10.1017/S0033291709992261

Tamatea, A. J. (this volume). Humanising psychopathy, or what it means to be diagnosed as a psychopath: Stigma, disempowerment, and scientifically-sanctioned alienation. In L. Malatesti, J. McMillan, & P. Šustar (Eds.), *Psychopathy. Its uses, validity, and status.* Springer.

van Voren, R. (2009). *On dissidents and madness: From the Soviet Union of Leonid Brezhnev to the "Soviet Union" of Vladimir Putin.* Rodopi.

Wakefield, J. C. (1992). The concept of mental disorder. On the boundary between biological facts and social values. *The American Psychologist, 47*(3), 373–388.

Wakefield, J. C. (2014). The biostatistical theory versus the harmful dysfunction analysis, part 1: Is part-dysfunction a sufficient condition for medical disorder? *Journal of Medicine and Philosophy, 39*(6), 648–682. https://doi.org/10.1093/jmp/jhu038

Wakefield, J. C., & Conrad, J. A. (2020). Harm as a necessary component of the concept of medical disorder: Reply to Muckler and Taylor. *The Journal of Medicine and Philosophy: A Forum for Bioethics and Philosophy of Medicine, 45*(3), 350–370. https://doi.org/10.1093/jmp/jhaa008

Wallinius, M., Nilsson, T., Hofvander, B., Anckarsäter, H., & Stålenheim, G. (2012). Facets of psychopathy among mentally disordered offenders: Clinical comorbidity patterns and prediction of violent and criminal behavior. *Psychiatry Research, 198*(2), 279–284.

Willoughby, C. D. E. (2018). Running away from drapetomania: Samuel a. cartwright, medicine, and race in the antebellum south. *Journal of Southern History, 84*, 579–614. https://doi.org/10.1353/soh.2018.0164

Yannoulidis, S. (2012). *Mental state defences in criminal law.* Ashgate.

Printed in the United States
by Baker & Taylor Publisher Services